# PHYTOHORMONES AND PATTERNING

### The Role of Hormones in Plant Architecture

T0350249

## Previous Books by Esra Galun

Frankel, R. and Galun, E. (1977) *Pollination Mechanisms, Reproduction and Plant Breeding*. Springer Verlag, 281 pp.

Galun, E. and Breiman, A. (1977) *Transgenic Plants*. Imperial College Press/ World Scientific, 376 pp.

Galun, E. and Galun, E. (2001) *The Manufacture of Medical and Health Products by Transgenic Plants*. Imperial College/World Scientific, 332 pp.

Galun, E. (2003) *Transposable Elements — A Guide to the Preplexed and the Novice*. Klumer Academic Publishers (Springer Verlag), 335 pp.

Galun, E. (2005) *RNA Silencing*. World Scientific, 456 pp.

Galun, E. (2007) *Plant Patterning* — Structural and Molecular Genetic Aspects. World Scientific, 499 pp.

# PHYTOHORMONES AND PATTERNING

## The Role of Hormones in Plant Architecture

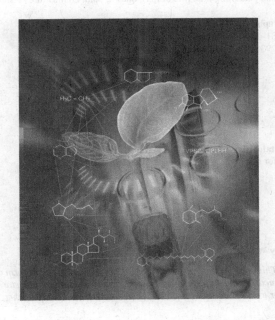

## Esra Galun

Weizmann Institute of Science, Israel

**World Scientific**

NEW JERSEY · LONDON · SINGAPORE · BEIJING · SHANGHAI · HONG KONG · TAIPEI · CHENNAI

*Published by*

World Scientific Publishing Co. Pte. Ltd.

5 Toh Tuck Link, Singapore 596224

*USA office:* 27 Warren Street, Suite 401-402, Hackensack, NJ 07601

*UK office:* 57 Shelton Street, Covent Garden, London WC2H 9HE

**Library of Congress Cataloging-in-Publication Data**
Galun, Esra, 1927–
    Phytohormones and patterning : the role of hormones in plant architecture / Esra Galun.
      p. cm.
    Includes bibliographical references and index.
    ISBN-13: 978-981-4293-60-0 (hardcover : alk. paper)
    ISBN-10: 981-4293-60-1 (hardcover : alk. paper)
    1. Plant hormones. 2. Plant pattern formation. 3. Plant regulators. I. Title.
    QK745.G33 2010
    571.7'42--dc22

                                                    2010023318

**British Library Cataloguing-in-Publication Data**
A catalogue record for this book is available from the British Library.

Typeset by Stallion Press
Email: enquiries@stallionpress.com

Printed in Singapore.

# Preface

*Tis evident that all reasonings concerning matter of fact are founded on the relation of cause and effect, and that we can never infer the existence of one object from another, unless they be connected together, either mediately or immediately. In order therefore to understand these reasonings, we must be perfectly acquainted with the idea of a cause; and in order to that, must look about us to find something that is the cause of another.*

David Hume (1711–1776)
from "An Abstract of a Treatise of Human Nature."

Hume pioneered the idea that getting intimately acquainted with one matter (e.g. a plant hormone) will not provide adequate information on this matter. To acquire such information, we should examine causality. In our case: What roles the plant hormones have in patterning and what are the impacts of interactions between plant hormones (phytohormones) on the shaping of plant organs?

I should apologize because the analogy of Hume's matter and plant hormones is far-fetched. Hume was interested in the nature of matter. He provided an example of a baby who has a ball. When he throws an elastic ball to the floor, it bounces. But if he throws a soft-cloth ball to the floor, it rests on the floor without bouncing. By that, the matter of the ball is characterized. In the present book we are less interested in the nature of the matter (phytohormones) and more in the impact of the matter on a phenomenon (plant patterning). We go further and are interested in the roles of interactions between plant hormones and the shaping of plant organs.

The role of phytohormones in the shaping of plant organs was mentioned in my previous book (Galun, 2007). During recent years, this subject was intensively studied and the understanding of the biosynthesis, signal transduction, and interactions of phytohormones improved considerably. Therefore, a comprehensive treatment of this subject is timely.

This book, like my previous books (Galun and Breiman, 1997; Galun and Galun, 2001; Galun, 2003, 2005, 2007), is aimed at a wide range of readers. A basic knowledge of botany, biochemistry and genetics should be helpful for the readers of the book, although I attempted to explain relevant concepts and processes to the novice. An example of such an explanation concerns *ubiquitination*, a process that takes place in (probably) all eukaryotic organisms and causes the degradation of specific target proteins. Because several phytohormones are involved in the ubiquitin-proteasome system, a section in the "Introduction" is devoted to this process.

Scholars engaged in the study of plant patterning commonly focus on one or very few plant organs. The book covers most of the plant organs and should provide these scholars with a broad view on the role of phytohormones in plant patterning.

I wrote this book in a form that will enable its use also as a companion textbook for students in advanced plant biology courses, especially courses that deal with plant hormones and patterning of plants. This text will thus supplement books that deal mainly with housekeeping biochemistry and molecular biology of plants (e.g. Buchanan *et al.*, 2000).

I am grateful to my wife, Professor Margalith Galun, who read parts of the draft of this book and provided constructive remarks. The moral support of other members of my family is very much appreciated. I also thank the staff members of my Department of Plant Sciences at the Weizmann Institute of Science in Rehovot: Professor Dan Atsmon, Professor Moshe Feldman, Professor Robert Fluhr, who discussed with me several issues of this book during my writing. I am especially grateful to Professor Yuval Eshed, an expert on the molecular aspects of plant differentiation, who provided me with ample advice. Professor Rudolf Hagemann of the Institute of Genetics of the Martin-Luther University in Halle/Saale is hereby acknowledged for detailing to me the contributions of various German scholars in choosing Arabidopsis (previously named *Stenophragma thalianum*) as a model plant for genetics investigators.

I very much appreciate the efforts of my publisher, World Scientific Publishing Co. and especially of Ms. Sook-Cheng Lim, scientific editor at WSPC, for editing and production of this book. I acknowledge the professional and good work of the Graphic Arts Department of the Weizmann Institute of Science for handling the figures. I further acknowledge the permission to use figures and tables that were granted by the respective publishers and authors. I provided the sources of these figures and tables below each figure and table, and the full citations are listed in the reference list. These citations should mean that I am grateful for the consent of the publishers to use their copyrighted material in this book. A "corpus acknowledgement" to the respective publishers is provided at the end of the book. Thanks are due to Mrs. Suzanne Trauffer Ramakrishna for her professional typing.

As in most of my previous books, Dr. Nillie Weinstein was a source of inspiration, especially for non-scientific aspects of this book, that were particularly relevant for writing the Preface, Introduction, and Epilogue.

**Esra Galun**
Rehovot, Israel
7th April 2010

# Contents

*Preface*                                                               v

Introduction                                                            1

**Part I    The Phytohormones**                                        **13**

Chapter 1      Auxin*                                                   15
Chapter 2      Ethylene*                                               49
Chapter 3      Gibberellins*                                           71
Chapter 4      Brassinosteroids*                                       91
Chapter 5      Cytokinins*                                            111
Chapter 6      Jasmonates*                                            131
Chapter 7      Abscisic Acid*                                         151
Chapter 8      Peptide Phytohormones*                                 165
Chapter 9      Strigolactones and Branching*                          173

**Part II    Plant Organs and Tissues**                               **189**

Chapter 10     Patterning of the Embryo                               207
Chapter 11     The Root Patterning and Phytohormones                  227

---

* In each of these chapters the main characteristics, biosynthesis pathways, local-ization of syntheses, transfer, signal transduction, and some of the interactions, with other hormones, are discussed.

Chapter 12    Patterning of the Shoot                247
Chapter 13    The Leaf                              297
Chapter 14    Patterning of Flowers: Genes and      325
              Phytohormones

*Epilogue*                                          367

*References*                                        373

*Corpus Acknowledgment*                             409

*Index*                                             411

# Introduction

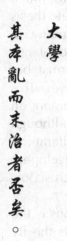

大學

其本亂而末治者否矣。

In the Introduction of the old Chinese text, *Da Xue* (*Great Learning*), there is a passage that is reproduced on the left side of this page. In a free translation, this passage says: "One who destroys his roots will never be able to keep order in his branches." This metaphor can be understood as the requirement to maintain good foundations in order to provide proper future development. It also may hint on the establishment of Bonsai trees. The Bonsai practice of obtaining miniature trees was started in Japan about 1000 years ago but it was based on an older Chinese practice to trim, continuously, the roots of young trees and by that, to obtain dwarf trees.

## Remarks on the Study of Plant Architecture

With respect to architecture of plants, there are two points that deserve some discussion. Some attention to these points, but from a different point of view, was already presented in my previous book (Galun, 2007). The first point is the relevance of "tools" in the history of the study of pattern formation. In a way, the history of understanding plant patterning is parallel to the history of emerging "tools" that enabled progress in this field of endeavor.

Looking back, the first decisive "tool," that was of benefit to developmental biology, was the optical lens. The use of the optical lens is well known in the history of astronomy: lenses provided Galileo

Galilei (1564–1642) with a much sharper vision of celestial bodies than looking at these bodies with naked eyes. Galilei was a member of the first post-Renaissance academy that was founded in Italy in 1603 by Federico Cesi: the Licean Academy (see Freedberg, 2002). Other members of this academy used optical lenses to probe into the morphology and the anatomy of various organisms, including plants. Such probings were then carried out also by investigators in other countries, such as England (e.g. Robert Hooke, 1635–1703) and the Netherlands (e.g. Antony van Leeuwenhoek, 1632–1723). Not much later (in 1759), Caspar Friedrich Wolff submitted his doctorate thesis ("Theoria Generationis") to the University of Halle in Germany, in which one part out of three was devoted to the development of plants. In his thesis, Wolff stated that all the parts of plants (angiosperms), but roots and shoots (i.e. including flowers), are modified leaves. This statement was made several years earlier than a similar statement on the metamorphosis of leaves made by J.W. von Goethe, although Goethe's statement was commonly cited in the botanical literature. Since about 150 years ago, several novel "tools" were developed, such as chemistry, biochemistry, genetics, electron microscopy, genetic transformation, system biology, etc.

An important "tool" for development biology investigations is the *organism* that serves these investigations. In animals, it was the fly *Drosophila melanogaster* and its related species that was employed since the early years of the 20th century by T.H. Morgan (1866–1945) and associates, and subsequently by many other geneticists and developmental biologists. *Drosophila* flies kept their place in the "Hall of Fame" till present, hence for about 100 years. In plant research, the favorite model plant is *Arabidopsis thaliana* (Arabidopsis). Arabidopsis has many features that render it a model plant of choice for genetic and developmental biology studies. Arabidopsis has an impressive "history" that was reviewed in detail by Meyerowitz (2001). The first botanical information on Arabidopsis is attributed to Johannes Thal (1542–1583), a physician from Thüringen, Germany. *Arabidopsis thaliana* changed names. Linnaeus termed it *Arabis thaliana*. It was later given the name *Stenophragma Thalianum*. When the German botanist/geneticist Friedrich Laibach (1885–1967) found (in 1907) that this plant has only five pairs of chromosomes, he still used the latter name. But in a later publication, in 1943, where Laibach recommended this plant for genetic studies, he used the name that is used presently: *Arabidopsis thaliana*. The species name, thaliana, was given in the honor of Johannes Thal, mentioned above. The torch of Arabidopsis investigations in Germany was passed over to Röbbelen, in Göttingen, who started to publish studies on this plant in 1956. The gospel of Arabidopsis as a model plant for genetic studies probably came to the USA from Hungary, when due to the Hungarian revolt in 1956, George P. Redei immigrated from Hungary to the USA. He brought the Arabidopsis to the University of Missouri in Columbia. In Germany, Andreas Müller of the Gatersleben Institute (then in the DDR) as well as other investigators performed Arabidopsis studies. As indicated above, Redei planted the "seeds" of Arabidopsis in the USA. Already in 1965, when the first International Symposium on Arabidopsis was held in Göttingen, in honor of Laibach's 80th birthday, that the US investigators attended. A few years after his arrival in Columbia, Redei began to publish a long list of papers on the genetics of Arabidopsis. The yield of the early years of Redei's work was published in two reviews (Redei, 1970, 1975). The advantages of Arabidopsis as a model plant appealed to several US investigators, such as Chris Somerville, then in East Lansing, Michigan (e.g. Somerville and Ogren, 1979) and Elliot Meyerowitz, at the California Institute of Technology. The latter, who came from the *Drosophila* world, set up a very prolific Arabidopsis laboratory. In his historical review, Meyerowitz (2001) detailed the advantages of Arabidopsis as a "tool" in genetic and development studies. He mentioned the ease of mutagenesis and the screening of mutants, the small genome that is convenient for gene cloning and the ease of genetic transformation. To that, one should add, now, one of the greatest advantages of Arabidopsis as a model plant: its genome is completely sequenced and the respective information can be accessed readily by every investigator. It can be claimed that other plants, such as tomato, petunia, pea, rice and maize, were successfully used in studies of plant pattern formation, but Arabidopsis had a central role as a "tool" in these studies.

With each additional "tool" that was added to the arsenal of investigations, the understanding of patterning was elevated to a higher level. The history of the evolution of knowledge of plant architecture was provided in Galun (2007), where several books on this subject were recommended (e.g. Sachs, 1991; Wolpert, 1998; Coen, 1999) as general references for studies on the development of patterns in plants.

The second point that I wish to bring up is about genes that are controlling a three-dimensional structure. The novice may infer that there are genes that are well defined (i.e. having known DNA sequences) that are leading to certain three-dimensional structures in plants, such as trichomes. This is obviously not so. Known genes may *modify* the three-dimensional structures (as reducing or increasing the number of cells in trichomes or affecting trichome branching), but we lack knowledge on genes that are causing the realization of the *idea* of the three-dimensional structure. In other words, in no case we know, in a cellular organism, how a linear information (the DNA sequence) is causing the formation of a three-dimensional form. We have to go "down" the phylogenetic ladder and reach the bacteriophages in order to find a system in which we know how such DNA sequences cause the final shape of a mature T4 bacteriophage. In T4, we are very close to a situation in which we can add, into a "soup", defined DNA sequences that cause the synthesis of specific proteins and by ordered sequence of these syntheses, will obtain viral "heads", viral base plates, viral "tails," etc. and then ultimately, mature and functional bacteriophages. What is known in T4 is not known even in bacteria, let alone in multicellular organisms.

There is another "unknown" in patterning: for correct patterning of a cell (or rather the genome in a given cell) it should receive and sense information on its *position*. Both the spatial and the temporal positions are vital. This means where in the space of a given organism and even the organ of an organism, the cell is located. Also vital is the information on the developmental stage of the organ or the organism, in which the cell is harbored. This spatial/temporal information, together with the influx of hormones, will then regulate the expression of specific genes required for correct patterning in the given cell. Only in very rare cases do we know what are the cues for position information. One example is the zygote of the brown alga *Fucus*; when illuminated by an unidirectional light beam, the embryo will initiate apical hairs towards the source of the light and rhizoids at the opposite side of the zygote (Galun and Torrey, 1969; Torrey and Galun, 1970). Another example is root epidermis cells that are destined to become root hairs (in Arabidopsis). In this case, if the epidermal cells are in contact with *two* cortex cells, genes that trigger the formation of root hairs will be activated in the epidermal cell, but if only *one* cortex cell contacts the epidermal cell, no root hair will be formed. We should keep in mind that the positioning of a cell in a plant is more complicated than position-ing a driver with the aid of a GPS instrument, in a specific location, because in the latter case, the localization is on a two-dimensional grid while in the former case, the localization should be in a three-dimensional grid. Nevertheless, there is already some progress in understanding positioning in angiosperms. This knowledge concerns the phenomenon known as "shade avoidance." It appears that plants have an efficient (but rather complicated) way in which they avoid the shade of another plant and the shoot then escapes the shade of another shoot. This set of plant responses was termed SAS for shade avoidance syndrome. The presence of a nearby plant or shoot causes a reduction of the ratio of red (R) to far-red (FR) light. The relative increase of FR light is perceived by the phytochromes of the leaves and in a cascade of regulatory processes (most of them not yet fully understood, but probably involving the phytohormone auxin), causes the SAS. More details on SAS and on light-regulated transcriptional networks in plants can be obtained from a recent research publication (Roig-Villanova *et al.*, 2007) and a review (Jiao *et al.*, 2007).

There is a further note on the establishment of specific architectures in plants that may be termed "The Pillars of the Mosque of Acre." I eluded on this metaphor in my previous book (Galun, 2007) but here is the essence of its meaning. The Mosque of Acre was built by the Turkish ruler of Lebanon and Syria, Achmad (al-Jazar — the beheader) Pecha. Achmad Pecha built the mosque in 1781 with an impressive court full of marble pillars. When one looks closely at the pillars, it is clear that the pillars differ considerably from each other. They are Doric, Ionic, Corinthian or Composite. No wonder: Achmad Pecha collected marble pillars from numerous Greek and Roman ruined temples of ancient cities along the Eastern Mediterranean shore. The pillars were modified to fit their setup in the court of the mosque. Well, nature acts in a way similar to Achmad Pecha: when a novel regulation of patterning is required, it does not invent completely *novel* regulatory components, but rather amends *existing* components for servicing the required novel regulations. This point should be remembered when we deal with the regulation of patterning. The regulatory systems frequently appear to us as excessively complicated; why were they not composed in a simpler way? However, we should recall that the various complex systems were formed from already available components (some of them already complex) that were further modified as required.

## Ubiquitination (Ubiquitylation)

The ubiquitin-proteasome system (UPS) is a widespread process that causes the regulated protein degradation in eukaryotes. It was revealed only a little over 10 years ago by Aaron Ciechanover, Avram Hershko and Irwin Rose (Hershko and Ciechanover, 1998), but since then it was found to have a major role in many processes. We shall see in Part I below that the activity of several phytohormones is involved with a UPS. Thus, a short description of the UPSs in plants will precede the handling of the individual phytohormones. In Arabidopsis, more than 1400 genes encode components of the ubiquitin-proteasome protein degradation. While the UPS has several variants, there is a similar overall strategy of the UPS that assures that the degradation will affect very specific proteins and will take place only in the presence of specific signals. For our deliberations, ubiquitination is of great interest because it is involved with phytohormones and plant patterning. Albeit, even if we focus on plants, the UPS is rather elaborated. To get an idea of how complicated it is, I shall note that in their review, Dreher and Callis (2007) had to list over 200 abbreviations that were used in their text! The UPS will therefore be described here only schematically and additional information will be provided in Part I, where the various phytohormones are handled. Readers interested in more details on UPS of plants are referred to reviews on this subject (e.g. Smalle and Vierstra, 2004; Moon *et al.*, 2004; Schwechheimer and Schwager, 2004; Dreher and Callis, 2007).

The main components of UPS are the ubiquitin (Ub), the proteasome and three E proteins: Ub activator (E1), Ub conjugating enzyme (E2) and Ub ligase (E3). In most eukaryotes, there is a small number of E1 proteins, more E2 proteins and a much greater number of E3 proteins. The latter many E3 ligases provide the final specificity: rendering a substrate to be degraded in the proteasome. The UPS is schematically shown in Fig. 1 and a scheme that includes the proteasome is shown in Fig. 2.

The Ub is a 76-amino acid protein and it is attached to the targeted substrate by a series of steps that involve the E proteins, until a "tail" of poly-Ub (commonly a chain of four Ub units connected by lysyl linkages) is attached to it. Then the poly-ubiquitylated target can enter the proteasome. The 26S proteasome is a 2 mD proteinous complex. The Ub is very similar in all eukaryotes. Plant Ubs differ from yeast Ub by only two amino acids and from animal Ub by three amino acids. Ubs are identical in all plants. The bulk of the Ub protein has a compact globulus shape. A flexible C-terminal extension protrudes from the Ub that includes an essential glycine. The glycine can interact with the E1, E2 and in some cases, with E3. Hence, this glycine serves to connect the Ub with its target protein that is destined for degradation. Interestingly, although Arabidopsis has 28 Ub-coding genes, they all encode the same Ub protein.

The modification of the Ub starts with the formation of an acyl phosphoanhydride bond between AMP and the above-mentioned glycine of the Ub, mediated by ATP. Then a thiol-ester is formed between the glycine of the Ub and a cysteine of E1. The activated Ub is then transferred to a cysteine in an E2 by trans-esterification. Finally, the Ub-E2 intermediate can bind to the substrate by E3. Commonly, more Ub units are added so that the target protein is

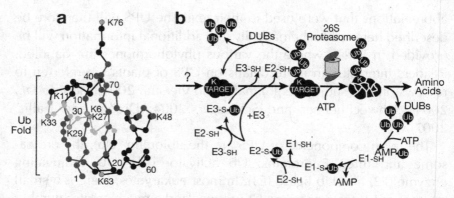

**Fig. 1.** Ub and the Ub/26S proteasome pathway. (a) The three-dimensional structure of plant Ub. The Ub fold is indicated with its mixed $\beta$ sheets and $\alpha$ helix. The lysines (K) at positions 6, 11, 27, 29, 48 and 63 that can participate in forming polyUb chains and the C-terminal glycine that forms the isopeptide bonds with targets. (b) Diagram of the Ub-26S proteasome pathway. The pathway begins with the adenosine triphosphate (ATP)-dependent activation of ubiquitin (Ub) by Ub-activating enzyme (E1), followed by the transfer of the Ub to a Ub conjugating enzyme (E2) and finally ends with the attachment of the Ub to a lysine in the target protein with the help of a Ub ligase (E3). Once the Ub-protein conjugate is formed bearing a poly-Ub chain, it is either recognized by the 26S proteasome and degraded in an ATP-dependent process with the concomitant release of Ub monomers, or the conjugate is disassembled by deubiquitinating enzymes (DUBs) to regenerate both the target protein and Ub intact. (From Smalle and Vierstra, 2004).

attached to a poly-Ub chain. In plants, the Ubs are usually linked to each other via their lysine48. The activating enzyme E1 is a relatively large protein (of about 1100 amino acids) but it is a very efficient enzyme; even a low concentration of it is sufficient for the UPS. The binding of E1 to Ub does not lead to specificity of target decomposition. The Arabidopsis genome encodes only two E1s. The Ub-conjugation enzyme E2 already leads to some specificity. Thus, there are many different E2s encoded by the genome of each organism. Arabidopsis produces at least 37 E2, and eight additional genes encode putative UEV (ubiquitin-conjugating E2 enzyme variants). The E3s are the members of the UPS that bestow specificity.

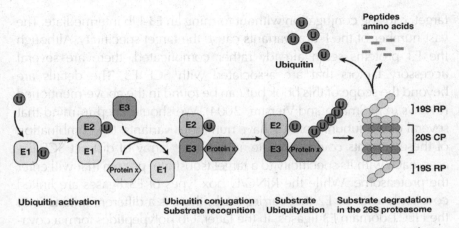

**Fig. 2.** Simplified scheme of the ubiquitin-proteasome pathway in eukaryotes. (From Schwechheimer and Schwager, 2004).

The E3 is actually a common name to a huge number of different proteins. Each variant of E3 binds to its specific target, causing the target to contain the Ub-tail. Only such "tailed" target proteins will enter the proteasome in order to be decomposed there. The Arabidopsis genome contains more than 1300 genes for the different E3s. These were grouped in families and four "types" of E3 were described in this plant, based on subunit composition and action. One was given the long name: Real Interesting New Gene (RING/U-box). This is a complex of three proteins: Skp1, CDC53 and F-box (together: SCF). They were divided into two groups: the RING group that has RING-fingers (about 480 in Arabidopsis) and proteins that contain a U-box (64 in Arabidopsis). The "fingers" are made of 70 amino acids that bind zinc. The U-box exploits electrostatic interaction instead of metal ion chelation to stabilize a RING-finger-like structure. There are also HECT E3s that are large proteins (over 100 kD); seven HECTs were identified in Arabidopsis. The RING/U-box serves as an Ub-E2 docking site, activating the transfer of Ub to the substrate (target) lysine(s).

SCF E3 consists of a complex of several polypeptides that function also as a scaffold to bring together the activated Ub-E2 complex and the target, causing conjugation without forming an E3-Ub intermediate. The vast number of the F-box variants cause the target specificity. Although the E3 proteins seem already rather complicated, there are several accessory factors that are associated with SCF E3. The details are beyond the scope of this book but can be found in the above-mentioned reviews (e.g. Smalle and Vierstra, 2004). We should keep in mind that several of the subunits of SCF have numerous variants; the combination of these variants could generate an "infinite" array of distinct SCF ligases, each with its specificity to a target (substrate) protein that will enter the proteasome. While the RING/U-box types of E3 ligases are linked covalently with an E2 that carries a Ub, there is a different linking with the HECT domain E3 ligases. In the latter, Ub polypeptides form a covalent thioester linkage with a cysteinyl sulfydryl group on HECT protein before being transferred to a lysine on the substrate (target protein). Again, as noted above, one may wonder why was the "palace" of the UPS built with such complicated building stones? The answer is that nature acts as Achmad Pecha (the al-Jazar of Acre): amending available building stones rather than using stones from the virgin rock.

A rough scheme of the 26S proteasome is shown in Fig. 2. It is an ATP-dependent proteolytic complex that degrades Ub-conjugates. Plants have several isoforms of proteasomes. The proteasomes are rather elaborate proteinous organelles. Again, the details of the structure of proteasomes will not be provided in this book. They were revealed mainly in yeast and animals. Albeit there is also some information on proteasomes of plants, based particularly on studies with Arabidopsis and rice. Here, only the main features of plant proteasomes will be given. The 26S proteasome contains about 30 principal subunits that are located in two subcomplexes: the 20S core protein (CP) and the 19S regulatory particle (RP). The CP has a wide range of proteases; it is a barrel-shaped organelle that has four "rings", each of which is composed of seven subunits. On both ends of this "barrel" are the $\alpha$-subunit rings and the middle two rings are composed of $\beta$-subunits. In the middle of the "barrel" is an active-site threonine. The "barrel" has trypsin-like and chemotrypsin-like activities, rendering the CP capable to clear most, if not all, peptide bonds. The "barrel" allows only unfolded proteins to enter the protease chamber.

The RPs associate with one or both ends of the CP and cause ATP-dependence as well as specificity to Lys48-linked poly-Ub chains to enter the proteasome. The RP has two subcomplexes: the "Lid" and the "Base." The RP assists in recognizing and unfolding the poly-Ub-tagged target proteins. The Ubs are removed, by the activity of the deubiquitylating enzyme (DUB), and released while the unfolded target protein then enters the proteolytic chamber of the CP. One may claim that the proteasome is an "environment-friendly" garbage disposal device — it separates the Ub particles from their complex with the target protein, digests the target protein into amino acids and allows the Ub particles to be reused as well as the amino acids to be polymerized into new proteins.

One final remark before we leave this general introduction to ubiquitylation and turn, in Part I, to UPSs related to specific hormones. The general strategy in many UPSs may be exemplified by the involvement of UPS with auxin (e.g. IAA) responsive genes (RG). The gene expression is regulated in a dual manner: there is a transcription activator (TA) that activates the expression of the auxin responsive gene (ARG); but commonly, this does not happen because the activity of TA is inhibited by another protein, the transcriptional repressor (TR) that binds to the TA. Only when the TR is eliminated, will the TA function and the ARG will be expressed. The elimination of the TR is mediated by the UPS. Before this happens, the TR has to undergo an elaborate process that includes the involvement of E1, E2 and E3. The E3 provides the final specificity; it is a multi-subunit complex and in the presence of IAA, will "mark" (attaching Ub) the TR for future degradation in the proteasome. For all this to happen correctly, a large number of genes are involved. Mutations in any of these may severely affect the UPS process. We are thus facing a situation that is problematic to men and plants — men are bewildered by the complexity of the process and plants have to cope with a very elaborate system that must flow without any faultiness.

As said above, the roles of UPS will be handled again, in the chapters dealing with the various types of phytohormones (e.g. auxins, gibberellins, abscisic acid, brassinosteroids, ethylene, jasmonates). A list of identified UPS components of Arabidopsis is updated on a website dedicated to this subject (http://plantsubq.genomics.purdue.edu).

# PART I

# The Phytohormones

This part of the book shall focus on several groups of phytohormones. The respective chapters shall start with a brief "history", followed by information on subjects such as biosynthesis, transport, signal transduction and involvement with ubiquitination. The interaction between phytohormones of different groups will be mentioned in this part only briefly as more detailed information will be provided in Part II, where the focus shall be on specific plant organs because hormone interactions have major roles in plant organ patterning.

In general terms, the early "history" of phytohormones (plant hormones or plant growth substances — I shall use all these three terms in this book) brings to mind the Cheshire Cat of "Alice's Adventures in Wonderland" by Lewis Carrol (C.L. Dodgson). However, the analogy between the Cheshire Cat and phytohormones is in a reverse sense ... In the story of Alice, the girl first saw the whole Cat but the cat gradually disappeared until only the grin of the cat was left. In the phytohormones (e.g. auxin), the early investigators (see Chapter 1) saw merely the effect of the hormones (the "grin") and only after many years, the whole Cat emerged (the auxin was chemically defined).

Even before most of the information on phytohormones was revealed, man utilized phytohormones or related synthetic components (e.g. dichlorophenoxyacetic acid — 2,4-D; naphthalene-1-acetic acid — NAA) for his needs: as herbicides or for the destruction of vegetation. Even earlier, one phytohormone (ethylene) was utilized in horticultural practice. The Bible tells us that the prophet Amos was

etching sycamore figs (The Hebrew Bible, Amos. Chapter 7, 14). He
did this to induce fruit setting. Such etchings cause the production of
high levels of ethylene and the latter helps in the setting of the figs.
Note that the figs of sycamore, as the common figs, are not regular
fruits, in botanical terms; they are "compound fruits" as each immature fig contains numerous individual flowers. A comprehensive
coverage of the utilization of phytohormones in agriculture and hor-
ticulture was provided by the book edited by Nickell (1983).
Although published over 25 years ago, the two volumes of the book
handle, in detail, 22 types of crops. Conspicuously, Nickell avoided
pineapple from his book!

Moreover, the parasitic bacterium *Agrobacterium tumefaciens* utilizes phytohormones for its parasitic life style. The bacterium utilizes
the same strategy as some cattle growers, who inject growth hormones into cattle to increase the yield of beef. The bacterium
"injects" into the genomes of its plant host genes that encode
enzymes for the synthesis of auxins and cytokinins. The latter induce
the formation of tumors in which the bacterium multiplies. The
*Agrobacterium* bacteria do even better … They introduce into their
host genes for the synthesis of metabolites (opines), forcing the host
to produce metabolites that the plant cannot use but are nourishing
the bacterium. The latter "trick" is not unique to *Agrobacterium*; it
can be found in several parasitic and symbiotic systems.

# CHAPTER 1

# Auxin

The clues that plants posses a transported entity that can cause responses of the plant to environmental effectors emerged already in the 19th century (Sachs, 1887). Experiments of Theophili Ciesielski (1872) and Sir Charles and Sir Francis Darwin (a team of father and son) studied phototropism and geotropism. Further studies on this theme were conducted by Went and by Cholodny as reviewed by Estelle (1996) for the gravitropic curvature in plants. All these pointed towards the existence in plants of a substance that can flow from one location to other locations. This substance, termed auxin, was then chemically identified as indoleacetic acid (IAA), as reviewed by Went and Thimann (1937). It was then revealed that treatment of plants with some synthetic chemicals, such as 2,4-D and NAA caused similar effects as IAA. Trichlorophenoxybutyric acid (2,4,5 TB) also has an auxin-like activity. This compound was mixed with 2,4-D to produce the "Agent Orange." During the production of 2,4,5 TB, dioxin was manufactured as a by-product, rendering the mixture toxic also to animals. The former auxin-like chemicals had the advantage that their degradation, in the plants, is slower than the degradation of IAA. This chapter will deal primarily with the natural auxins (IAA and indole butyric acid, IBA).

## Synthesis and Conjugation — Main Pathways

An update on auxin synthesis was provided in my previous book (Galun, 2007). It was based primarily on the review of Woodward and Bartel (2005). Figure 10 in my book featured a scheme of auxin

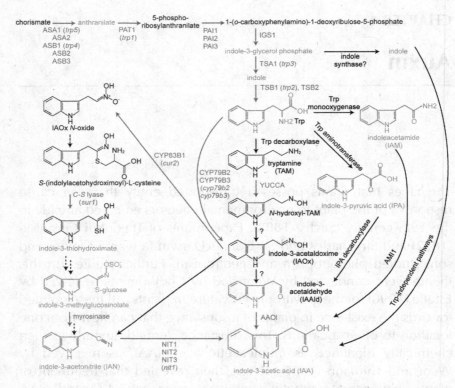

**Fig. 3.** Potential pathways of IAA biosynthesis in Arabidopsis. *De novo* IAA biosynthetic pathways initiate from Trp or Trp precursors. Compounds quantified in Arabidopsis are in blue, enzymes for which the Arabidopsis genes are identified are in red, and Arabidopsis mutants are in lower-case italics. Suggested conversions for which genes are not identified are indicated with question marks. Trp biosynthesis and the P450-catalyzed conversion of Trp to IAOx are chloroplastic, whereas many Trp-dependent IAA biosynthetic enzymes are apparently cytoplasmic. (From Woodward and Bartel, 2005).

synthesis. The figure is provided here as Fig. 3. It is primarily based on information from Arabidopsis.

Revealing the pathways that result in IAA and IBA is a difficult task because of several problems. There is a problem with enzymes involved in auxin biosynthesis, because there can be more than one pathway. The isolation of the respective mutants is thus difficult.

Moreover, because of the many vital tasks of auxins, plants that are unable to produce IAA or IBA, do not exist.

Due to the many compounds involved in IAA and IBA synthesis, this list of abbreviations should be helpful.

| | |
|---|---|
| IAA | Indole-3-acetic acid |
| IAAld | Indole-3-acetaldehyde |
| IAM | Indole-3-acetamide |
| IAN | Indole-3-acetonitrile |
| IAOx | Indole-3-acetaldoxime |
| IBA | Indole-3-butyric acid |
| IPA | Indole-3-pyruvic acid |
| oxIAA | Oxidized IAA |

Plants use two routes for IAA biosynthesis: Trp-dependent and Trp-independent, that are not mutually exclusive. The same plant, such as Arabidopsis, may produce IAA by both main pathways. The relative rate of synthesis in these pathways may vary considerably in different organs (or stages of development) in the very same plant species. However, it should be remembered that the information on these pathways is limited to a tiny fraction of angiosperms.

## Trp-dependent IAA biosynthesis

The potential pathways of IAA biosynthesis are shown in Fig. 3 and a simplified scheme is provided in Fig. 4. Both are based on the review of Woodward and Bartel (2005), who summarized the experimental results of their own work as well as publications by others. Several metabolic conversions are marked by question marks and await verification. However, most probably, there are four main routes for the trp-dependent IAA biosynthesis:

1. The IPA pathway that starts with Trp and goes through IPA, IAAld to IAA, exists in microorganisms and probably also in angiosperms, although some of the enzymes required for this pathway were not detected in plants (e.g. IPA decarboxylase).

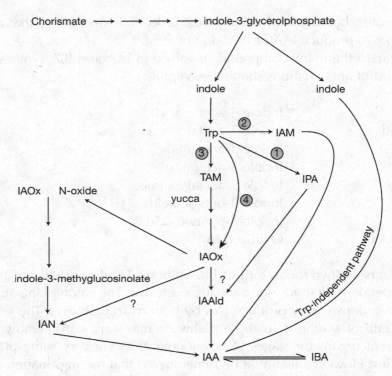

**Fig. 4.** A simplified scheme of the synthesis of IAA and IBA based on Woodward and Bartel (2005). See the abbreviations of the compounds found in the main text. The encircled numbers denote four main Trp-dependent pathways of IAA synthesis.

2. The IAM pathway converts Trp to IAM and from the latter, IAA is synthesized. This pathway also exists in microorganisms. One of the enzymes required for this pathway is trp monooxygenase (IaaM) that converts Trp to IAM. This enzyme is crucial for the parasitic relationship between the *Agrobacterium tumefaciens* parasite and its angiosperm host. The gene that encodes this enzyme is transferred from the bacterial plasmid into the plant's genome, causing a localized high level of IAA and together with locally increased cytokinin, promotes active cell division in the root of the host and consequently, the formation of a tumor

where the parasite multiplies. IAM was detected at high levels in Arabidopsis, suggesting that the Trp-IAA route is taking place in this plant.

3. The TAM pathway converts Trp to TAM and then to N-hydroxyl-TAM; the latter is probably converted to IAOx and to IAAld. IAAld is then converted to IAA. It is not clear if the pathway takes place in Arabidopsis because TAM was not detected in this plant. However, there are indications that the TAM pathway does exist in plants. It was suggested that the *YUCCA* gene (in Arabidopsis) or its homologue in other plants, that encode enzymes for the conversion of TAM to N-hydroxyl-TAM, play a role in IAA biosynthesis. Moreover, *YUCCA* may be rate-limiting in this biosynthesis pathway. The verification of this possibility is hampered by the existence of a family of ten *YUCCA*-like genes in Arabidopsis. Also, the conversion by dehydrogeneration of N-hydroxyl-TAM to IAOx as well as the dehydrogenation and hydroxylation to IAAld is not substantiated by the existence of respective enzymes in Arabidopsis.

4. There is a possible pathway in which IAOx is not converted to IAAld but is rather converted first to IAOx N-oxide and then from IAOx N-oxide (in several steps) to IAN. IAN may then be converted to IAA. This possibility points to IAOx as a potential branch-point from where the metabolic pathways can go in three directions: to IAAld, to IAOx N-oxide, or to IAN. The last compound in the IAOx N-oxide conversion to IAN is indole-3-methylglucosinolate. It appears that this compound serves as a "buffer" and is accumulated when IAA-precursors reach a very high level. Arabidopsis plants, in which the level of indolic glucosinolate is very low, are unable to adequately compensate for high IAA-precursor levels.

The conversion of IAN to IAA is performed by nitrilases. In Arabidopsis, these nitrilases are encoded by several NIT genes; such genes were also found in other plants (such as maize). Although the four nitrilase genes of Arabidopsis encode proteins with similar metabolic roles, they are differentially regulated. *NIT1* is highly expressed, while *NIT2* expression is enhanced under specific condition, and *NIT3* is induced by sulfur starvation. In maize, one nitrilase gene, *ZmNIT1*, is highly expressed in embryonic tissues.

## Trp-independent IAA biosynthesis

The existence of this pathway for IAA biosynthesis (see Figs. 3 and 4) in plants was revealed by analyzing Trp mutants. In spite of low Trp levels, these mutants accumulated IAA conjugates. This was found in Arabidopsis Trp mutants as well as in maize Trp mutants. The bifurcation from the Trp-dependent IAA biosynthesis pathway was located to indole-3-glycerol phosphate. It should be noted that the alkaline conditions under which the various IAA-related compounds are extracted from plant tissue can cause hydrolysis and thus blur the real levels of these compounds in plant tissue. When plants were fed with isotopically labelled substrates, the analyzed labelled IAA-related compounds indicated that plants indeed are able to produce IAA in a Trp-independent pathway. Is the same plant tissue capable to shift from Trp-independent to Trp-dependent IAA biosynthesis (e.g. under stress) or is either of these pathways preferred at specific phases of the plant's development (as embryonal tissue versus mature plant tissue) and in specific plant organs (such as seeds versus young leaves)? The answer seems to be in the affirmative.

## Conjugation of IAA and IBA

Angiosperms can conjugate IAA and IBA, and some of these conjugates of IAA and IBA can be hydrolyzed or β-oxidized back to free IAA and free IBA, respectively. Woodward and Bartel (2005) summarized the relevant information on the various pathways based on the publications by numerous authors. This is shown in Fig. 5. A summarized version of these potential pathways is provided in Fig. 6. There are still quite a lot of question marks on these figures but grossly, there are conjugates that can be converted back to free IAA

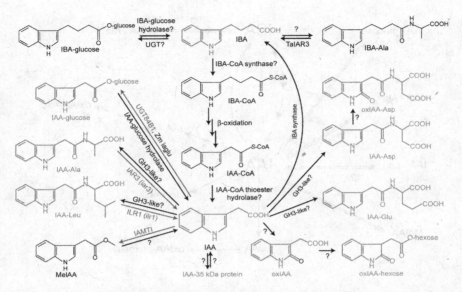

**Fig. 5.** Potential pathways of IAA metabolism. Compounds quantified in Arabidopsis are in blue, enzymes for which the Arabidopsis genes are cloned are in red. Suggested conversions for which plant genes are not identified are indicated with question marks. A family of amidohydrolases that apparently reside in the ER lumen can release IAA from IAA conjugates. ILR1 has specificity for IAA-Leu whereas IAR3 prefers IAA-Ala. Maize (Zm) iaglu and Arabidopsis UGT84B1 esterify IAA to glucose; the enzymes that form the hydrolyse IAA-peptides have not been identified. IBA is likely to be converted to IAA-CoA in a peroxisomal process that parallels fatty acid β-oxidation to acetyl-CoA. IAA can be inactivated by oxidation (oxIAA) or by formation of non-hydrolysable conjugates (IAA-Asp and IAA-Glu). IAA-amino acid conjugates can be formed by members of the GH3/JAR1 family. OxIAA can be conjugated to hexose and IAA-Asp can be further oxidized. IAMTI can methylate IAA but whether this activates or inactivates IAA is not known. IBA and hydrolysable IAA conjugates are presumably derived from IAA; biosynthesis of these compounds may contribute to IAA inactivation. Formation and hydrolysis of IBA conjugates may also contribute to IAA homeostasis; the wheat (Ta) enzyme TaIAR3 hydrolyses IBA-Ala. (From Woodward and Bartel, 2005).

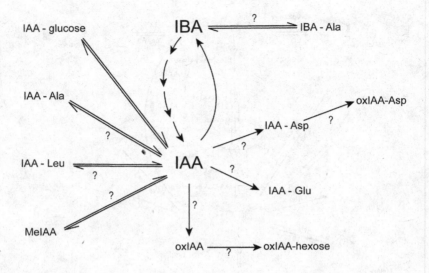

IAA - glucose

IBA

IBA - Ala ?

IAA - Ala

oxIAA-Asp

?

IAA - Asp ?

IAA - Leu

IAA

?

?

?

IAA - Glu

?

MeIAA

oxIAA

oxIAA-hexose

?

Reversible conjugates: IAA - glucose
(possible storage of IAA) IAA - Ala
IAA - Leu
Non-reversible conjugates: IAA - Asp
(destined for IAA degradation) IAA - Glu

**Fig. 6.** Reversible and non-reversible IAA conjugates. (According to Woodward and Bartel, 2005).

while others are not converted. Those that can be reversed to free IAA, such as IAA-glucose, IAA-Ala and IAA-Leu, may serve as IAA storage (e.g. in seeds), from where free IAA can be made readily available while keeping the level of free IAA at a height that is required by the plant tissue. These conjugates may also have roles in protecting IAA from peroxidative degradation and they may be useful in IAA transport. The IAA conjugates are abundant in plant tissues: in some estimates, about 90% of the IAA was in amide linkages and 10% in ester-linked conjugates and only about 1% was free IAA, but the levels of conjugates vary considerably and only low levels of IA-Ala, IAA-Asp, IAA-Glu and IAA-Leu were detected in seeds of Arabidopsis (Rampey *et al.*, 2004). It should be noted that IAA conjugates are not unique to angiosperms; they were revealed also in lower plants such as ferns, mosses and liverworts. As clearly shown in Fig. 6, some IAA conjugates are not reversible and thus do not serve as storage for free IAA. These are considered as destined for degradation, thus reducing IAA levels. The laboratory of B. Bartel intensively studied the genes encoding enzymes that cause cleavage of IAA conjugates, indicating that some conjugates, such as IAA-Ala have a role in addition to generating free IAA. In their review, Woodward and Bartel (2005) listed 21 genes that are implicated in IAA conjugate metabolism. Part of these genes encode enzymes for conjugation while others encode amidohydralases. Most were revealed in Arabidopsis but some were detected in monocots (wheat and maize).

There is a strong metabolic link between IAA and IBA. The latter is also considered a natural auxin, at least in some plants. Is IBA's activity as auxin based on its conversion to IAA? At least some experiments indicate such a possibility. The location of the IBA-to-IAA conversion is probably in peroxisomes, and this conversion is parallel to $\beta$-oxidation of fatty acids; defects in this oxidation reduces the energy supply of lipids in germinating seeds as well as renders the seedling resistant to high IBA; no high level of IAA is reached. Mutants that have defects in $\beta$-oxidation, or in peroxisome integrity, require sugar as energy during germination. From various experimental approaches, it appeared that some mutants that are peroxisome defective, and their IBA to IAA conversion is abolished, also have specific morphological defects in adventitious root formation. Hence, more information may provide a better understanding of the impact of IBA-to-IAA conversion and root patterning. It is possible that IBA itself has a role in root architecture. I would like to note that many years before the molecular genetics of IBA metabolism was clarified, horticulturists dipped shoot sections in IBA to improve their rooting. Practical wisdom can precede scientific knowledge... .

As indicated in Figs. 5 and 6, some IAA conjugates are reversible, while others, such as IAA-Asp and IA-Glu, are destined for degradation. Arabidopsis seedlings showed a low level of IAA-Asp and IAA-Glu hydrolyzation. Rather, IAA-Asp can be oxidized to oxIAA-Asp on the way to further degradation. As noted in the above mentioned figures, there are still a lot of question marks in the scheme for reversible and non-reversible auxin conjugation. However, the available information suggests that plants are equipped with mechanisms that tend to keep the level of free IAA at bay. Two additional phenomena are noteworthy. One is that the type of IAA conjugates that are formed after IAA challenge, are in part, dependent on the amino acid pool in the respective plant tissue. The other phenomenon is the reversible conversion of IAA to IBA. If, indeed, high levels of free IAA are toxic while IBA is less toxic, the conversion of IAA to IBA may have a similar consequence as the conjugation of IAA to Ala or to Leu.

## Specific Expression of IAA Synthesis Genes in Plant Organs

The spatial and temporal expression of *YUCCA* genes was intensively, studied in Arabidopsis (e.g. Zhao *et al.*, 2002; Cheng *et al.*, 2006, 2007). There are 11 genes in the YUC family in Arabidopsis, four of which were analyzed by the Cheng/Zhao team of the University of California at San Diego. The enzymes encoded in the *YUC* genes encode flavin monooxygenases that are active in the conversion of TAM to N-hydroxyl-TAM in the Trp-dependent IAA synthesis pathway. Four of the 11 *YUC* genes were used in detailed analyses. It was revealed that these four genes (*YUC1, YUC2, YUC4, YUC6*) are not ubiquitously expressed in the Arabidopsis plants. Thus, loss of function in specific genes can cause developmental phenotypes in specific plant organs. For example, *YUC1* and *YUC4* are strongly expressed in apical meristems and floral primordia. Moreover, *YUC1* expression is later localized to define groups of cells in the stamens and carpel while in mature flowers, this expression is shut down. A different expression pattern was revealed for *YUC2*. Also, there are typical expression patterns of *YUC1* and *YUC6* in developing leaves. On the other hand, *YUC6* was apparently not expressed at all in leaves. Different combination of YUC mutants lead to different phenotypic defects in plant architecture, pointing to the spatial and temporal specificities in the expression of the various *YUC* genes (and their homologues in *Petunia*). On the other hand, there is a redundancy by which these genes can at least partially "cover" each other, hence the four studied genes seem to have overlapping functions. Overall, it became very clear that the *YUC* genes have a major role in IAA biosynthesis. When the synthesis of IAA mediated by *YUC* genes is impaired (e.g. in *yuc1 yuc2 yuc6* plants), then the Trp-independent IAA biosynthesis and the bacterial gene *iaaM* can rescue the mutant Arabidopsis plants.

In a recent study, Cheng *et al.* (2007) investigated, in detail, the roles of several *YUC* genes as well as genes involved in the efflux and influx of auxin (out and into cells; *PIN1* and *AUX1*, respectively) in Arabidopsis (see next section on IAA transport). They found that the *YUC* genes as well as the combination of *YUC* genes with efflux and influx genes affect embryogenesis, leaf formation and flower development. The genes *YUC1* and *YUC4* were found to be expressed in discrete groups of embryo cells and their expression pattern overlaps the expression pattern of *YUC10* and *YUC11*. When all these four genes were mutated, the quadruple mutants failed to develop a hypocotyl and a root during germination. Also, the triple mutants of *yuc1 yuc4 pin1* failed to form leaves and flowers. When *YUC* mutants were combined with a mutated carrier gene (*aux1* for IAA influx), leaf and flower formation was compromised. These investigations thus clearly indicated that auxin synthesis mediated by the YUCCA flavin monooxygenase is essential, in Arabidopsis, for normal embryogenesis and postembryogenic organ formation.

## IAA Transport

In his review on auxin transport and its role in shaping the plant, Friml (2003) noted that "short distance activity" is related to the role of auxin as a *morphogen*. We should remember that the term morphogen was used (probably for the first time) by the British mathematician Alan Mathison Turing (1912–1954), when he used mathematical approaches to understand the differentiation of Hydra. Morphogens, by dictionary definition, are chemicals that regulate morphogenesis, but by biologists, morphogens are regarded as substances that form a concentration gradient, and can conceptionally be viewed as *flow-ing* substances. Turning to plants and their cells, we face a problematic issue. If we look at a plant cell, we can assume that this cell has the capacity to evaluate the level of auxin (i.e. of IAA) in its cytoplasm as well as in other intracellular locations (a capability that botanists are still lacking). However, in order to sense *flow*, the cell has to be able to sense the difference, in level, of auxin at its two poles on the line of the flow. Intuitively, we can assume that a flow can be sensed by the cell, if the influx facilitator of auxin, the AUX1 protein (see Swarup *et al.*, 2001) is reduced, causing a reduction of the auxin level in the cell, or if the efflux of auxin (by PIN1 and/or other proteins with a similar effect) is reduced — causing an increase in the level of auxin. Such changes in PIN1 or/and AUX1 can be imposed by investigators — but do plant cells use this approach to sense flow? As noted above, there is a technical problem. The sensitivities of the presently available methods for the analysis of IAA levels are too low to provide information on IAA levels in individual cells, let alone IAA levels in intracellular locations. For example, Friml *et al.* (2002) used the "microscale" technique of Edung *et al.* (1995) to evaluate IAA levels in root sections. The former had to extract fifty root sections, of 1 mm each, to obtain information on how many ng IAA were in a root section of 1 mm. The localization of proteins involved in either influx or in efflux of IAA can be performed by immuno-staining of tissue section, providing semi-quantitative estimates of the respective proteins. Thus, in a file of cells that have AUX1 protein on one side and PIN1 protein on the other side (of each cell), one can assume that auxin flows in and out of these cells. Also, when a small group of cells have in their periphery PIN proteins, one can assume that these cells constitute a source of auxin. Figure 7 demonstrates such an immuno-staining of a root tip of Arabidopsis. While this and similar demonstrations provide a reasonable explanation for IAA flow, the immuno-staining image furnishes a circumstantial evidence for flow. It shows that the "tools" for flow do exist, but is there also a flow? Probably there is.

There are great differences in the distance of the transport of auxin. In adult plants, where the source of auxin is presumed to be in the

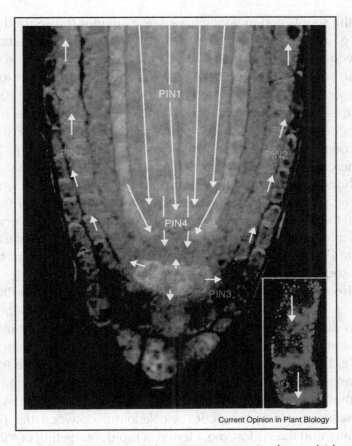

Current Opinion in Plant Biology

**Fig. 7.** Immunolocalization of the PIN3 protein in the Arabidopsis root apex and probable routes of polar auxin transport (white arrows). PIN3 (in green) is localized symmetrically in columella cells and apparently mediates lateral auxin distribution to all sides of the root cap. After the root is turned 90 degrees away from vertical (i.e. after a gravistimulus is applied), PIN3 rapidly relocates to the bottom side of the columella cells (inset), and thus probably regulates auxin flux to the lower side of the root. Auxin is further transported through lateral root cap and epidermis cells basipetally by a PIN2-dependent route (polar localization of PIN2 at the upper side of cells is depicted in blue). This basipetal transport also requires AUX1-dependent auxin influx. AUX1 is present in the same cells as PIN3 and PIN2. Auxin is supplied to the root cap by the PIN1- and PIN4-dependent acropetal route (which is depicted in white). (From Friml, 2003).

tips of the shoots and in young leaves, the auxin is transported a long way (e.g. in the phloem) and is causing apical dominance (reduced outgrowth of axilary shoots). The vascular tissues of plants provide a transport route all the way from the upper regions of shoots to close to the root tips. Clearly, in roots, the source of IAA and the location of the affected tissue, as the elongation zone, where the vascular tissue is being differentiated, are remote from each other. Still, the root has its own source of auxin, in the columella (green in Fig. 7) and from there the auxin is transported to a relatively small distance. An even shorter transport of IAA occurs in the early globular embryo. However, there may even be a zero distance of transport. This takes place in the formation of *Agrobacterium* tumors where the very same cells that initiate the tumor are "seduced" by the bacterial plasmid to integrate IAA synthesis genes into the plant genome and start to produce a source of auxin in the tumor cells.

As also shown in Fig. 7, Arabidopsis has genes for several PIN proteins (e.g. PIN1, PIN2, PIN3 and PIN4). The various PIN proteins seem to differ in their role in different cell types. The exact intracellular location of PIN protein is an elaborated issue. Hence, for example, PIN3 changes its location in root cells in response to gravity changes, that are sensed by starch-containing organelles (statolithes). It was also found that PIN1 and PIN3 can cycle in membrane vesicles, along the actin cytoskeleton, between the plasma membrane and the endosome. Hence, a rapid retargeting of PIN protein is replacing the much more elaborated process of protein degradation followed by new protein synthesis targeted to new sites. How is the sedimentation of statolithes connected to the relocation of PIN3? There is no clear answer, yet, to this question.

A subsequent study of Friml, Jürgens and associates (Paciorek *et al.*, 2005) indicated that auxin regulates PIN abundance and activity at the cell surface, providing a mechanism of feedback regulation of auxin transport.

Various PIN proteins were found to interact with the expression of another gene, *PLETHORA* (*PLT*). The study of Blilou *et al.* (2005) indicated that specific PIN proteins restrict *PLT* expression in the basal end of the young Arabidopsis embryo, initiating the root primordium of this embryo. In parallel, *PLTs* maintain *PIN* transcription and by that, stabilize the position of the distal stem cell niche. At a distance from the auxin maximum *PLTs* maintain PIN3 and PIN7, which reinforce provascular acropetal auxin flux. By that, a reflux-loop is established that controls auxin transport. But other auxin regulation is not excluded; local biosyntheses and local degradation may also have roles in the very young root.

A detailed review on the *PIN* genes in Arabidopsis as well as in other plant species (dicots and monocots) was published by Paponov *et al.* (2005). Eight different *PIN* genes were revealed in Arabidopsis that differ considerably in their structure and are located in four of the Arabidopsis chromosomes. They all encode transmembrane domains. The authors also constituted phylogenetic trees for *PIN* genes from Arabidopsis, soybean, potato, alfalfa, rice and wheat. However, a major question was not yet answered: are the PIN proteins the actual *transporters* of auxin out of the cell or are they *regulators* of this transport? So the authors gave PIN protein a noncommuting name: "auxin efflux facilitators." Blakeslee *et al.* (2005) added another component to the mechanism of IAA efflux: plant orthologues of the mammalian multi drug-resistance/P-glycoproteins (MDR-PGP). The latter apparently interact with PIN to perform the efflux. Blakeslee *et al.* suggested that the mechanism of auxin efflux from long and mature cells may differ from the mechanism in small meristematic cells. The authors argued that in the latter cells, there may be a diffusive re-entry of auxin, impeding polar transport. Thus, an energy-dependent efflux mechanism, provided by PGPs, could counteract the re-diffusion. The p-glycoprotein (*PGP*) subfamily of the ATP-binding cassette (*ABC*) transporter genes deserve special attention. The *ABC* genes were revealed in Arabidopsis when the genome was fully sequenced: more than 120 putative *ABC* transporter genes were revealed (compared to much less in non-plant multicellular organisms). Since the mammalian orthologues of *PGP* were found to be associated with multiple drug resistance, plant (Arabidopsis) *PGPs* (22 genes) were initially also presumed to function in detoxification. But soon, the *PGPs* were found to have developmental roles. Angus Murphy and associates focused on the involvement of *PGP* genes in auxin transport (e.g. Blakeslee *et al.*, 2005; Geisler and Murphy, 2006; Bandyopadhyay *et al.*, 2007). Vieten *et al.* (2007) came up with a model for PGP-, PIN- and AUX1/LAX-mediated auxin transport in plants (Fig. 8), in long cells

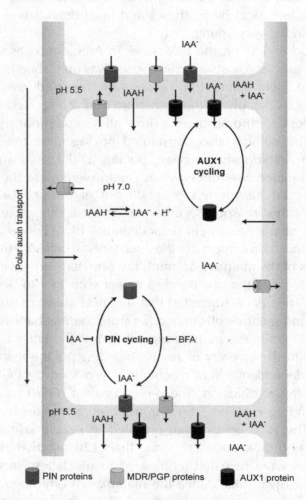

**Fig. 8.** Cellular model for polar, cell-to-cell auxin transport. According to the chemiosmotic hypothesis, a pH gradient across the plasma membrane leads to the accumulation of IAA in the cell. A higher pH inside the cell causes auxin molecules (IAAH) to dissociate, making them unable to pass passively

of mature tissue and in small meristematic cells, taking into account also diffusion of auxin, into the cells and outside of the cells. They also listed PIN, PGP and AUX1 proteins according to the tissues and the locations where they were detected (Table 1). Bandyopadhyay *et al.* (2007) suggested the following scheme for the auxin transport in Arabidopsis hypocotyl tissue. Accordingly, PGP19 and PIN1 function coordinated in auxin transport through the hypocotyl. PIN1 transports auxin primarily through the vascular parenchyma and PGP19 transmits the auxin through the bundle sheath cells, while PIN1 exhibits basal localization in non-apical cells. PGP19 is present all around bundle sheath cells, but is more abundant at the bottom of the cells. In addition to transporting auxin polarly through bundle sheath cells, PGP19 appears to enhance polar transport by preventing lipophilic auxin uptake from the vascular tissue. Expanded expression and abundance of PGP19 and PIN1 in apical tissues suggests that the two proteins interact to enhance loading of apical auxin into the vascular stream. Enhanced auxin transport capacity observed in these tissues is consistent with this function. Co-localization of PIN3 and PGP19 on the centripetal surface of bundle sheath cells suggests that these proteins interact to re-direct auxin back to the vascular stream. In the dark-grown upper hypocotyls, PIN1 distribution expands into bundle sheath cells, suggesting a direct interaction between PGP19 and PIN1 in these tissues. In (the mutant) *pin1*, auxin transport is reduced approximately 30% in light-grown seedlings, but

---

**Fig. 8.** (*Continued*) back through the membrane. Hence, the IAA⁻ becomes trapped inside the cell. Auxin efflux carriers (PINs, some MDR/PGPs) are needed to transport auxin out of the cell. In addition to the simple diffusion of non-dissociated IAA molecules into the cell, auxin influx carriers (AUX1) transport auxin anions (IAA⁻) into the cell. The polar subcellular localization of PIN proteins is important for directional auxin transport, and is accompanied by constitutive endocytic cycling of the PIN proteins. Auxin itself inhibits endocytosis of PINs, increasing their levels at the cell surface. The inhibitor of vesicle trafficking, BFA, represses the exocytosis of PINs but not of AUX1. The subcellular dynamics of PGP proteins is as yet unclear. (From Vieten *et al.*, 2007).

**Table 1.   Tissue-Specific and Subcellular Localization of PINs, PGPs and AUX1**

| Tissue | Protein | Localization |
|---|---|---|
| Cotyledons | PIN3 | Bundle sheath, polar |
| | PIN7 | Epidermis, apolar |
| | PGP19 | Bundle sheath, polar and apolar; epidermis, apolar |
| Shoot apex | AUX1 | Epidermis, apolar/polar (top) |
| | PIN1 | Epidermis, apolar/polar (top) |
| | PGP1 | All tissues except primary meristem, apolar |
| | PGP19 | Provasculature, apolar; cortical, bundle sheath, apolar/polar (bottom) |
| Hypocotyl | PIN1 | Vascular parenchyma, polar (bottom) (light) |
| | | Vascular parenchyma, bundle sheath, cortex, polar (bottom) (dark) |
| | PIN3 | Bundle sheath, polar (lateral and bottom) |
| | PIN7 | Epidermis, apolar |
| | PGP19 | Bundle sheath, polar/apolar (light) |
| | | Bundle sheath, cortex, polar/apolar (dark) |
| Mature root | PIN1 | Bundle sheath, polar (bottom) |
| (>2 mm from | PGP1 | Cortex, epidermis, polar (top) |
| root tip) | PGP4 | Epidermis, polar (bottom) |
| | PGP19 | Endodermis, apolar; vascular bundle, polar (bottom) |
| CEZ/DEZ | PIN1 | Vascular bundle, polar (bottom) |
| boundary | PGP4 | Epidermis, polar (top) |
| (0.30–0.45 mm) | PGP19 | Endodermis, pericycle, apolar; vascular bundle, polar (bottom) |
| DEZ | AUX1 | Epidermis, polar (top) |
| (0.15–0.30 mm) | PIN1 | Vascular bundle, cortex, epidermis, polar (bottom) |
| | PIN2 | Epidermis, polar (lateral and top) |
| | PIN3 | Pericycle, polar (lateral) |
| | PIN7 | Vascular bundle, polar (lateral and bottom) |
| | PGP1 | Cortex, epidermis, polar (bottom) |
| | PGP4 | Epidermis, polar (top) |
| | PGP19 | Endodermis, pericycle, apolar; vascular bundle, polar (bottom); cortex, polar (top); epidermis, polar |

*(Continued)*

**Table 1.** (*Continued*)

| Tissue | Protein | Localization |
|--------|---------|--------------|
| Root apex | AUX1 | Columella, LRC, protophloem, polar/apolar |
| (tip 0.15 mm) | PIN1 | Provasculature, polar (bottom) |
| | PIN2 | Epidermis, polar (lateral and top); cortex, polar (bottom) |
| | PIN3 | Columella apolar |
| | PIN4 | QC, surrounding cells, apolar; provasculature, polar (bottom) |
| | PIN7 | Columella, apolar |
| | PGP1 | Stele, apolar; epidermis (bottom); cortex (top) |
| | PGP4 | LRC, apolar |

CEZ, central elongation zone; DEZ, distal elongation zone; LRC, lateral root cap; QC, quiescent centre. (From Bandyopadhyay *et al.*, 2007).

is reduced by an additional 10% in dark-grown seedlings, whereas the *pgp19* (mutant) exhibits a 60% decrease in auxin transport in both light- and dark-grown seedlings. Further, in *pin1*, there is a small but significant reduction in transport specificity as seen by a slight increase in bundle sheath transport. This is consistent with the increased substrate specificity exhibited when PGP19 and PIN1 are co-expressed in heterologous systems. In short, interaction between PIN1 and PGP19 seems to enhance efflux of auxin.

Ottoline Leyser, of the University of York in England, whom we shall meet, while handling suppression of branching, started her review on auxin distribution and plant patterning (Leyser, 2005) with a theological question: "How many Angels can dance on the point of a PIN?" These angels are regarded, in Christian theology, as messengers and as codified, in the 5th century by St. Dionysius, the Areopagite, there are several kinds of angels, in descending hierarchy; the upper ones are Seraphim and Cherubim. Some of these angels have wings (e.g. Cherubim), others lack wings. Leyser attributed the "messenger" task to auxin (IAA). If so, there are two kinds of "angels," the "winged Cherubim" that are protonated IAA and can "fly," by diffusion, into cells and the "wingless angels" that are ionized and require an influx carrier in order to enter a plant cell. As the auxin is ionized in the cell cytosplasm, due to higher pH, this "wingless" auxin requires an active efflux complex to exit the cell as shown schematically in Figs. 9(a) and 9(b). Figure 9(b) provides the relatively simple transports of IAA in the tip of a young root. Note that it appears that auxin in the root can be "recycled" from the "elongation zone" to the polar transport stream, down the center of the root, towards the tip, at the stem cells of the root (above the columella root cap, where there is a "fountain" of auxin biosynthesis). Thus, an auxin maximum is established from where the auxin flow is directed up, again, at the root periphery in the "cell division zone". The efflux is an active, carrier-dependent process facilitated by PIN protein and

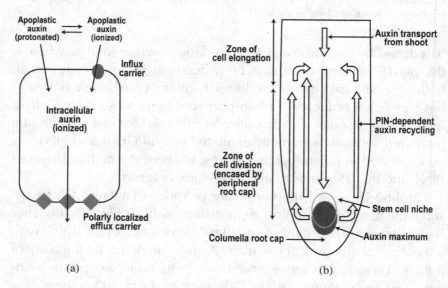

**Fig. 9.** Auxin movement through a cell and a root. (a) Auxin movement through plant cells. Auxin entry into plant cells is mediated by diffusion or active uptake via an uptake carrier. Auxin export depends on an active efflux carrier. (b) Auxin movement in a root. Auxin accumulates in a specific pattern in root tips and is required for cell specification around the stem cell niche. Auxin enters the auxin maximum from above and exits it through the root cap. Auxin from the root cap is channeled back up the root and can be recycled into the auxin maximum from above. (From Leyser, 2005).

at least in some cases by PGP proteins, as noted above. The PIN proteins cycle between the plasma membrane and intracellular vesicular compartment. On the plasma membranes, PIN proteins are commonly located polarly to a specific cell surface according to the direction of the auxin flow (Fig. 9a). While the efflux facilitators (PIN, PGP) are always required to drive auxin outside the cell, the influx facilitator, AUX1, is apparently critical only when rapid uptake is required. In the latter case, the dissipation of the auxin in the apoplast, by diffusion, is prevented. The acropetal/basipetal movement of auxin in the root has its origin in the early embryo and is further elaborated in the young seedling, when this movement affects tissue differentiation, as shall be noted below when the plant organs shall be handled. In animal systems, the *gradient* of morphogen is decisive in differentiation along the *flow*. While the movement of auxin in plants has regulators (e.g. influx and efflux facilitators), thus the auxin movement is much less dependent on the source of auxin and its diffusion. Also, auxin can originate from more than one source and can be recycled as in the root (Fig. 9b) and may "pave" its own "road" (in leaf venation and the induction of vascular tissue) as reviewed by Berleth and Sachs (2001) (see also Galun, 2007). "Paving its own way" is probably executed by the ability of auxin to facilitate PIN formation. This "road-pavement" takes place when the polar transport from the shoot apex towards the root is interrupted by wounding. Then, auxin accumulates at the upper edge of the cut, initiating cell division to bridge the cut and reestablish polar transport. Animal morphogens generally lack such characteristics. Plants use the efflux facilitators, such as PIN protein, to direct auxin flow to a specific direction. Moreover, at least in some tissues, the efflux is driven by specific PIN proteins. For example, PIN1 directs auxin in the shoot apical meristem toward a small location, at the flank of the apical dome, so that surrounding locations are depleted of auxin. The PIN7 is active in the very young embryo where the auxin level in the small apical cell is increased while the auxin in the basal cell is decreased. PIN7 is found in the upper side of the basal cell, apparently "pumping" auxin from the basal to the apical cell. On the other hand, at least some PIN proteins can compensate each other so that a mutation in one specific *PIN* gene may not result in a strong phenotype.

Leyser (2006) compared the system of auxin flows in plants to a complex and dynamic road network through which the traffic (the auxin) flows. She suggested two central challenges for auxin biologists: to decipher the auxin "highway code" and to determine how the behavior of auxin in this system is read out to drive specific patterning and growth processes. Clearly, read-outs from auxin distribution network require local *sensing* of auxin. For this sensing, auxin signal transduction is a key issue. This signal transduction will be handled in a section below.

Kramer and Bennett (2006) focused their review on auxin transport in plant roots. They updated the information on transport by diffusion versus transport by the influx and efflux facilitating proteins (AUX1, PGP and PIN). As mentioned above, the main plant auxin (IAA) can have either a protonated form (IAAH) or an anionic form (IAA$^-$). Being a weak acid with a dissociation constant of pH 4.8, IAA will be mostly (99.4%) in the IAA$^-$ form in neutral or basic pH, but in acidic compartment (at pH 2), IAA will be mostly (99.8%) protonated (IAAH). Because the apoplast space in Arabidopsis is commonly about pH $\simeq$ 5.8, about 20% of this apoplast-IAA will be protonated. The IAAH can diffuse into plant cells through their lipid plasma membrane but at a low speed and the influx by specific protein facilitator seems to be, in epidermal cells of the root elongation zone cells, much greater than diffusion. The estimated rate of facilitator-mediated transfer/diffusion transfer, in these cells, was about 15. The importance of the influx facilitator AUX1 is revealed by *aux1* mutants where the influx is drastically reduced. This led to an estimate that in these root cells, the AUX1 import is about 10 times more efficient than diffusion. Thus, Kramer and Bennett (2006) suggested that influx diffusion in these root cells is only "supplemental" for root developmental processes. It should be noted that the facilitator of influx of IAA causes also influx of 2,4-D but not of NAA. As for the aforementioned PGP, it is not yet clear if it can also play a role in influx of IAA (the role in efflux of IAA by proteins, encoded in *AtPGP1* and *AtPGP19*, was established). It is also noteworthy that the main route of IAA transport appears to change with the maturation of the Arabidopsis seedling (Ljung *et al.*, 2005). In the root of the very young seedling, polar auxin transport predominates, while after 5 days, phloem-mediated transport takes over. There are large zones of PIN1 and PGP1 expression in the roots stele, while the expression of the influx facilitator AUX1 is limited to two files of protophloem cells that connect mature phloem with the center of the root apex. This suggests that the protophloem files unload auxin distally from the conducting phloem. Not only PIN1 but also four additional PIN proteins (e.g., PIN2, PIN3, PIN4 and PIN7) participate in the fine-tuning of the root apical auxin gradient. This fine-tuning affects the expression of developmental determinants, such as the PLETHORA transcription factor. There is thus a probably additional interaction involved in the stabilization of auxin gradient at the root tip.

In a recent review, J. Friml and associates (Vieten *et al.*, 2007) were wondering how a *single hormone* (IAA) can be involved in a plethora of patterning schemes in various plant organs and responses to environmental effectors (e.g., root meristem maintenance, vascular tissue differentiation, hypocotyl and root elongation, apical hook formation, apical dominance, fruit ripening, growth responses to environmental stimuli, as well as several others). Vieten *et al.* (2007) looked for an answer in the mechanisms of auxin transport, and the "position awareness" of the affected cells. By "position awareness" I mean, for example, that cells in the flanks of the shoot apical dome sense their position and are "ready" to respond (directly or indirectly) to IAA in a very specific manner that is different from the response of cells at the periphery of young leaves, where vascular tissue is required for veination of these leaves. In the former case, IAA will be involved in lateral outgrowth while in the latter case, IAA will be involved in vein formation. The "position awareness" cells should also gain the capability to sponge IAA or to remove it. For this capability, an effective transport system is required. Transport can be long range (i.e. via the phloem) or transport between adjacent cells. Vieten *et al.* (2007) focused on the latter transport. As already noted above, the plasma membrane location, where PIN protein is located, is providing the *direction* of IAA efflux. In *gnom* mutant embryos, the coordinated polar localization of PIN1 protein is not established, resulting in embryos with poorly defined apical basal axes and lack of bilateral symmetry. *GNOM* encodes a factor which mediates formation of exocytic vesicles at endosomes responsible for carrying PIN proteins. We should recall that PIN proteins are capable of "constitutive cycling" between endosomes and plasma membrane. Hence, the GNOM protein probably has a major role in placing PINs at the required site on the plasma membrane. GNOM is not the only factor that affects the correct localization of PIN proteins. Moreover, it could be that each factor affects specific PIN proteins (i.e. PIN2 but not PIN1). PIN2 is involved in the gravitropic response. When roots of Arabidopsis are turned sideways, there is an increase of auxin concentration in the epidermis and the lateral root cap on the new lower side, and a decrease on the new upper side. This decrease coincides with a decrease of PIN2 on the upper side. It seems that the decrease of PIN2 is caused by ubiquitination. One additional factor identified to mediate decisions about the polarity of PIN targeting is the Ser/Thr kinase PINOID (PID). The *pid* mutants have severe phenotypes: bare inflorescence shoot with much reduced lateral outgrowths. Overexpression of PID also has severe morphological consequences as primary roots collapse, due, probably to auxin depletion. It seems that PID determines apical to basal localization of PIN in a given cell, and by that can control the direction of IAA flow. It was pointed out above that auxin is probably capable to *pave* its own road. Due to an indication that auxin is involved in PIN expression and possibly also its location on the plasma membrane, Vieten *et al.* (2007) suggested that auxin could influence both the rate (paving its own road) and the directionality of its own flow.

Boutte *et al.* (2007) of the Swedish Agricultural University in Umeå also updated the advances towards an understanding of cellular auxin transport activities of presumptive carrier proteins, as well as the intercellular signaling of auxin-dependent cell and tissue polarity. One of the questions that was recently asked was whether the auxin influx and efflux proteins (AUX1, PIN and PGP) can mediate the auxin transfer only by complexing with other (not yet revealed) proteins or they do not require such "helpers." To answer such questions, the influx and efflux facilitators were introduced into heterologous cells, such as yeast, mammalian cells and *Xenopus* oocytes. For AUX1, a simple answer was obtained. This influx facilitator can import auxin across the plasma membrane of yeast and human HeLa cells. Hence, AUX1 can be regarded as a *bona fide* auxin influx carrier. Surprisingly, the proteins PGP1, PGP19 and PGP4 can also cause influx of auxin into HeLa cells. Such an influx activity was attributed in the past to PGP4 but not to PGP1 and PGP19 that were considered efflux facilitators. Do PGP1 and PGP19 reverse the direction of their transport in the heterologous cells? Clearly, in heterologous cells, each of the four presumed transport facilitators, PGP1, PGP19, PIN2 and PIN7, act as efflux carrier. However, this does not exclude the possibility that in Arabidopsis cells (or in some of them), *in planta* PIN proteins require a complex (that includes PGP) for efficient efflux activity. As for PIN2 and root gravitropism, it was found that a specific directionality of polar PIN2 localization is required for the realization of the bending of the root. Not only transport facilitators such as PIN1 and PIN2 can affect IAA transport, but also IAA itself can affect the intracellular localization of these facilitators: IAA can affect its own transfer by preventing endocytosis of PIN1 and PIN2 and retain them in the plasma membrane. However, the endocytosis of PIN proteins can be mediated by other cellular factors. Moreover, evidence exists that at least PIN2 can undergo ubiquitination. It is not only the efflux facilitators that can move (in both ways) from the plasma membrane to the endocytic compartment; such a movement was revealed also for AUX1, an influx facilitator. However, the route of this movement by AUX1 differs from the route of PIN proteins.

A rather elaborate issue is the involvement of IAA transfer and root hair formation in the root of young seedlings. Root hair formation was studied in detail in Arabidopsis (see Galun, 2007). Files of epidermal cells have the potential to develop hairs. Only those files of which the epidermal cells face two (rather than one) cortex cells will produce hairs. But not all the cells in such a file will produce hairs. Also, the distance from the root tip to the hair initiation can be changed. Triple mutants of the genes *AUX1*, *EIN2* and *GNOM* (i.e. *aux1*, *ein2*, *gnom*) perturb the coordination of hair positioning and also perturb the auxin level gradient. Precise localization of applied IAA (by the application of an IAA-containing sephadex bead) will strongly promote hair formation near the IAA application site. The actual control of root hair localization is probably more complicated because another family of proteins, Rho-of-plant (ROP2 and ROP4) also seem to be involved. We shall not elaborate this subject but shall point out that an external application of IAA (by IAA-coated sephadex beads) can change the polarity of PIN proteins in cortical, endodermal and vascular cells — in line of the above mentioned presumption that IAA can affect its own transport. An IAA effect on the polarity of PIN, that is followed by PIN regulating the direction of IAA flow, is apparently also happening in the early vascular differentiation in young leaves as well as in the replacement of the vasculature (e.g. in pea epicotyls) that were cut by wounding. The latter system was morphologically and histologically described in the classical work of the late Tsvi Sachs (Sachs, 1991) of the Hebrew University of Jerusalem. There is another link between efflux facilitators and auxin flow. Under regular conditions, PIN4 is located at the basal plasma membrane (PM) of the quiescent center (QC) cells. But after ablating (by a laser beam) the QC, and in mutants termed *shr* and *scr*, the location of the PIN4 is reversed and this protein is relocated to the PM, away from the ablated QC. Whatever the measure of this relocation of PIN4, it is clear that PIN4 can undergo such a relocation in certain conditions and consequently reverse the flow of auxin.

## Perception, Signaling and Ubiquitination

As an introduction to the perception and signaling of auxin (we shall focus on IAA), let us start from the "end" and work our way back to the start (the perception). Albeit we should recall that the "end" is still enigmatic. Meaning, we do not know how specific genes impose a three-dimensional structure (an organ) in a plant (or in any other multicellular organism). We do have considerable information on how genes *change* the differentiation of an organ. This apparently strange statement will become clear by an example that was already mentioned in the Introduction. Trichomes are formed on the surface of leaves in Arabidopsis and in many other plants. We have information on genes that suppress or promote the differentiation of trichomes from epidermal cells. Also, there are known genes that will affect the branching of trichomes. However, we have no knowledge on how the conceptual blueprint of a normal mature trichome is translated, by the plant's genetic information, that is stored in the plant's genome (in the form of linear sequences of nucleotides), to create the typical three-dimensional trichome. In other words, when we speak of genes that have a role in patterning, we mean a role in *amending* (changing) that pattern rather than fundamentally shaping plant organs. These considerations apply also to those genes, in which the level of expression is regulated by auxin and the changes in this level have morphogenetic impact. These are the auxin responsive genes (ARG). Auxins were found to have many effects in plant patterning such as in affecting lateral outgrowth from the shoot, initiation of secondary roots from the main root, causing vein production in leaves, differentiation of vascular tissue, etc. The impact on these phenomena is mediated by specific ARGs as shall be mentioned in Part II of this book, where the roles of phytohormones in shaping specific plant organs will be discussed. This means that the expression of these genes (commonly encoding proteins) affects these phenomena. Clearly, the level of expression is regulated by various means (at the levels of chromatin, DNA, messenger RNA and proteins, as discussed in some detail in Chapter 5 of Galun, 2007). One important mean of regulating the expressions of ARGs is by controlling their transcription that is mediated by transcription factors, termed auxin responsive factors (ARF) that bind to the promoters of ARGs. Although the regulation of ARGs transcription is an important mean of controlling the expression of these genes, and much work was conducted with this regulation, there are other regulatory mechanisms that control the expression of ARGs. Many ARGs were characterized in Arabidopsis and other angiosperms, and most of them have a similar basic composition: (1) an amino-terminal DNA-binding domain (DBD); (2) a middle region that functions as an activation domain (AD) or as a repression domain (RD); (3) a carboxy-terminal dimerization domain (CTD). The DBD binds to a consensus DNA sequence: TGTCTC of the promoter region of the *ARGs*.

Guilfoyle and Hagen (2007) updated the research on ARFs in Arabidopsis and other plants. The details are outside the scope of this book but some of this information shall be noted here. Twenty-two full-length ARFs and one partial-length ARF were found in Arabidopsis and the respective genes are distributed in all five chromosomes. In rice, there are 25 ARF loci distributed in 10 out of 12 chromosomes. The expression of the different ARFs in Arabidopsis is apparently different in different organs. For example, the ARFs of chromosome 1 appear to be restricted in their expression to embryogenesis/seed development. The *ARF1* gene was found to be expressed in developing flowers and *ARF2* in developing floral organs. Other organ-specific expressions were revealed for *ARF3*, *ARF4*, *ARF5*, *ARF6*, *ARF7*, *ARF8*, *ARF12*, *ARF16* and *ARF19*. Environmental conditions, light and even hormone signals, do affect the expression of some Arabidopsis and rice *ARF* genes. The level of expression of the *ARF* genes (i.e. the level of resulting protein) is also regulated, at least in some of the genes, by RNA silencing (see Galun, 2005, for a detailed discussion). Thus, for example, *ARF6* and *ARF8* are targets for miR167 and *ARF10*, *ARF16* and *ARF17* are targets for miR160. Other *ARFs*, such as *ARF3* and *ARF4* are targeted by *trans*-acting-small interfering RNA (ta-siRNA). The various regulations have different effects; for example, the regulation of *ARF6* and *ARF8* has effects on anther and ovule development. *ARFs* were mutated by several methods, such as T-DNA insertion in the *ARF* genes. Such mutagenesis yielded 18 or more *ARF* mutations. The respective plants had typical altered phenotypes. We face the situation that ARGs are regulated by *ARFs* but the expression of the *ARF* genes themselves is also regulated and can be changed by several mechanisms.... Complicated? Yes, it probably is, but we shall see that the further IAA signaling is far more complicated. The 33rd US president, Harry S. Truman (1884–1972) in an address to fellow politicians said: "Those who cannot endure the heat, should not enter the kitchen." In our case, those who shunt from complicated biological systems, should not study the role of phytohormones on plant patterning.... .

Double *ARF* mutants have frequently much stronger phenotypes than single mutants. This was revealed for *arf1 arf2*, for *arf3 arf4* and *arf5 arf7*. The *arf5 arf7* are severely defective in embryo patterning and in vasculature. Also, single mutants *arf6* and *arf8* show delayed flower maturation and reduced fertility but the double mutant plants *arf6 arf8* have early arrest of flower formation, before bud opening, and are completely infertile (Nagpal *et al.*, 2005). Moreover, no phenotypes were revealed in either *arf10* or *arf16* but the double mutant *arf10 arf16* has a severe root cap defect. These stronger phenotypes of double mutants suggest that some of the *ARF* genes are at least partially redundant. Let us turn to the *ARGs* and how they can be identified. By *ARG*, we define a gene that in response to an increased level of auxin will change the patterning of a plant tissue or a plant organ. We commonly mean the *terminal* link of genes between the perception of the auxin and the morphogenetic manifestation. One indication that a gene is indeed an *ARG*, is that in the DNA of its promoter region is a sequence of nucleotides that have the potential to bind regulatory proteins of the ARF family; hence, the AUXRE sequences TGTCTC or GAGACA. There is a method to test whether or not a given *ARG* is itself the target for an ARF. One measures the transcript level in the presence and in the absence of the proper ARF and repeats these measurements with and without an inhibitor of protein synthesis. When the transcript is changed (usually increased) even without protein synthesis, this indicates that the changed transcript is from the target *ARG*. If for a change of transcript level protein synthesis is required, the analyzed gene is not the terminal link in a chain of genes that is regulated by the ARF used in such a test.

The review of Guilfoyle and Hagen (2007) lists many analyses on the roles of Arabidopsis ARFs in patterning. Many of these focused on ARF7 alone or on ARF7 in combination with one other ARF (e.g. ARF19). The details will not be presented here but it should be noted that ARFs commonly activated *ARG* genes but in some cases, ARFs were found to repress *ARGs*. It is also noteworthy that a given *ARG* may be a target of more than one ARF. In the case of the *GH3* genes (involved in conjugating IAA to Ala and

Leu, see Fig. 5), there are indications for the three ARF activators and one ARF repressor.

As we are going backwards in looking at the auxin signal transduction, we should now know that the ARFs are *not* the direct targets of auxin. For auxin response, the ARF has to interact with the AUX/IAA repressor. This interaction is most probably through the CTD domain (at the carboxy-terminal) of the ARF rather than with either the amino-terminal domain (DBD) or the AD or RD domains. However, what affects the association or the dissociation of ARFs from ARGs? In most (or all?) cases, these are proteins termed AUX/IAA repressors. The latter can form a complex with the ARFs, on the promoters of *ARGs* and consequently repress the transcription of the *ARGs*. But the AUX/IAA repressors are labile: they can be degraded by a ubiquitination process that requires binding of IAA to the TIR1 protein of the SCF$^{TIR1}$ complex, as shall be narrated below. There is one further issue that should be remembered before we leave the interactions between *ARG* genes, ARFs and AUX/IAA suppressors, and it is the following: The family of ARF proteins is large, even in Arabidopsis, and the various family members have to compete among themselves for a target on an *ARG*. There is also a large family of AUX/IAA repressors (29 family members were revealed in Arabidopsis). It was estimated that there are, in Arabidopsis, five ARF activators that could interact with 29 AUX/IAA repressors. Again, there may be a competition among the AUX/IAA repressors for this interaction.

The molecular mechanisms of the activation and repression of *ARG* genes by ARFs and the repression imposed by AUX/IAA is still not clear. On the other hand, there is evidence for "co-regulators" that also participate in regulating the transcription level of *ARG* genes. One of these is the PICKLE (PKL) protein that is predicted to be a component of an ATP-dependent chromatin-remoduling complex, involved in histone deacetylation (causing the reduction of the transcription from the deacetylated region). Another protein, SEUSS (SEU) appears to function by directly interacting with an ARF (e.g. ARF3) and promotes floral organ development.

After dealing with the AUX/IAA repressor/ARF/ARG interaction, we can move "upstream" towards the ubiquitination and the perception of auxin (which *in planta* is IAA and in some cases IBA, only there is very little information on IBA perception and ubiquitination).

The overall picture, after neglecting some exceptions, is that when the level of IAA is low, the AUX/IAA complexed with the ARF represses the transcription of the *ARG*. However, when the level of IAA is increased, the AUX/IAA repressor protein (that commonly exists in low levels), is degraded, enabling the free ARF to promote the transcription from the *ARG* gene.

Now we can ask how the increased level of IAA is perceived. To answer this question, we have to look again at the SCF$^{TIR1}$ complex, this complex was mentioned in the section of ubiquitination in the Introduction. In brief, the complex has the following components: (1) an $\underline{S}$kp1 protein; (2) a $\underline{C}$ullin protein; (3) an $\underline{F}$-box protein that in the IAA ubiquitination process is termed TIR1. These three components comprise the SCF$^{TIR1}$. There are two additional proteins that complex with the Cullin: RUB1 and Rbx1.

A breakthrough in our understanding of the IAA signal transduction was the verification that the TIR1 F-box is a receptor of IAA. This verification was published, sequentially, in two articles in *Nature* (Dharmasiri *et al.*, 2005a; Kapinski and Leyser, 2005). The two articles had the same title (only Kapinski and Leyser added "Arabidopsis" to their title). Both teams were careful with the phrasing of their title and claimed that the F-box protein TIR1 is *an* auxin receptor (not *the* auxin receptor). By an *in vitro* assay, Dharmasiri *et al.* (2005a,b) found that the interaction between TIR1 and AUX/IAA proteins does not require stable modification of either protein. Instead, auxin promotes AUX/IAA-SCF$^{TIR1}$ through its TIR1. Also, TIR1 binds to AUX/IAA proteins in an auxin-dependent manner. The study of Kapinski and Leyser likewise provided evidence that there is a direct binding of auxin to TIR1, showing that TIR1 is an auxin receptor that mediates the transcriptional response to auxin.

An important further step towards understanding the regulation of plant development imposed by auxin was made by a team of eight investigators from the University of Cambridge (Tan *et al.*, 2007). They described the crystal structure of a TIR1 complex and revealed how auxins fit into a "pocket" of TIR1 and stabilize the binding of the

AUX/IAA repressor to this pocket. The AUX/IAA repressors have four conserved domains. One of these domains has an amino acid motif of Gly-Trp-Pro-Pro-Val (GWPPV). This domain is recognized by the TIR1. However, this "recognition" is not sufficient for binding. To enhance the binding (or stabilize it), auxin is recruited. The study of Tan *et al.* (2007) was based on crystalizing a TIR1-ASK1 (adaptor) complex. They found that the complex had a mushroom shape with the Leu-rich-repeat domain of the TIR1 forming a cap. A pocket on the top of the Leu-rich-repeat domain functions in both the auxin binding and the AUX/IAA binding. The auxin occupies the lower part of the pocket and the AUX/IAA is above it. The GWPPV motif of AUX/IAA is packed directly against the auxin and covers the auxin binding site. This suggests that the auxin is trapped in the pocket until the AUX/IAA is removed from it. The presence of auxin at the bottom of the pocket probably also assures the binding of AUX/IAA. The investigators tested the binding of three auxins to TIR1: the natural auxin, IAA and the two synthetic auxins 1-NAA and 2,4-D. All these three bound auxins enhanced the recruitment of AUX/IAA to TIR1, but the affinity of IAA was higher than the affinities of the synthetic auxins. One may ask, then, why does at least in some cases, 2,4-D have a stronger auxin effect than IAA? The answer, by the present author, is that it probably results from the quicker degradation of IAA in the plant, than 2,4-D.

The study of the crystal structures performed by Tan *et al.* (2007) revealed that an inositol hexakisphosphate ($InsP_6$) is tightly bound to TIR1 and the association of $InsP_6$ may be essential for auxin binding and the function of the receptor.

The studies on the perception of auxin, its role in the ubiquitination of the transcription inhibitors (e.g. AUX/IAA) and the impact of these issues on plant patterning are being performed currently by several research teams. It is expected that in future, research will focus on very specific ARGs in which the coding sequence as well as the sequence of the promoter region are known, and the impact of this ARG on morphogenesis is direct and well characterized.

There was a fundamental question: How can a very specific protein, such as one of the AUX/IAA proteins, be degraded by a process

**A**

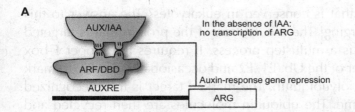

In the absence of IAA:
no transcription of ARG

Auxin-response gene repression

**B**

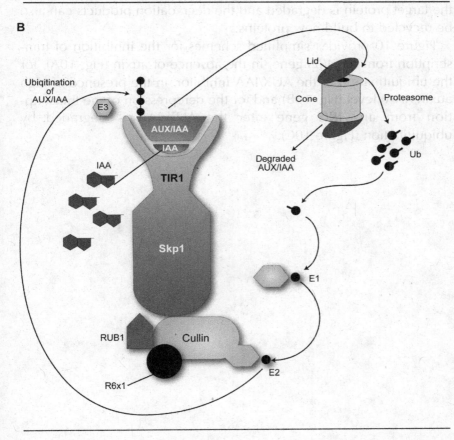

**C**

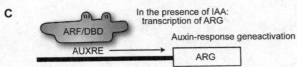

In the presence of IAA:
transcription of ARG

Auxin-response geneactivation

**Fig. 10.**   Ubiquitination of auxin/IAA.

(ubiquitination) that is conserved in eukaryotes? The answer to this question is emerging! The recognition of the protein that is targeted for degradation is a multistep process. It requires the proper F-box (TIR1) and carrier of the Ub (E1, E2 and occasionally also E3) to mark the target with polyubiquitin. The marked target is then recognized by the proteasome. The ubiquitin molecules are then recycled and the target protein is degraded and the degradation products can also be recycled to build new proteins.

Figure 10 provides simplified schemes for the inhibition of transcription from an *ARG* gene, in the absence of auxin (Fig. 10A), for the ubiquitination of the AUX/IAA inhibitor, in the presence of high auxin (IAA) level (Fig. 10B) and for the derepression of the transcription from an *ARG* gene after the AUX/IAA is degraded by ubiquitination (Figure 10C).

# CHAPTER 2
# Ethylene

Ethylene (ET) is rather unique among the phytohormones examined in this book. Its chemical structure enables a swift movement among plant cells, as well as uninhibited entrance from the outside into the plant, and exit from the plant to the outside. ET is produced only in organisms that belong to the plant kingdom. However, it also does affect animals. ET produced by phytochemical reactions in seawater induced ET-inducible genes in sponges. Also, when mammalian cell cultures were treated with the ET-releasing ethephon, there was a dramatic increase in cytosolic calcium influx (see Wang *et al.*, 2002).

Similar to auxin (as mentioned in Chapter 1), the effects of ET on plants and the practical utilization of these effects by horticulturalists were known many years before the biosynthesis of ET and its signal transduction were revealed.

The early history of ET was narrated in a study by Crozier *et al.* (2000). You can also read my own personal observations on its history (Galun, 2007, Chapter 2). The practice of etching figs (wounding) — that is now known to induce ET biosynthesis — goes back to thousands of years; it was noted in the Hebrew Bible in the book of the Prophet Amos. Already in 1886, Dimitry Nikolayevich Neljubow (1879–1926), as a graduate student in St Petersburg at the age of 17, observed that in laboratory settings, etiolated pea seedlings grew horizontally, while in the outside air, they grew vertically. Neljubow demonstrated that ET, used for the illumination of the laboratory, induced the horizontal growth. About 50 years later, in Kiev, in 1938, E.G. Minina reported that the air of heated greenhouses that contained ET, caused the femalization of cucumber flowers. Unfortunately, further studies by Minina on ET and sex expression were curtailed by T. Lysenco. Minina and her husband refused to follow Lysenco's dogma on the inheritance of acquired characters and were subsequently sent to Siberia in 1948, together with some other Soviet geneticists. Lysenco blamed them saying that "while Soviet heroes died in the defense of Stalingrad, they played with fruit flies." Once in northern Siberia, Minina was assigned to study the sex of pine trees. When Lysenco failed to deliver cold-resistant wheat, he was removed from the office and died in 1979. Yet, the dissident geneticists had to remain in Siberia since their crime was not bad science but dissidence against the authorities. In the early 1960s, a method was developed to increase ET levels in plants by the application of 2-chloroethylphosphonic acid ("ethephon"). The compound decomposes in the plant to ET. Also, an antidote against ET was revealed: aminoethoxyvinyl glycine (AVG). Being interested in the sex expression of cucumbers during the late 1950s and early 1960s, I, together with Dan Atsmon (also from the Department of Plant Sciences at the Weizmann Institute of Science), found that ethephon had a femalization effect on male cucumber flowers, while AVG initiated stamens in female flowers (see details in Frankel and Galun, 1977). Our results were helpful in the establishment of seed- and pollen-parents for commercial hybrid cucumber seed production. Again, the horticultural practice preceded by many years the information on the biosynthesis and signal transduction of ET.

The utilization of ET, antidotes of ET (such as AVG and silver thiosulfate), as well as ET-releasing compounds (ethephon) in the horticultural and agricultural practice are not covered in this book. A very good update on this subject can be found in the two volumes of the book edited by Nickell (1983), where the information is presented according to crops. The effects of ET on the physiology and growth of plants were revealed in a vast number of reports. These include studies that showed ET involvement in response to biotic and abiotic stresses, involvement in seed germination, root and shoot elongation, flower development, senescence and involvement in the abscission of flowers and leaves, as well as in the ripening of fruits.

In some of these involvements, ET plays a role in major economical issues, such as ripening and inhibition of ripening in major fruit crops and submergence tolerance in rice. These effects will not be detailed in this book. Readers interested in them will find the relevant citations in several reviews and articles (e.g. Kende, 1993; Kende and Zeevaart, 1997; Wang *et al.*, 2002; Voesenek *et al.*, 2005; Perata and Voesenek, 2007; Etheridge *et al.*, 2006; Ortega-Martinez *et al.*, 2007). In the present chapter, I am dealing with three issues of ET: (1) the biosynthesis; (2) the signal transduction; and (3) a few interactions between ET and other phytohormones. The involvement of ET in plant organ patterning and additional interactions between ET and other phytohormones — that affect patterning — will be handled in Part II, where specific plant tissues and organs are discussed.

As noted in the Introduction of this book, progress in the acquisition of scientific knowledge is achieved by the availability of new tools. This was also the case in the accumulation of knowledge on ET as a phytohormone. Two such "tools" were especially helpful for research on ET. One "tool" was the development of gas chromatography and flame ionization detectors for the quantitative analysis of ET that replaced the previous "quantitative" bioassays that were based on leaf-epinasty or growth of ethiolated legume seedlings. This "tool" was developed in the early 1960s and increased the quantitation of ET more than a 1000-fold. The other "tool" is the "Triple Response Assay." In this assay, ethiolated seedlings (i.e. of Arabidopsis) are exposed to ET, and the inhibition of the hypocotyl and horizontal growth is recorded. This assay is performed with seedlings of plants that were mutated in order to isolate mutants that were defective in the "Triple Response." When ethiolated Arabidopsis seedlings treated with ET showed no — or only minor — phenotypic responses, they were termed "ethylene-insensitive" (*ein*) or "ethylene-resistant" (*etr*) mutants. There were also mutants that had a constitutive triple response in the absence of ET. A vast number of such mutants were thus isolated, and these were very useful in understanding the molecular genetics of ET signal transduction. Interestingly and as noted above, the phenomenon of the

"Triple Response" was already observed and reported (in legume seedlings) by Dimitry Neljubow over 100 years ago; however, it became useful as a "tool" for ET investigations only after it was combined with the swift procedure to induce and isolate Arabidopsis mutants.

## Biosynthesis

The resolution of the metabolic pathway that leads to the biosynthesis of ET was initiated by two pioneering teams. Lürssen *et al.* (1979) established that 1-aminocyclopropane-1-carboxylic acid (ACC) is a major intermediate of ET synthesis in plants. In the very same year, the team of S.F. Yang of U.C. Davis, USA, published the same finding (Adams *et al.*, 1979) and also reported that ACC synthase is a key enzyme in ET biosynthesis. The molecular cloning of ACC synthase was published by Sato and Theologis (1989). The biosynthesis of ET was comprehensively reviewed by Kende (1993) and handled in some detail in the chapter on the biosynthesis of plant hormones by Crozier *et al.* (2000) in the book edited by Buchanan *et al.* (2000). It was recently reviewed by Argueso *et al.* (2007), and a brief account on ET biosynthesis can also be found in my previous book (Galun, 2007).

We can begin to understand the biosynthesis of ET by looking at the Methionine Cycle (or the Yang Cycle as it is also known, named after S.F. Yang who led much of the early work on the methionine pathway). Methionine is an essential sulfur-containing amino acid. It can be synthesized by plants but not by humans. As seen in Fig. 11, it can be converted, in the presence of ATP and adenosine, to S-adenosyl-L-methionine (SAM) by SAM synthetase. SAM is a potent donor of the methyl group and is utilized for the synthesis of several important substances. In the pathway that leads to ET, SAM releases 1-aminocyclopropane-1-carboxylic acid (ACC). The release of ACC from SAM is catalyzed by ACC synthase. Subsequently, ACC is oxidized to ET in a reaction that releases cyanide (HCN) and $CO_2$. ACC can also be converted to N-malonyl-ACC by ACC-malonyl

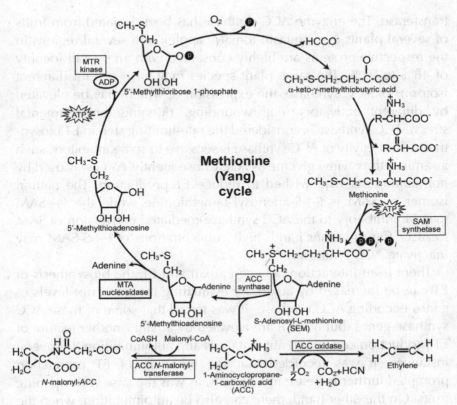

**Fig. 11.** The methionine cycle and ET biosynthesis. ET is synthesized from methionine by SAM and ACC. The enzymes that catalyze these three steps are ATP:methionine S-adenosyl-transferase (SAM synthase), S-adenosyl-L-methionine methyl-thioadenosine-lyase (ACC synthase) and ACC oxidase. 5'-methyl-thioadenosine, a product of the ACC oxidase reaction, is salvaged for the resynthesis of methionine through the methionine cycle. If the methylthio-group from SAM were not recycled, methionine availability and ethylene biosynthesis would probably be restricted by sulfur availability. By converting ACC to N-malonyl-ACC instead of converting it to ethylene, plants can deplete the ACC pool and thereby reduce the rate of ethylene production. (From Crozier *et al.*, 2000).

transferase. The enzyme ACC synthase has been isolated from fruits of several plants (e.g. squash, tomato, apple), and several regions in the respective proteins are highly conserved with an overall identity of 48% to 97%. The same plant species may have several different isoforms of ACC synthase, the expression of which may be elevated by different activators (e.g. wounding, ripening, environmental stress). ACC synthase is considered the rate-limiting step in ET biosynthesis. The activity of ACC synthase is sensitive to some inhibitors, such as aminoethoxyvinyl glycine (AVG). Consequently, AVG was used by horticulturalists who wished to reduce ET production. The natural isomer of SAM is (−)-S-adenosyl-L-methionine while the (+)-SAM may be inhibitory to the ACC synthase-mediated conversion of SAM to ACC. On the other hand, high concentrations of (−)-S-SAM may inactivate ACC synthase.

There is an interaction between auxin and ET: the biosynthesis of ET can be increased by auxin. By evaluating the transcript levels of genes encoding ACC synthase, it was found that some of these ACC synthase genes (but not all) are elevated by auxin. Another feature of ET production concerned autocatalysis and autoinhibition. In several instances, it was recorded that the initiation of ET production prompted further increase in ET levels, as was the case with ripening fruits. On the other hand, there can also be autoinhibition: when the outer peel of grapefruit was treated with ET, there was a marked reduction of ET production by this peel. These increases and decreases in ET levels were attributed to the respective increase and decrease in ACC synthase activities.

The final step in the biosynthesis of ET (Fig. 11) is the ACC oxidase-mediated conversion of ACC to ET, which requires the presence of $O_2$, $Fe^{2+}$ and ascorbate (as mentioned above). The ACC oxidase is a very unstable protein and it inevitably caused difficulties in its direct isolation from plant tissue, but it was eventually identified by isolating the respective cDNA of ACC oxidase genes (by Hamilton *et al.*, 1991 and by Spanu *et al.*, 1991). The proof that the identified mRNA is indeed encoding ACC oxidase came from a heterologous experiment in which the cDNA was introduced into yeast cells, and the transgenic yeast was able to convert ACC to ET. Regular untransformed

yeast is not capable to perform this conversion. A different presentation of the ET biosynthesis, in which Methionine and SAM are shown
as providers of other metabolic needs, is presented in Fig. 12.

An important difference between ET and other phytohormones is
that the levels of other phytohormones have to be reduced by specific
catabolitic processes. This is not the case with ET; ET is volatile and
can exit plant tissue. However, ET was also found to be oxidized or
converted to ethylene oxide and/or ethylene glycol. Nevertheless, it
is assumed that the main way of reducing ET levels in plant tissue is
through diffusion into the surrounding air. This swift diffusion across
aqueous and lipid phases has a two-pronged consequence: ET can
be released, but in an environment of high ET level, ET can also enter
plant tissue. When potato seedlings are densely planted in a closed
plastic box, the high levels of ET in the closed box strongly affect the
seedlings, stunting them and severely reducing the size of their
leaves. This can be reversed by opening the box. Interestingly, when
tomato seedlings are planted in closed boxes, they are much less
affected than potato seedlings. The great difference in what appears
to be the effect of high ET levels between two species of the same
genus (i.e. *S. tuberosum* and *S. lycopersicum*) is noteworthy (unpub-
lished observations of A. Perl, D. Aviv, and E. Galun). However, this
difference should not surprise us because the genes of different isoforms of ACC synthase are differentially controlled. For example,
Nakatsuka *et al.* (1998) found that during fruit maturation, *Le-ACS2*
and *Le-ACS4* in tomatoes are positively regulated, while *Le-ACS6* is
negatively regulated by the ET that is synthesized during this fruit
maturation. The names of these genes, Le-ACS, are a bit confusing.
"Le" stands for the old Latin name of tomato: *Lycopersicum esculen-
tum* and "ACS" stands for ACC synthase. As different ACC synthase
genes — even in the same organ of a given species — can have different regulatory mechanisms, such as different mechanisms in
different ACC synthase genes, different species may vary in their
regulatory mechanisms. The existence of many ACC synthase isoforms and the various control mechanisms in Arabidopsis (where
seven ACC synthase genes were characterized) and in several other
plants prompted several research teams to study this subject. The

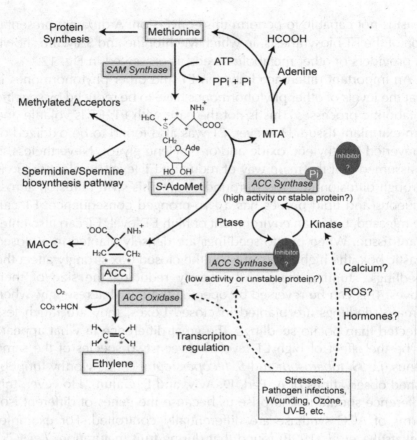

**Fig. 12.** Biosynthetic pathway and regulation of ET. The formation of S-AdoMet (SAM) is catalyzed by SAM synthetase from the methionine at the expense of one molecule of ATP per molecule of S-AdoMet synthesized. S-AdoMet is the methyl group donor for many cellular molecules (methylated acceptors), including nucleic acids, proteins and lipids. In addition, S-AdoMet is the precursor of the polyamine synthesis pathway (Spermidine/Spermine biosynthesis pathway). ACC is the immediate precursor of ET. The rate-limiting step of ET synthesis is the conversion of S-AdoMet to ACC by ACC synthase under most conditions. Methylthioadenosine (MTA) is the by-product generated along with ACC production by ACC synthase. The recycling of MTA back to methionine conserves the methylthio-group and is able to maintain a constant concentration of cellular methionine even when ET is rapidly synthesized. Malonylation of ACC to malonyl-ACC

details of these studies are beyond the scope of this book, but can be found in Wang *et al.* (2002). One point is worth mentioning. It appears that at least some ACC synthases are active as dimers. If so, dimers between two isoforms may furnish an additional control of ACC synthase activity.

While ACC synthase activity is controlled by the regulation of gene transcription, there is evidence (for *Le-ACS2*) for post-transcriptional regulation of ACC synthase activity, presumably by phosphorylation, by the level of another phytohormone (cytokinin) and even by wounding. It is reasonable to assume that when a change in ET levels is induced within seconds of a given treatment, ACC synthase is regulated post-transcriptionally rather than by a change at transcription level. Further details on the regulation of ET biosynthesis were supplied by Argueso *et al.* (2007).

## Perception and Signal Transduction

The phytohormone ET essentially serves as a releaser of the inhibition of gene expression in various ET-responsive genes. This means that in the absence of ET, these genes are repressed. The processes of ET perception and signal transduction are rather elaborate, and the full picture is emerging gradually as a result of a vast number of studies. The Triple Response of etiolated seedlings maintained in an ET

---

**Fig. 12. (*Continued*)**   (MACC) deprives the ACC pool and reduces the ET production. ACC oxidase catalyzes the final step of ET synthesis using ACC as substrate, and generates carbon dioxide and cyanide. Transcriptional regulation of both ACC synthase and ACC oxidase is indicated by dashed arrows. Reversible phosphorylation of ACC synthase is hypothesized and may be induced by unknown phosphatases (Ptase) and kinases, the latter presumably activated by stresses. Both native and phosphorylated forms (ACC synthase-Pi) of ACC synthase are functional, although the native ACC synthase may be less stable or active *in vivo*. A hypothetical inhibitor is associated with ACC synthase at the carboxyl end and may be dissociated from the enzyme if it is modified by phosphorylation at the vicinity. (From Wang *et al.*, 2002).

atmosphere — as well as other methods using mainly mutagenized Arabidopsis seedlings, but also mutagenized seedlings of other plant species (e.g. tomato) — resulted in numerous mutants that were defective in ET response. These yielded proper tools for the investigation of ET perception and signal transduction. The respective proteins that were identified in these studies may cause some confusion. As these proteins will be mentioned in the text and figures that follow, here is a list of commonly used acronyms:

| | |
|---|---|
| CTR | Constitutive Triple Response |
| EBF | Ethylene F-box Protein |
| EDF | Ethylene Defense Factor |
| EIL | EIN-Like |
| EIN | Ethylene Insensitive |
| ERF | Ethylene Response Factor |
| ERS | Ethylene Response Sensor |
| ETR | Ethylene Receptor |
| MAPK | Mitogen-Activated Protein Kinase |
| MAPKK | MAPK Kinase |
| MAPKKK | MAPK Kinase Kinase |

First, we will take a look at the overall picture of ET perception and signal transduction that leads to differential growth and to ET responses (i.e. the external impacts such as biotic and abiotic stresses). A somewhat surprising finding, published relatively recently (Chen *et al.*, 2002), was that the receptor/s of ET is/are located in the endoplasic reticulum (ER) membrane. Receptors for other phytohormones are located on the plasma membrane. The identification of a receptor protein, ETR1, in Arabidopsis (Chang *et al.*, 1993; Hua and Meyerowitz, 1998), preceded the localization of the receptors by only a few years. Several studies mentioned in the reviews listed below suggested that the evolution of ET as a phytohormone started with the presumed engulfment of a cyanobacterial-like organism into a eukaryotic cell ("endosymbiosis") that resulted in the very early plant cells. In the presence of ET, the signal transmission of the receptor proteins (in Arabidopsis, ETR1 and other receptors) is turned off and the CTR1-inhibition ceases, thus enabling the downstream activation of the positive regulator EIN2. EIN2 can then activate EIN3 and ERF1, causing the ET responses. While the sequence of processes from the receptor to EIN2 takes place in the ER, the induction of expression of ET responsive factors, that are induced by EIN3/EILs and ERF1, take place in the nucleus. In the absence of ET, the receptors (ETR1 and others) transmit a signal that keeps the CTR1 active as an inhibitor of further steps: EIN2 is repressed and consequently, EIN3 and EILs are also not activated; there is thus, no activation of transcription of the ET-responsive genes. The overall picture described above is a vast simplification of the process. Many additional details of the perception and signal transduction of ET were revealed (especially in Arabidopsis) during the last 16 years, and these were reported in numerous reviews and research reports. Here is a partial list of these reviews and reports: Kieber *et al.*, 1993; Chang *et al.*, 1993; Kende and Zeevaart, 1997; Kieber, 1997; Hua *et al.*, 1998; Bleecker, 1999; McCourt, 1999; Alonso and Stepanova, 2004; Guo and Ecker, 2004; Chen *et al.*, 2005; Etheridge *et al.*, 2006; Chow and McCourt, 2006; Binder *et al.*, 2007; Robles *et al.*, 2007; Hall *et al.*, 2007. The recent and rather thorough review by Binder (2007) covers the perception of ET in Arabidopsis and in numerous other plants, especially crop plants, such as the fruit trees apple, peach, pear, plum, persimmon and coffee, as well as cucumber, tomato, wheat, maize, sugar cane and several ornamental plants, such as *Petunia*, *Pelargonium* and *Oncidium*.

## Perception of ET

As noted above, the receptors of ET are located in the ER. Studies with Arabidopsis and tomato showed that these receptors could be divided into two subfamilies as demonstrated in Fig. 13. All these receptors share three domains: (1) an ET-binding domain, (2) a GAF domain, and (3) a kinase domain. Subfamily II receptors contain a signal sequence at their N-end, and most receptors (but not all) also contain a receiver domain at their C-end. These receptors share sequence similarity with the bacterial two-component histidine kinases. Five

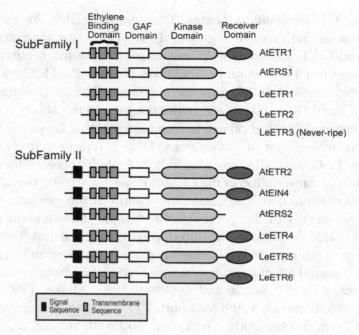

**Fig. 13.** ET receptor structure and families in Arabidopsis (At) and tomato (Le). Based on sequence comparisons of the ET binding domains, the different ET receptor isoforms in plants separate into two subfamilies. All the receptors share similar structures containing an ET-binding domain, a GAF domain and a kinase domain. A subset also contains a receiver domain. All members of subfamily II contain an extra putative N-terminal signal sequence. (From Binder, 2008).

such receptors were found in Arabidopsis, and several receptors were revealed in other plants: in tomato, there are six receptors; five of them are found to be functional. Such receptors were also revealed outside the angiosperms as in the alga *Chara*) and even in a group of cyanobacteria, but they were not found in Protozoa and Metazoa. The receptors form homodimers that are connected at their N-termini by two disulfide bonds. While all of the angiosperms' ET receptors that were studied have overall structures as shown in Fig. 13, the ET receptors of cyanobacteria differ from these structures. Moreover, it is not excluded that there are ET receptors that are not localized on the ER.

Specific gain-of-function mutants in the binding domain of Arabidopsis receptor isoforms confer dominant *ET insensitivity*, while the loss of several receptors causes a *constitutive ET response*: in the latter case, there seems to be no activation of the CTR1 inhibitor and the downstream EIN2 transmits its activation further down (to EIN3/ EIL1). Hence, the receptors serve as negative regulators in the pathway; ET itself restrains this inhibitory effect of the receptors.

Looking again at the structure of receptors indicates that the ET-binding domain is at the N-terminal, transmembrane domain. It was found that Cu ions are required for ET binding and there is one copper ion per receptor dimer (e.g. two AtETR1 receptors). While the copper ion is essential for the functionality of receptor dimers, the disulfide bonds between the two receptors are apparently not essential. Other metals (such as gold) can also bind between the homodimers but they may affect the functionality of the receptors as negative regulators of CTR1. Silver ions were employed by horticulturalists to negate ET effects for more than 30 years (later, horticulturalists used AVG for the same purpose). The Ag ions probably interfere with the transmission of the ET signal in the receptors. The binding of the copper ions and ET is in the transmembrane (ET-binding domain) of the receptors. There are indications that suggest a three-stage model. In the first stage, there is no binding to ET and the receptors do not inhibit the CTR1 inhibitor. In the second (intermediate) stage, ET binds to the receptors but the receptors are still signaling. In the third stage, ET abolishes the signaling of the receptors and inhibits the CTR1 inhibitor enabling the activation of EIN2.

The receptors of Arabidopsis have a functional His-kinase, but it is still not clear what function this kinase has in the ET signaling. The Ser/Thr kinase was also found in some receptor isoforms of Arabidopsis, but again, the exact role of this kinase is not yet clear. The CTR1, in its non-mutated form, is also associated with the ER membrane, and there are indications that suggest a physical contact between AtETR1 and CTR1. It is probable that the association of CTR1 with the ER membrane is receptor-dependent. The receptor function may even be more complicated. Another protein that is associated with ER, as well as with Golgi membranes, may modify reception. Moreover, the receptor/signaling of ET may differ in different plant organs (such as root patterning, fruit maturation and mutational bending of Arabidopsis hypocotyls).

While all the studied receptors contribute to the signaling, and there is a functional overlap between the receptor isoforms, the various isoforms are not entirely redundant in their function. Moreover, isoforms may replace each other in one function but not in another: AtETR1 and AtEIN4 do not replace each other in the repression of a root hair mutant (*rhd1*) by ET, but these two isoforms may replace each other in other functions. In general, members of subfamily I receptors (Fig. 13) appear to have a greater role in ET signaling of Arabidopsis. Interestingly, the Ag ion appears to mainly target subfamily I in replacing copper ions and abolishing the receptor's role. It was, therefore, proposed that the members of subfamily I are the isoforms that more strongly physically interact with CTR1, whereas isoforms from subfamily II receptors interact less so. However, this greater role of subfamily I may not be shared in plants other than Arabidopsis: in tomato subfamily I, isoforms do not appear to have a greater role than subfamily II isoforms, at least not with respect to fruit ripening. In short, one may conclude (Binder, 2007) that receptors have overlapping but distinct roles in ET signaling.

Studies with ET in Arabidopsis indicated that this plant responds to a very wide range of ET concentrations: from 0.2 nl per L to 100 µl per L. No major differences in ET affinity were revealed between receptor isoforms, and it was calculated that an ET response can occur when about 0.1% of the receptors bind ET. The ET response is not an all-or-none process. Plants can sense small changes in ET concentrations. To explain this capability as well as other phenomena of ET receptor function, the team of Bleacker and Binder from the Department of Botany at the University of Wisconsin, Madison, came up with a receptor-clustering model (see Binder, 2007 for details and references). This receptor-clustering model (Fig. 14) was also intended to explain the phenomenon of quick growth recovery after the removal of ET. It was assumed that a small number of receptors that did not bind ET converted their neighboring

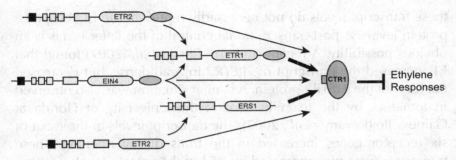

**Fig. 14.** Receptor signaling model in Arabidopsis. In this model, the subfamily II receptors predominantly signal through the subfamily I receptors. However, subfamily II receptors also interact with — and stimulate — CTR1, but at reduced levels compared to subfamily I receptors. See Fig. 13 for receptor domains. (From Binder, 2008).

ET-bound receptors to the signaling state, thus facilitating the recovery of growth. It was discovered that a truncated *etr-1* receptor caused high levels of *ET-insensitivity*. Also, specific missense mutations in Arabidopsis receptors—such as EIN4 that are expressed at low levels — result in an *ET-insensitivity*. This receptor-clustering model suggests interactions between the receptors but also implies that the subfamily I receptors (ETR1 and ERS1) are the primary signaling receptors that antagonize CTR1, thus derepressing EIN2 and further downstream processes. The subfamily II receptors (ETR2, EIN4 and ERS2) act, according to this model, cooperatively to enhance the signaling of subfamily I receptors. Such an interaction would explain why a poorly expressed EIN4 would still cause a dominant insensitivity since it is keeping ETR1 and ERS1 in the signaling state, in the presence of ET. The model also suggests that subfamily II receptors can act independently from subfamily I receptors at a reduced level. While the receptor-clustering model can be used to explain several phenomena, yet it still fails to provide direct proof.

There is another problem with understanding the receptors' role in the signal transduction of ET. Much data was assembled on the changes of transcript levels of genes encoding the ET receptors, but these transcript levels do not necessarily represent the ET receptors' protein levels. A post-transcriptional control of the latter levels is an obvious possibility. A recent study by Chen *et al.* (2007) found that ET increased the transcript of *AtETR2* in Arabidopsis but decreased the level of the AtETR2 protein. A similar situation was also observed in tomatoes by the team of Klee at the University of Florida at Gainesville (Kevany *et al.*, 2007). The transcript levels of three out of six receptor genes increased in the fruits during ripening. These transcripts were also increased by ET but the protein levels of these receptors decreased with maturation of the fruits. Also, ET applications reduced the levels of these proteins. Clearly, the levels of LeETR4 and LeETR6 are controlled post-transcriptionally.

## From receptors to ethylene-responsive genes

The overall pathway of the ET perception and signal transduction can be schematized as follows:

This very schematic presentation of the ET pathways, adopted from Chen *et al.* (2005), illustrates that *in air*, the receptors activate the inhibitor, CTR1. This inhibitor inhibits EIN2, and by that, the

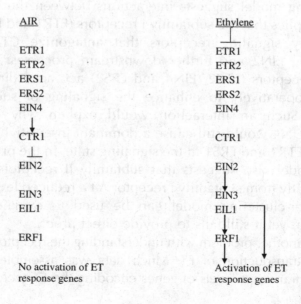

pathway is stopped and the ET response genes are not activated. In the presence of ET, the receptors do not activate the inhibitor, CTR1, thus allowing the EIN2 to activate EIN3/EIL1 and possibly other EIN-like transcription activators, causing the induction of transcription activation of ET responsive genes (such as EDF1, EDF2, EDF3, EDF4, as well as ERF1). ERF1 can then activate additional ET responsive genes. It appears that while the receptor-CTR1-EIN2 components reside in the ER, EIN3/FILs and ERF1 control the transcription of ET responsive genes in the nucleus.

In recent years (e.g. Guo and Ecker, 2003), another process of the ET pathway came to light: EIN3 BINDING F-BOX1 (EB-F1) and EBF2 were found to function in this process. In the absence of ET, EIN3 and possibly other EIL proteins are targeted for ubiquitination. Binder *et al.* (2007) found that EIN3 and EIL1 are the main targets of EBF1 and EBF2. A ubiquitination complex composed of $SCF^{EBF1/2}$ is formed, as shown in the model in Fig. 15 (left side). Ub units are attached to EIN3/EIL1 and the ubiquitinated complex is entered into a proteasome. The EIN3/EIL1 is thus degradated to amino acids and the Ub units are released for re-use. However, there is a "twist": EBF1 and EBF2 have distinct roles. While their roles overlap, EBF1 plays the main role in the absence of ET (i.e. in air) and EBF2 has a more prominent role during the initial phase of signaling (in the condition that there was first an exposure to ET and the ET was then removed, causing a recovery from ET effect). In other words, the EBF1 and EBF2 have a partial redundant role in fine-tuning the ET effect.

It should be noted that in Arabidopsis, there are five EIL proteins. The abovementioned information concerns EIL1; the exact functions of the other four EIL proteins (EIL2–5) are not yet clear. The level of EIN3 can be changed by several affectors: it rises in the presence of ET, it is stabilized by glucose and can be degraded as noted above. Intact EBF1 and EBF2 are most likely vital for Arabidopsis seedling development. The phenotypes of double-null mutants (ebf1, ebf2) are the following. Such seedlings emerge slowly from their seed coats; the seedlings are stunted and they fail to proceed with normal growth; there is an increase in root hair number, enhanced

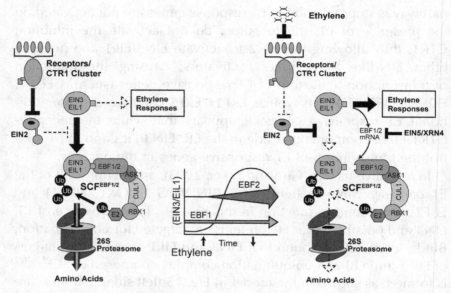

**Fig. 15.** Model for the action of EBF1 and 2 in the ET signal transduction pathway. In the absence of ET, the receptor family (ATR1, ETR2, ERS1, ERS2, EIN4) activates the negative regulator CTR1, which leads to the inhibition of EIN2. EIN3 and EIL1 levels are kept low by selective ubiquitination of the proteins by SCF^EBF1^ and SCF^EBF2^, which induces their subsequent breakdown by the 26S proteasome. In the presence of ET, the receptors are inhibited, thus reducing the output of CTR1 and its subsequent inhibition of EIN2. EIN2 acts in part to directly — or indirectly — block the interaction of EIN3 and EIL1 with the SCF E3s containing EBF1 and EBF2. The reduction in ubiquitination allows EIN3 and EIL1 levels to rise to mediate ET responses. Over a slower time course, EIN2 activation also leads to an increase in EBF1 and EBF2 mRNA and presumably protein levels (shown in inset), which further dampens the accumulation of EIN3 and EIL1. Via an unknown mechanism, the exoribonuclease EIN5/XRN4 dampens the accumulation of the EBF1 and EBF2 transcripts. During ET signaling, EBF1 plays a special role at no — or low — hormone levels to maintain low basal levels of EIN3/EIL1. By contrast, EBF2 accumulates during ET signaling to prevent excess accumulation of EIN3/EIL1 and to remove EIN3/EIL1 after ET levels dissipate. (From Binder *et al.*, 2007).

anthocyanin accumulation and a failure to develop normal leaves, stems and flowers. Such seedlings have an elevated level of EIN3 (and possibly also of EILs). The experimental work of Binder *et al.* (2007) clearly indicated that the SCF$^{EBF1}$ and SCF$^{EBF2}$ E3 complexes (see Fig. 15) work in concert to fine-tune the abundance of EIN3/EIL1 in response to ambient ET levels.

## Interactions between the components of the ET pathway: from receptors to regulators of the expression of ET responsive genes

While genetic studies, especially in Arabidopsis, revealed the sequence of signaling in the ET pathway, the biochemistry of the interaction between these components is much less clear. For example, genetic studies clearly placed EIN2 between CTR1 and EIN3/EIL1, but what chemical changes occur in CTR1 (in the presence and absence of ET), and how these changes affect the chemistry of EIN2 is not yet clear. Information on this chemistry may follow revealing the crystaline structures of these proteins; however, this has not yet happened. What we do know is that in air (no ET) the receptors activate the kinase CTR1 and after this activation, CTR1 suppresses the downstream component (EIN2). The suppressed EIN2 does not activate EIN3/EIL1 and the pathway is stopped. In the presence of ET, the receptors (e.g. ETR1, ETR2, ERS1, ERS2, EIN4 of Arabidopsis) do not activate the kinase of CTR1 and thus, the latter protein does not inhibit EIN2. The active EIN2 then activates the EIN3/EIL1. A slightly different scheme for the ET signal transduction pathway was put forward by Alonso and Stepanova (2004) as shown in Fig. 16.

It should be noted that plants of different species have a different number of receptors. Information on this was mainly obtained from crop plants. Here is a sample list:

| | |
|---|---|
| Six receptors | Tomato |
| Five receptors | Arabidopsis |
| Four receptors | Tobacco |

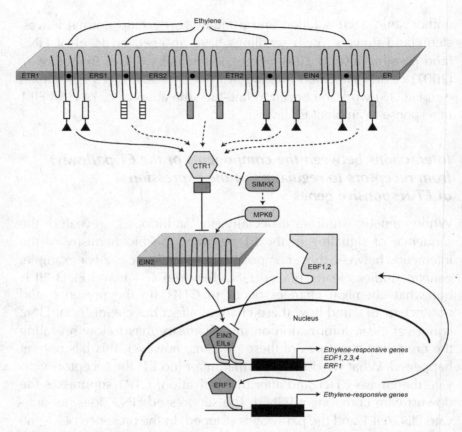

**Fig. 16.** Representation of the ET signal transduction pathway. ET is perceived by a family of receptors located in the ER membrane. The binding of ET to the hydrophobic pocket of the receptors is mediated by a copper cofactor. The receptors physically interact with the Raf-like kinase, CTR1. The binding of ET to the receptors results in the inactivation of both receptors and CTR1, causing derepression of a positive regulatory molecule, EIN2. A MAPK cascade may be involved in the signal transduction between CTR1 and EIN2. By an unknown mechanism, a positive signal is then transmitted from EIN2 to the transcription factors EIN3/EILs, resulting in the stabilization and consequently, accumulation of the EIN3/EIL proteins in the nucleus, where they induce the transcription of *ERF1*, *EDF1, 2, 3, 4* and other ET-regulated genes; this is the first step in a transcriptional cascade that unleashes the downstream ET responses. The levels of EIN3 are regulated by two F-box proteins, EBF1 and 2, whose transcription is inducible by ET. (From Alonso and Stepanova, 2004).

| | |
|---|---|
| Three receptors | Cucumber, carnation, pear |
| Two receptors | Peach, passion fruit, melon, maize |
| One receptor | Wheat |

These receptors have common characteristics. All have similarity to the bacterial two-component sensors. The various receptors of a given species seem to be (at least in part) redundant. All these receptors act as negative regulators of the downstream component, CTR. The receptors also seem to show differential expressions during plant development. For example, the LeETR4 of a tomato is specifically involved in fruit ripening.

While Arabidopsis has only one *CTR* (*CTR1*) gene, there are probably three *CTR1-like* genes in tomatoes (e.g. Klee, 2004). It should be noted that there are phenomena in tomatoes (like fruit ripening) that do not exist in Arabidopsis. Thus, it should not surprise us that the components of the ET pathway are not identical in these two species.

Additional and detailed reviews were recently written by several experts of signal transduction of ET: Chang, 2007; Li and Guo, 2007; Binder, 2007 and Hall *et al.*, 2007.

# CHAPTER 3

# Gibberellins

The awareness of gibberellins (GA) first emerged in Japan more than 80 years ago, in 1926, when Eiichi Kurosawa studied the *bakanae* disease of rice that was responsible for causing severe damage to rice yields. The affected plants were taller than normal but their flowers were malformed and did not yield rice grains. It was found that these plants were infected with a fungus: *Gibberella fujikuroi*. This fungus produced diterpenoid compounds that were later purified and even crystallized, and these were given the name gibberellins. The fungus produces several GA compounds, such as $GA_1$ and $GA_3$ (gibberellic acid). Surprisingly, no publications on GA studies appeared outside Japan until the early 1950s. Then, numerous publications on GA appeared in both the UK and the USA. These publications mainly concerned angiosperms and were primarily dealing with three subjects: (1) the discovery of GA in various plant species; (2) the effect of treating plants with GAs; and (3) the biosynthesis of GAs in plants. Only much later were subjects such as perception and signal transduction studied.

The application of GA compounds to plants commonly had a very clear effect: it caused the elongation of shoots and this elongation was conspicuous, especially in some dwarf mutants, indicating that dwarf mutants could be defective in GA biosynthesis. However, it became clear at an early stage that not all plant species reacted to GA; moreover, a given plant species could react to a GA compound, for which there was no sensitivity in other species. Furthermore, elongation was not the only result of GA application. These phenomena became evident in my own research during the late 1950s

71

and 1960s (summarized in Frankel and Galun, 1977). I found that $GA_4$ and $GA_7$ strongly suppressed pistillate (female) flowers in gynoecious and monoecious cucumbers (*Cucumis sativa*) but $GA_3$ had almost no effect on these flowers. In melon plants (*Cucumis melo*) — which belong to the same genus as cucumbers — the sex expression of flowers is not affected at all by GA compounds. Moreover, in some plants, GAs have an opposite effect: suppression of stamen development. Many other effects on growth and development are caused by the application of GAs to plants. For example, GAs can induce formation of cones in conifers and cause flowering in angiosperms (e.g. Jerusalem artichoke) that normally require specific photoperiods or exposure to cold. Also, there were reports on the retardation of leaf and fruit senescence and the promotion of plant germination. The effect of GAs on the *de novo* synthesis of $\alpha$-amylase and other enzymes in the aleurone layer of barley is extensively utilized in the beer industry, because GAs can regulate (and facilitate) the production of malt. In Part II of this book, where the patterning of plant organs will be examined, many specific effects of GA and GAs in combination with other phytohormones, will be provided.

## The Occurrence of Gibberellins

Certain GAs probably exist in all vascular plants, but their existence varies greatly among these plants. Jake MacMillan from the UK, a pioneer of GA studies, collected the information up until 2002 (MacMillan, 2002) and compiled the information in two lists. In the first list, he arranged all the GAs by their number, from $GA_1$ to $GA_{126}$. Since 2002, additional GAs were detected and the total number of GAs now approaches 200. The plant list did not include certain GAs (such as $GA_2$, $GA_{10}$ and $GA_{14}$) despite the fact these GAs were found in *Gibberella fujikuroi*. Only seven bacterial species were listed by MacMillan and these contained the "common" GAs of plants (e.g. $GA_1$, $GA_3$, $GA_4$ and $GA_{20}$). In another table, MacMillan listed numerous plants alphabetically and indicated which GAs were found in specific tissues. Up-to-date records of these lists are probably kept at the University of Bristol in the UK. We are already faced with a vast number of GAs in plants, and this is contrary to ET, where there is only a single compound and only one or two auxins (IAA and IBA) in the whole plant kingdom.

All the GAs can be divided into two main groups: those that contain a $C_{20}$ and those that lack it. The latter contain a $\gamma$-lactone ring, connecting $C_{10}$ with $C_{19}$. One group of GAs contains this hydroxyl (e.g. $GA_1$ and $GA_3$) while the other group lacks it (e.g. $GA_4$ and $GA_7$).

## Biosynthesis and Inactivation

The history of discovering GAs' biosynthesis in the fungus *G. fujikuroi* and in plants has been extensively reviewed by Hedden *et al.* (2002). Studies on this biosynthesis were first conducted with *G. fujikuroi*, following the revealing of the structure of $GA_3$. Investigators then used plant homogenates to start the research on the pathway for GA synthesis in angiosperms. As we will see below, there is a considerable difference between GA biosynthesis in the fungus and in vascular plants. This should not surprise us because in the latter, the early stages of GA biosynthesis (up to *ent*-kaurene) take place in the plastids, intracellular organelles that do not exist in fungi. Was there, in the early evolution of fungi, a close association between the photosynthesizing organelles and fungi? There is no experimental evidence for such an association but I think it is important that we should recall that there were no multicellular plants that existed about one billion years ago. Fungal organisms are either saprophytes, symbionts (lichens) or parasites of plants (including algae and cyanobacteria). So what was the "food" of fungi at about a billion years ago? A plausible answer is that at that period, fungi had to live in close association (a kind of symbiosis similar to present-day lichens) with photosynthesizing cells. It seems likely that during that early period, there was a transfer of metabolic capabilities from plastids into the ancestors of the *Gibberella* species. There is a finding that supports my guess: in the deep gold mines of South Africa, investigators revealed an association of photosynthesizing cells with fungal-like hyphae. Were the fungal-like hyphae instrumental in the concentration of the gold?

There is no answer to this question but surely, present-day fungal hyphae are capable in absorbing heavy metals from their environment (see Galun *et al.*, 1983a, b; Siegel *et al.*, 1986).

There are very good reviews on the early studies of GA biosyntheis in *G. fujikuroi* and in plants (e.g. Graebe, 1987; Hedden and Kamiya, 1997; MacMillan, 1997; Lange, 1998; Crozier *et al.*, 2000). First, the biosynthesis in the fungus was studied; thereafter, homogenates of plants were also covered in these studies, and later, this biosynthesis was studied in whole plants, too. Interestingly, while some steps in the metabolic pathway of the three systems are identical, there are steps in which the three systems differ. In addition, differences in the metabolic pathway were found between plant species and even in different tissues of the same species. This was clearly indicated in the review of MacMillan (1997) and also emerged from the list of GA contents in plant species and their tissues (MacMillan, 2002); for example: $GA_1$, $GA_3$, $GA_4$ and $GA_9$ were found in Arabidopsis seeds; $GA_3$ was *not* reported in shoots, but 17 other GAs not found in seeds *were* detected in Arabidopsis shoots.

The plant GAs can be divided into two main groups (MacMillan, 1997) as shown in Fig. 17: $C_{20}$ and $C_{19}$ GAs. The many types of plant GAs are all oxidized variants of $C_{20}$ and $C_{19}$. The first biosynthetic steps of GAs, up to the formation of *ent*-kaurene — e.g. the conversion of geranylgeranyl diphospate (or geranylgeranyl pyrophosphate) by *ent*-copalyl diphosphate synthase (CPS), as well as the conversion of GGDP (or GGPP) to copalyl diphosphate (CDP or CPP), and also the conversion of CDP to *ent*-kaurene by *ent*-kaurene synthase (KS) — take place in the plastids. The pathway then continues in the cytoplasm until the formation of $GA_{12}$-aldehyde (Fig. 18) and from the latter compound to the first gibberellin $GA_{12}$. Although fungi lack plastids, the biosynthesis pathway until $GA_{12}$-aldehyde in *G. fujikuroi* and in plants is probably along the same metabolic pathways (see Fig. 19). However, after $GA_{12}$-aldehyde, the pathway of the fungus is different from the one in the plant. Nevertheless, plants as well as the fungus can produce both kinds of GAs: those that are hydroxylated in $C_{13}$ (e.g. $GA_1$, $GA_3$) and GAs that lack this

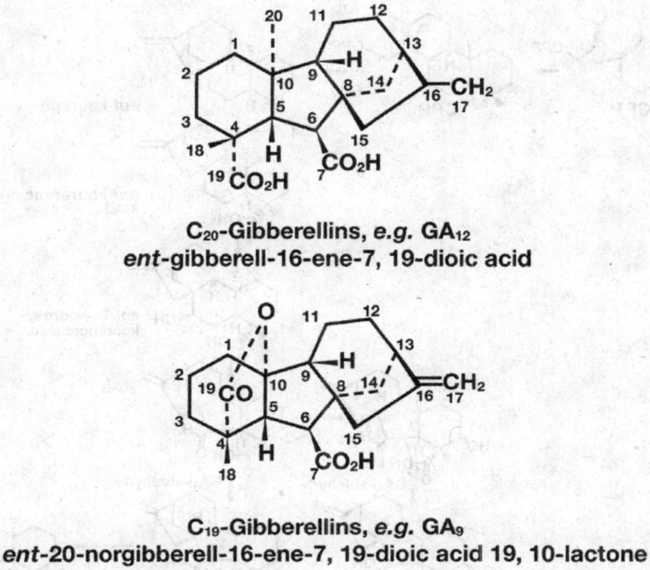

**C₂₀-Gibberellins, *e.g.* GA₁₂**
***ent*-gibberell-16-ene-7, 19-dioic acid**

**C₁₉-Gibberellins, *e.g.* GA₉**
***ent*-20-norgibberell-16-ene-7, 19-dioic acid 19, 10-lactone**

**Fig. 17.** Gibberellin structures and nomenclature. (From McMillan, 1997).

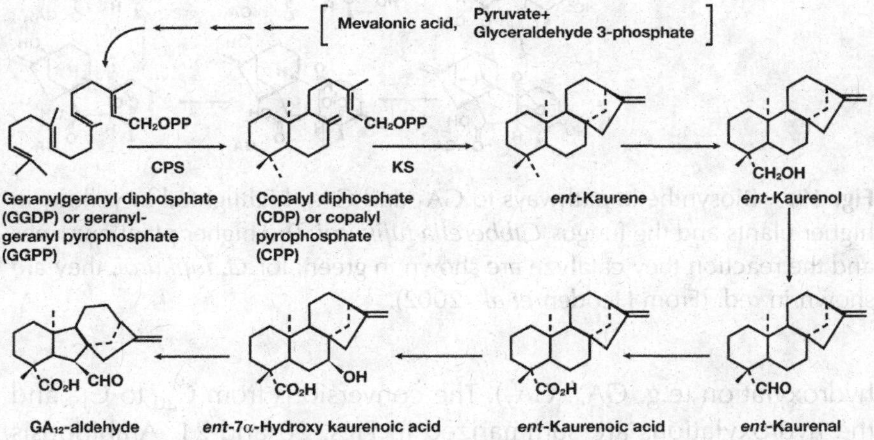

**Fig. 18.** Early GA biosynthetic pathway to $GA_{12}$-aldehyde. GGDP is produced in plastids by the isoprenoid pathway, originating from mevalonic acid, or possibly from pyruvate/glyceraldehyde 3-phosphate. (From Hedden *et al.*, 2002).

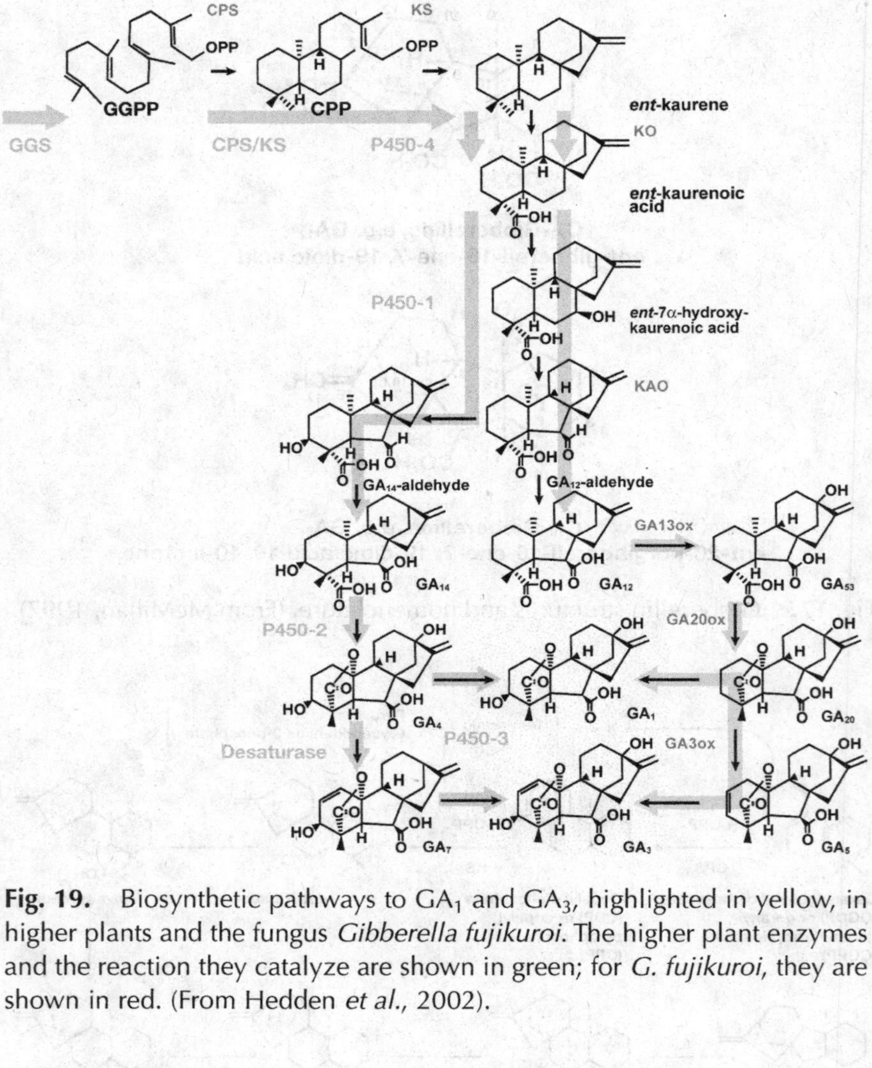

**Fig. 19.** Biosynthetic pathways to $GA_1$ and $GA_3$, highlighted in yellow, in higher plants and the fungus *Gibberella fujikuroi*. The higher plant enzymes and the reaction they catalyze are shown in green; for *G. fujikuroi*, they are shown in red. (From Hedden *et al.*, 2002).

hydroxylation (e.g. $GA_4$, $GA_7$). The conversions from $C_{20}$ to $C_{19}$ and the hydroxylations are summarized in Figs. 20 and 21. Arabidopsis genes that are involved in GA biosynthesis, as well as the functions of their encoded enzymes, are shown in Table 2 (updated until the year 2002).

It should be noted that there are two groups of important enzymes in the biosynthesis of GAs. One is the GA20ox group and the other

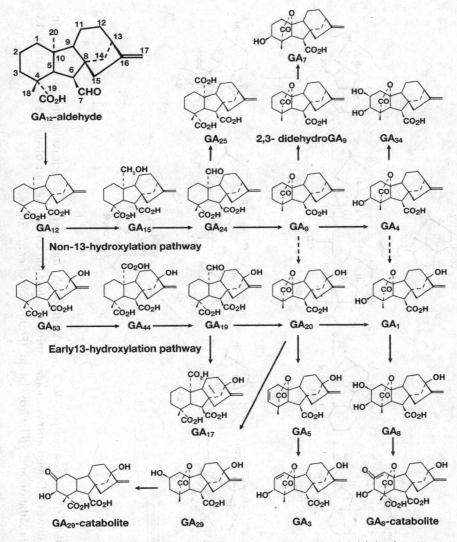

**Fig. 20.** Gibberellin biosynthesis pathway from $GA_{12}$-aldehyde. (From Hedden and Kamiya, 1997).

is the GA3ox group. The gibberellin 3-oxigenases catalyze a final step in GA biosynthesis — the hydroxylation of $GA_9$, at $C_3$ — to form $GA_4$. The other group, the GA20ox, causes the conversion of $C_{20}$ GAs (as $GA_{12}$) to $C_{19}$ GAs (as $GA_{20}$ and $GA_1$). Recently, Hu *et al.*

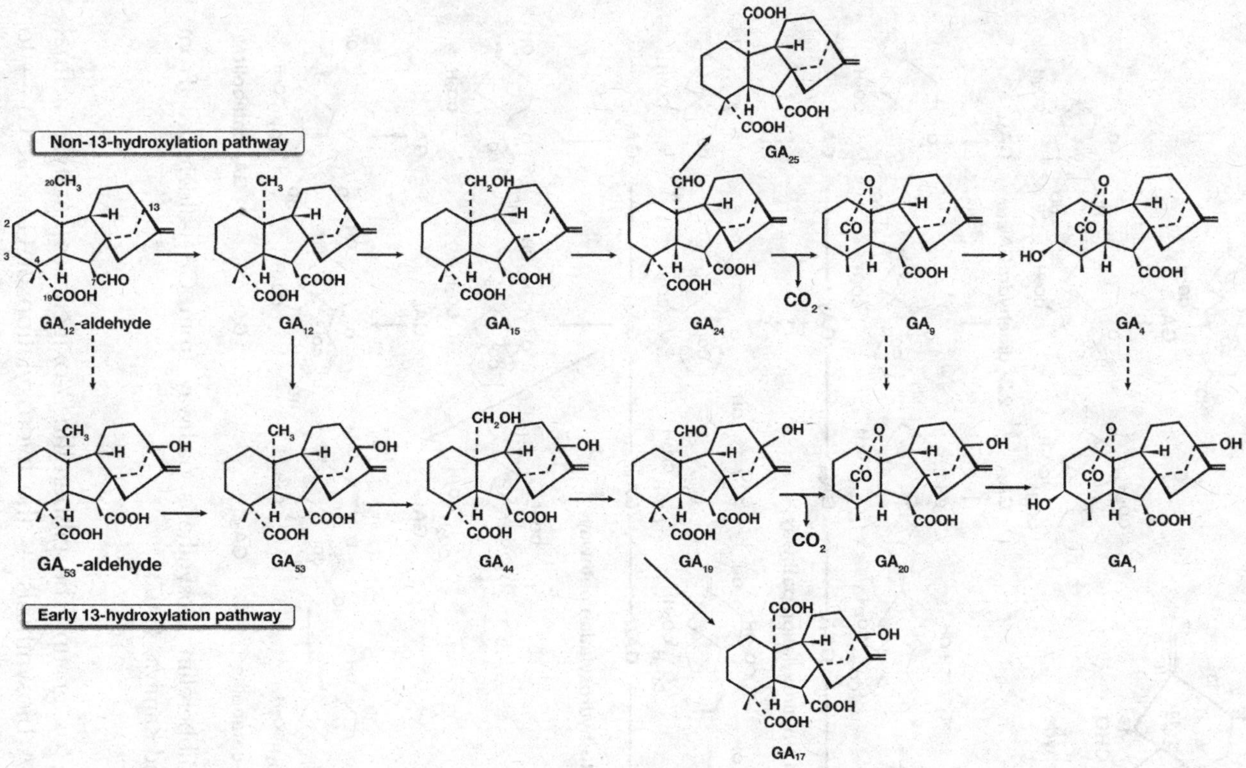

**Fig. 21.** Metabolism of $GA_{12}$-aldehyde to $C_{19}$-GAs through the early 13-hydroxylation and the non-13-hydroxylation pathways. Dotted arrows indicate minor routes. The conversions involved are 7-oxidation, 13-hydroxylation, C-20 oxidation and 3β-hydroxylation. (From Crozier et al., 2000).

**Table 2. *Arabidopsis* Genes Involved in GA Biosynthesis, Including the Function of Their Encoded Proteins and Mutants**

| Gene | Enzyme Function | Mutant |
|------|----------------|--------|
| *CPS* | CPP synthase (GGPP to CPP) | ga1 |
| *KS* | *ent*-Kaurene synthase (CPP to *ent*-kaurene) | ga2 |
| *KO* | *ent*-Kaurene oxidase (*ent*-kaurene to *ent*-kaurenoic acid) | ga3 |
| *KAO1* | *ent*-Kaurenoic acid oxidase (*ent*-kaurenoic acid to $GA_{12}$) | |
| *KAO2* | *ent*-Kaurenoic acid oxidase | |
| *GA20ox1* | GA 20-oxidase ($GA_{12}/GA_{53}$ to $GA_9/GA_{20}$) | ga5 |
| *GA20ox2* | GA 20-oxidase | |
| *GA20ox3* | GA 20-oxidase | |
| *GA20ox4* | Undetermined[a] | |
| *GA20ox5* | Undetermined[a] | |
| *GA3ox1* | GA 3$\beta$-hydroxylase ($GA_9/GA_{20}$ to $GA_4/GA_1$) | ga4 |
| *GA3ox2* | GA 3$\beta$-hydroxylase | |
| *GA3ox3* | GA 3$\beta$-hydroxylase | |
| *GA3ox4* | GA 3$\beta$-hydroxylase | |
| *GA2ox1* | GA 2-oxidase ($GA_1/GA_4/GA_9/GA_{20}$ to $GA_8/GA_{34}/GA_{51}/GA_{29}$ and corresponding catabolites) | |
| *GA2ox2* | GA 2-oxidase | |
| *GA2ox3* | GA 2-oxidase | |
| *GA2ox4* | GA 2-oxidase | |
| *GA2ox5* | Probably pseudogene[b] | |
| *GA2ox6* | GA 2-oxidase | |

[a]Classified as GA 20-oxidases on the basis of predicted amino acid sequences. [b]The gene contains a large insertion and is apparently not expressed. From Hedden *et al.* (2002), where references were provided.

(2008) focused on the Arabidopsis *GA3ox* genes while Rieu *et al.* (2008) focused on the Arabidopsis *GA20ox* genes. As for the latter, five *GA20ox* genes were revealed in Arabidopsis; two of them, *GA20ox1* and *GA20ox2*, were the most highly expressed during vegetative and early reproductive development. Both appear to be redundant with respect to hypocotyl and internode elongation as well as to other developmental phenomena; however, the two genes do differ: *GA20ox1* showed a greater impact on internode and filament elongation than *GA20ox2*, while *GA20ox2* had a greater effect on flowering time and silique length.

Other *GA20ox* genes had lesser impact than *GA20ox1* and *GA20ox2*, and also showed somewhat different effects on elongation. It can be assumed that all these five genes (that probably originated from a single ancestral gene) contribute to the normal development of Arabidopsis.

Hu *et al.* (2008) followed the expression of four Arabidopsis gibberellin 3-oxydase (*GA3ox*) genes that encode the enzyme that causes the hydroxylation on GAs at $C_3$ (e.g. $GA_9$ to $GA_4$ and $GA_{20}$ to $GA_1$). They found that the *GA3ox* genes are only expressed in stamen filaments, anthers, and flower receptacles. Plants that lack *GA3ox1* and *GA3ox3* functions had stamen and petal defects showing that these two genes are important for GA production in the flower. Hence, *de novo* synthesis of active GAs is necessary in stamen development in early flowers. It was suggested that bioactive GAs produced in the stamens and/or flower receptacles are transported to petals, where they promote petal development. In the developing siliques, *GA3ox1* is mainly expressed in the replums, funiculi, and the silique receptacles. The other *GA3ox* genes of Arabidopsis are only expressed in developing seeds. It was suggested that active GAs are transferred from the endosperm into the surrounding maternal tissues where they promote growth, possibly through interaction with other phytohormones.

While most of the research activity in GA metabolism was directed to the biosynthesis of active GAs, there were also some interesting studies on the conversion of active GAs into inactive GAs. One metabolic step to convert an active GA into an inactive one is by hydroxylation of $C_2$ as the conversion of $GA_4$ to $GA_{34}$ or the conversion of $GA_{20}$ into $GA_{29}$ (the latter can be further metabolized to the $GA_{29}$-catabolite). An interesting study on the inactivation of a GA by GA2ox was recently reported by Kloosterman *et al.* (2007). These authors investigated the early stage of potato tuber formation and the subapical stolon region. Under normal tuber formation conditions, this subapical region swells into a tuber. The authors found that an increase in expression of the potato *GA2ox1* gene will inhibit the swelling of the subapical stolon region and thus suppress tuberization. Although the mechanism controlling the levels of active GA (such as $GA_1$) is complex and probably includes feedback inhibition of further active GAs, the GA2ox1 enzyme seems to play an important role in regulating the level of active GAs in the potato stolon. Hence, the StGAox1 fulfills a central role in the transition from longitudinal stolon growth to tuber initiation by regulating the active GA levels in the subapical stolon region, facilitating radial growth.

## Perception and Signal Transduction

The overall mechanism of GA perception and signal transduction is the following. Active GAs bind to a receptor that is termed GID1. Once bound, a repressor of GA-regulated gene enters the scene. This repressor protein contains a 17-amino-acid sequence in which the 5 amino acids (Asp, Glu, Len, Leu and Ala) are situated, and it is therefore termed DELLA (according to the five amino acids, by the single letter designation). In the absence of an active GA, GID1 does not bind firmly to DELLA so that DELLA is free to repress the GA-regulated genes. In the presence of an active GA, this GA binds to the soluble GA receptor, GID1 in the nucleus, and this complex binds DELLA; a triple complex is then formed with $SCF^{GID2}$. In this complex, DELLA is connected to several units of ubiquitin (polyubiquitinated) and is then entered into the 26S proteasome. The units of ubiquitin are released but the DELLA is degraded into amino acids in the proteasome in a similar manner to the degradation described above for the auxin signal transduction. In addition to soluble receptors of GA, there also seem to exist membrane-bound receptors, as for example, in the oat aleurone.

The receptor GID1 that binds to active GA was discovered by Ueguchi-Tanaka *et al.* in 2005. It was discovered in rice where a dwarf mutant (GA-insensitive Dwarf1) was found to be insensitive to GA and consequently, the respective gene, causing this suppression of shoot elongation, which was termed *GID1*. A detailed review on GA receptors and their role in GA signaling was later provided by the same Japanese investigators (Ueguchi-Tanaka *et al.*, 2007b). While one GA receptor was revealed in rice, there are three *GID1* homologues in Arabidopsis: AtGID1a, AtGID1b, and AtGID1c, all of which interact with the five DELLA proteins of Arabidopsis. The revealing of the *GID1* GA receptor in rice was a major step in the understanding of the perception and signal transduction of GA in plants. Interestingly, the awareness of GAs started in rice by Japanese investigators, and about 80 years later, it was Japanese investigators again who used rice for the discovery of a GA receptor. Arabidopsis served as a main tool for the investigation of another component of GA perception and signal transduction: the DELLA subfamily proteins of the GRAS family. These proteins have a sequence of five amino acids (hence the name as indicated above), as well as another motif of six amino acids, TVHYNP, near their N-terminal region. The DELLA concern different motifs of this protein, affecting the capability of DELLA to suppress GA-responsive genes; or, when the TVHYNP motif is defective, DELLA is unable to be bound with the GA–GID1 complex. Such mutations may thus cause constitutive DELLA effects or insensitivity to GA, respectively. The activity suppression of GA-responsive genes (e.g. binding to the promoters of these genes) is probably caused by motifs in the C-terminal region of DELLA termed VHIID, PFYRE and SAW domains.

As noted above, the GA–GID1–DELLA complex may enter a ubiquitination during which DELLA is degraded in proteasomes. For this to happen, F-box proteins are required. These are encoded by the *OsGID2* of rice and *AtSLY1* of Arabidopsis. Models of GA signaling are provided in Figs. 22 and 23. There is a problem of nomenclature in the literature. Although the five amino acids were termed DELLA, in many cases, the name DELLA was given to the protein that also has the TVHYNP motif in the N-terminal, and the VHIID, PFYRE and SAW motifs in the C-terminal. The *OsGID2* and *AtSLY1* genes are orthologous and encode F-box domain containing proteins. The F-box protein is a component of the ubiquitination complex SCF (having Skp1, cullin and F-box protein subunits). The SCF complex catalyzes the transfer of ubiquitin from E2 to the target protein (the DELLA proteins). The degradation of the DELLA in the proteasome takes place only after the DELLA is fully polyubiquitinated. OsGID2 and AtSLY1 share conserved amino acid motifs in the N-termini as well

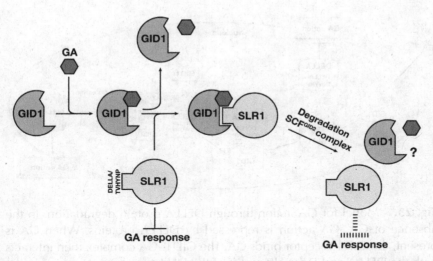

**Fig. 22.** Model of GA signaling in rice. Under low GA concentrations, SLR1 represses the GA responses. Under high GA concentrations, a soluble receptor, GID1, binds to GA; however, the binding is unstable and easily dissociates from the other. The GID1–GA complex specifically interacts with SLR1 at the site of DELLA and TVHYNP domains. The triple complex composed of GID1–GA–SLR1 is stable and does not easily dissociate. The triple complex is in turn targeted by the SCF$^{GID2}$ complex and the SLR1 protein is degraded by the 25S proteosome, which releases the repressive state of GA responses. (From Ueguchi-Tanaka *et al.*, 2007a).

as in the C-termini. Deletions in the C-termini cause loss of function. Several lines of evidence indicated that the DELLA proteins are the targets of complexing by SCF$^{GID2}$ and SCF$^{SLY1}$. It is not yet clear whether or not the phosphorylation of DELLA is a prerequisite for DELLA degradation that is initiated by active GAs.

Generally, there is a positive correlation between the level of activity of a given GA and the ligand specificity of this GA to GSI–GID1. However, there are exceptions: the GA$_4$-binding affinity to GID1 is about 20 times higher than the GA$_3$-binding to GID, but GA$_4$ is frequently lower in physiological activity than GA$_3$. This paradox was explained by the quicker degradation of GA$_4$. Based on my own observations, I have noted that some plants — such as

*Phytohormones & Patterning*

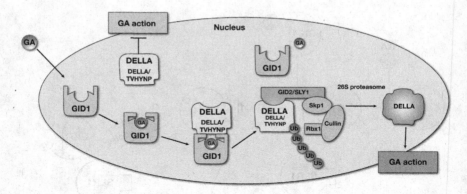

**Fig. 23.** Model for GA action through DELLA protein degradation. In the absence of GA, GA action is repressed by DELLA proteins. When GA is present, the GID1 receptor binds GA. The GID1–GA complex then interacts with the DELLA and TVHYNP motifs of the DELLA proteins, resulting in the recognition of DELLA proteins by the SCF$^{GID2/SLV1}$ complex (consisting of Skp1, Cullin, F-box protein, and Rbx1). After DELLA proteins are polyubiquitinated by the SCF$^{GID2/SLV1}$ complex, DELLA proteins are degraded through the 26S proteasome pathway, and as a consequence, GA action is activated. This consecutive reaction is predicted to occur in the nucleus. Ub, ubiquitin. (From Hirano *et al.*, 2008).

cucumbers — have a strong physiological response to GA$_4$ and only a weak response to GA$_3$. Indeed, in mutants that are defective in GA$_4$ degradation, GA$_4$ is more active than GA$_3$. It should be noted that the DELLA/TVHYNP domains of the rice SLR1 and the conserved HSL (hormone-sensitive lipase-like) regions of GID1 are essential for the interaction between GID1 and SLR1.

The association and dissociation of GID1 and active GA occur frequently and also in the absence of SLR1, but once the GID1–GA complex binds to SLR1, the GID1–GA complex is stabilized. The "trio" complex can then serve as a target of GID2, causing the degradation of SLR1, through the ubiquitination of the SCF$^{GID2}$ complex.

While the rice GID1 was the first reported GA receptor, three similar GA receptors (AtGID1a, AtGID1b and AtGID1c) were soon detected by Nakajima *et al.* (2006) in Arabidopsis. As in the rice receptors, the

strength of binding of the Arabidopsis receptors was positively correlated to the physiological activity of the various GAs in Arabidopsis.
Since Arabidopsis has five types of DELLA proteins and three GA
receptors, 15 combinations are possible. It appears that all these
combinations are effective in the signal transduction. Clearly, GAs
have different physiological and structural effects on the various
Arabidopsis organs, and it is reasonable to assume that the 15 combinations have various roles in these effects. Information on this
possibility is still insufficient (see below for additional information).
Interestingly, AtGID1 can replace the rice receptor for GA in rice
mutants that are defective in *gid1*.

While in most analyzed cases, GID1 is a soluble GA receptor, this
may not always be the case. In the system of $\alpha$-amylase expression in
the aleurone cells of cereals (e.g. barley), GA binds to a receptor in
the plasma membrane. The case of induction of $\alpha$-amylase in the
aleurone cells has some exceptional features. The aleurone cells of
cereals are exceptional with respect to GA biosynthesis: they do not
produce GAs and for the $\alpha$-amylase induction, the GA has to be
transported into these cells from the grain embryo. Interestingly, all
cells in rice, except for the aleurone ones, can produce GAs. Also,
while the aleurone GA receptor is localized in the plasma membrane,
the GA-dependent $\alpha$-amylase induction is extremely low in *gid1*
aleurone cells.

The perception described above together with the signal transduction of GA should be supplemented with several other observations as
noted below.

Studies by Hartweck and Olszewski (2006) showed that GA
inhibits the activity of cytokinins. They also showed that in cereal
grains with the *gid1* mutation (lack of functional GA receptor), even a
100-fold higher level than regularly optimal would not induce
the synthesis of $\alpha$-amylase in the aleurones. In such mutants, there is
also no elongation of the sheath of the second leaf of seedlings by GA.
On the other hand, in these *gid1* mutants, there is an accumulation of
GA. In *GID1*, there is a low synthesis of GA and a mechanism to
degrade GA. While most of the receptor (GID1) is soluble and located
in the cell nuclei, some *GID1* is also located in the cytosol.

The GID1 is a protein that has similarity to hormone-sensitive lipases (in animals), but the plant GID1 lacks lipase activity. This is obviously another example of the "The Pillars of the Mosque of Acre": an existing protein was changed during evolution in order to serve a very different purpose.

From the study of Hirano *et al.* (2008), it is evident that while the main function of the DELLA proteins is to repress the expression of GA-responsive protein, there are two additional activities of the DELLA proteins: they maintain the GA homeostasis (by affecting the expression of genes involved in GA synthesis) and they regulate the "cross-talk" between GA and other phytohormones, such as abscisic acid (see Fig. 24).

Schwechheimer (2008) further reviewed the GA–GID1–DELLA system and noted that contrary to the information about other proteins, the degradation of DELLA does not require phosphorylation. There is an apparently paradoxical situation: while GA is instrumental for the degradation of the DELLA proteins, GA facilitates the expression of DELLA. There is also a feedback inhibition of GA on enzymatic activities that are in the metabolic pathway of GA biosynthesis: both the activities of GA3ox ($GA_{20}$ to $GA_1$) and of GA20ox (e.g. $GA_{12}$ to $GA_9$) are reduced by active GA. On the other hand, there is an enhancement, by active GA of GA20ox, that converts active GA into inactive GA (as $GA_1$ to $GA_8$). While there appears to be no good evidence for the need of phosphorylation for DELLA degradation, there is proof that inhibitors of kinases may inhibit the degradation of the DELLA proteins. Another important information concerns the effect of light on germination of Arabidopsis seeds. Light and GA facilitate this germination. It seems that the light effect is conveyed by a factor termed PIL5 (phytochrome-interacting factor-3 like5). PIL5 reduces the GA activity, the light degrades PIL5 and thus elevates the GA level, consequently promoting the degradation of DELLA.

In order to reveal which protein expressions serve as targets to the DELLA proteins, Zentella *et al.* (2007) used microarray gene expression analysis in Arabidopsis. For their experiments, these investigators used the *ga1-3* mutant that is deficient in GAs. They then performed their

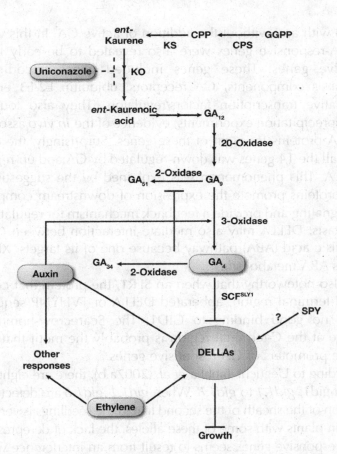

**Fig. 24.** GA biosynthesis and DELLA proteins. GAs are synthesized via a complex pathway involving multiple enzymes and cellular components. The main, and possibly only, consequence of the perception of active GAs (such as $GA_4$ in Arabidopsis) is the degradation of various DELLA proteins (such as GAI) via the $SCF^{SLV1}$ E3 ubiquitin ligase complex. DELLA protein levels are also regulated by auxin and ET, raising the possibility that the DELLA proteins are general regulators of plant growth that are regulated by several plant hormones rather than being specific components of GA signal transduction. CPP: copalyl diphosphate; CPS: copalyl diphosphate synthase; GGPP: geranylgeranyl diphosphate; KO: *ent*-kaurene oxidase; KS: *ent*-kaurene synthase; SPY: SPINDLY. (From Swain and Singh, 2005).

analyses with and without the addition of active GA. In this way, 14 early GA-responsive genes were also revealed to be early DELLA-responsive genes. These genes included genes encoding GA biosynthesis components, GA receptors, ubiquitin E2/E3 enzymes and putative transcription factors/regulators. They also found, by immunoprecipitation experiments, evidence of the *in vivo* association of DELLA proteins for eight of these genes. Surprisingly, the expressions of all the 14 genes was down-regulated by GA and up-regulated by DELLA. This phenomenon was explained by the suggestion that DELLA proteins promote the expression of downstream components of GA signaling and provide a feedback mechanism for regulating GA homeostasis. DELLA may also mediate interaction between GA and the abscisic acid (ABA) pathway because one of its targets, XERICO, regulates ABA metabolism.

It is also noteworthy that when an SLR1, the protein that contains at the N-terminal region, abberated DELLA or TVHYNP sequences, there is no good binding to GID1. The Scarecrow-homologous sequence at the C-terminal region is probably the motif that recognizes the promoters of GA-responsive genes.

According to Ueguchi-Tanaka *et al.* (2007a,b), there are eight alleles in rice of gid1: *gid1.1* to *gid1.8*. When *gid1.1–gid1.6* are defective, the elongation of the sheath of the second leaf in the seedlings is very short. At least in plants with some of these alleles, the lack of derepression of the GA-responsive genes seems to result from an interference with the SLR1–GID2 complex.

According to Griffiths *et al.* (2006) and Nakahima *et al.* (2006), there are three GA receptor genes in Arabidopsis: *AtGID1a*, *AtGID1b* and *AtGID1c*. All three genes (*atgid1a*, *atgid1b*, *atgid1c*) bind strongly to $GA_4$. The binding of GA is strongest by the AtGID1b protein, but its good binding is only in a small range of pH 6.4 to pH 7.5. When all three genes are mutated, the triple mutant has a very clear phenotype: there is hardly any germination of seedlings and the triple mutants have no sensitivity to GA. When the seed coat is removed, the seeds begin to germinate. The removal of seed coat in order to provide the seeds with the ability to germinate is a well-known practice of horticulturalists and geneticists, especially when they attempt to germinate seeds that result from interspecific crosses. Incidentally, several Solanaceae seeds will have a much better germination after treatment with active GA. Arabidopsis with the double mutation *atgid1a*, *atgid1c* is dwarfed. In the double mutant *atgid1a*, *atgid1b*, there is a strong reduction of male fertility. The three *AtGID1* genes are expressed in all the Arabidopsis tissues but at different levels. It seems that the proteins (receptors) of the three genes can compensate each other up to a certain level, but under normal (wild-type) conditions, these three genes differ in their impact on the signaling of derepression for specific GA-responsive genes.

The application of GA reduced the expression of all three *AtGID1* genes. Transgenic transfer of any of the three Arabidopsis *GID1* genes will restore the normal phenotype to the rice mutant *osgid1-1*. This indicates that the GA receptor is very conserved even among plants that have a large phylogenetic distance.

The signal transduction of GA, as described above, was found in all angiosperms that were analyzed. It was also found in the lycophyte *Selaginella moellendorffi* which belongs to the lowest lineage of vascular plants. However, the moss *Physcomitrella patens*, a bryophyte that diverted from the vascular plants about 430 million years ago, does not harbor this GA signal transduction. No traces of the GID1/DELLA-mediated GA perception and transduction were found in this moss. This provides us with an estimate for when the GID1/DELLA system evolved: probably when the aquatic plants emerged on land. In terms of years, this system evolved between the diversion of *Physcomitrella* (430 million years ago) and the emergence of *Selaginella* (400 million years ago).

# CHAPTER 4

# Brassinosteroids

## Short History and Occurrence

The brassinosteroids (BRs) are phytohormones that were discovered later than auxins, ethylene and gibberellins. The "discovery" is attributed to Japanese investigators who revealed a novel growth promoter in the pollen of *Distylium racemosum.* Their methods of evaluating growth-promoting capability were by using a rice bioassay. This bioassay was based on the angle formed between the leaf-blade and the sheath in the leaves of rice seedlings. This method is sensitive but has no specificity to BRs. An update on the "history" of BR research was provided in Chapter 2 of my previous book (Galun, 2007), and will not be further examined here. The bulk of the early BR research was by investigators of the United States Department of Agriculture (USDA) and affiliated investigators. An early significant publication was by Mitchell *et al.* (1970), in which four USDA investigators and one investigator from the University of Guelph (Canada) participated. They collected 1.5 g of pollen from oilseed rape (*Brassica napus*) and through several steps of extraction and separation on silica gel chromatographic plates (TLC) that were accompanied by a bioassay, they managed to isolate an "oily" product that was assumed to be a phytohormone. Their bioassay was based on the elongation of the second internode of bean (*Phaseolus vulgaris*) seedlings. They used several chemical methods to identify the "oily" product, but did not arrive at a real identification. They suggested that the product had a "glyceride structure" that represented a new family of plant hormones and termed the product as "Brassins." The USDA investigators did not give up, and in collaboration with other investigators,

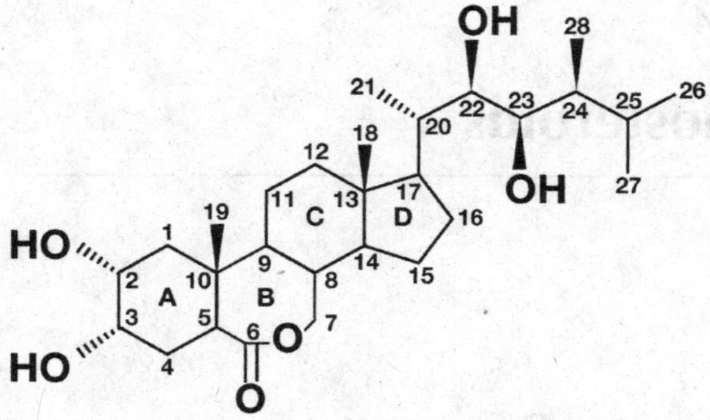

**Fig. 25.**   The structure of brassinolide, a commonly occurring BR with high biological activity, showing numbered positions mentioned in the text. In natural BRs, hydroxylation can occur in ring A at positions 3-, and/or 2-, and/or 1-; also found is epoxidation at 2,3-, or a 3-oxo-group. In ring B, alternatives are 6-oxo- and 6-deoxo-forms. In the side chain methyl, ethyl, methylene-, ethylidene, or *nor*-alkyl groups can occur at 24-methyl, and 25-methyl series is also represented. (From Clause and Sasse, 1998).

opted to start their extraction, purification and chemical identification with a huge quantity of pollen (about 230 kg). This resulted in a product that was chrystallized and a chemically defined compound was obtained (Fig. 25). The compound was termed brassinolide ($BR_1$) as a trivial name. The compound is a steroid with four "rings" and a side chain of seven carbon atoms carrying two hydroxylation sites and a lactone between $C_6$ and $C_7$ of ring B. The full chemical name is rather long: (22R, 23R, 24R)-2$\alpha$,3$\alpha$,22,23-tetrahydroxy-24-methyl-$\beta$-homo-7-oxa-5$\alpha$-cholestan-6-one.

   In this chapter, the trivial names of BRs will be mainly used for ease of reference. Changes in the hydroxylations, especially in the A-ring and in the side chain, as well as lack in the $C_6$–$C_7$ lactone will yield numerous BRs (see Clause and Sasse, 1998). When updated in Fujioka and Sakurai (1997), and Clause and Sasse (1998), there were already over 40 members in the BR family that are hydroxylated derivatives of cholestan. Every single angiosperm species seems to contain its own members of the BR family. In

Arabidopsis, Fujioka *et al.* (1996) identified four BRs in the shoots: Catasterone, 6-deoxocastasterone, typhasterol and 6-deoxotyphasterol. The level of content of the BRs was estimated as 0.1–1.0 ng per g (FW). They did not reveal BR in these shoots and suggested that BR is an active BR in Arabidopsis, but due to its low level, could not be detected in their analysis. There is an obvious difference in BR kinds and levels in the different components of plants. In general, the BRs appear to be much higher in pollen and immature seeds, and lower in other plant organs. For example, pollen and immature seeds contain 1–100 ng/g (FW) of BR, while shoots and leaves contain only 0.01–0.1 ng/g of this compound. An uncommonly high level of BR and castasterone (about 30 ng/g) was detected in crown gall cells of *Catharanthus roseus*. Young growing tissues generally have higher levels of BRs than mature tissues.

Several groups of investigators attempted to obtain general rules for structure-activity relationships among the BRs. However, no clear rules were obtained, although the 7-oxalactone and 6-keto forms seem to generally be the most active forms. As "activity" was evaluated by bioassays and these may differ in their response to specific BRs; the situation is complicated.

## Biosynthesis and Metabolism

Much of the early work on the biosynthesis of BRs was performed by several teams of Japanese investigators. These investigators used plant cell cultures and young seedlings for their research. These studies were reviewed by Yokota (1997), and by Clouse and Sasse (1998). Mainly based on Arabidopsis and pea mutants, feeding experiments, biological activities and the identification of the chemical structure, it was agreed that during the biosynthesis of sterol leads from squalene-2,3-epoxide to lanosterol, squalene in plants is converted to cycloartenol and from there (by some steps), to campesterol. There is now a consensus that campesterol in plants is the progenitor of brassinolide. Compounds such as teasterone, typhasterol and castasterone were suggested as intermediates. Yokota (1997) thus proposed

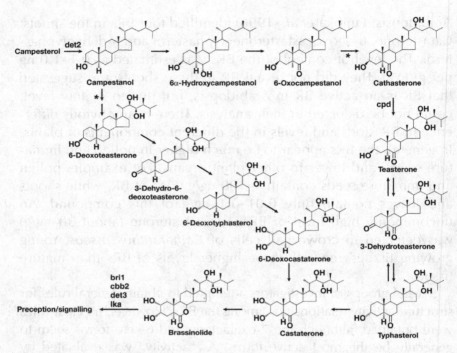

**Fig. 26.** Biosynthesis of brassinolide from campesterol. Campesterol is converted to campestanol, which is then metabolized to castasterone through bifurcation of the early or late C6 oxidation pathway. Castasterone is finally converted to brassinolide. Bold letters indicate putative lesions in the biosynthesis or sensitivity mutants of Arabidopsis. Pathways indicated by asterisks are hypothetical, due to insufficient metabolic evidence. (From Yokota, 1997).

the metabolic pathways from campesterol to brassinolide (Fig. 26). Note that from campesterol, the pathway is split: in one "branch," starting from $6\alpha$-hydroxy campastanol, $C_6$ is oxidized while in the other "branch," there are several metabolic steps, in which the compounds are not oxidized at $C_6$ — until castasterone, where $C_6$ is oxidized. The latter branch was termed the "late oxidation" pathway.

The earlier steps of BR biosynthesis, starting from mevalonate up to campesterol and from there, by the $C_6$ oxidation of campestanol

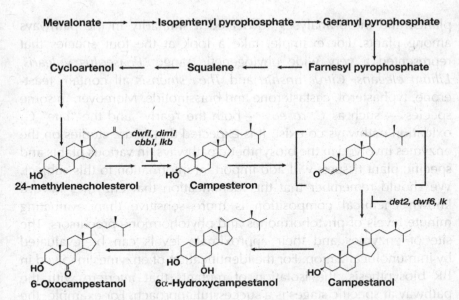

**Fig. 27.** Biosynthesis of early members of the BR biosynthetic pathway via the mevalonate and isoprenoid pathway. Campesterol is a bulk sterol, also found in membranes, while campestanol and later derivatives are considered committed to BR biosynthesis. Proposed blocks in BR biosynthesis in Arabidopsis (*dwf1, din1, cbb1, det2, dwf6*) and pea (*lkb, lk*) mutants are indicated. The structure in brackets is a probable intermediate based on molecular genetic and biochemical studies. (From Clause and Sasse, 1998).

to 6α-hydroxycampestanol (the "early oxidation pathway") were presented by Clouse and Sasse (1998) as shown in Fig. 27.

It should be noted that the scheme of BR biosynthesis is not applicable for all angiosperms. For example, castasterone is converted to brassinolide in *Catharanthus roseus* but not in tobacco or rice. Moreover, the pool level of intermediate compounds may be very different in different plants: before a quick conversion, the pool may be extremely low, while in front of a very slow conversion, there can be a build-up of a pool. When the pool is extremely low, it may not be detected suggesting differences in the biosynthesis pathway. Clearly, analyses of intermediates were performed only with some plants, and these analyses advocated a similarity in the pathways among plants. For example, take a look at the four species that represented a very wide phylogenetic range: *Phaseolus vulgaris*, *Lilium elegans*, *Citrus unshiu* and *Thea sinensis* all contain teasterone, typhasterol, castasterone and brassinolide. Moreover, in some species — such as *C. roseus* — both the "early" and the "late" $C_6$ oxidation pathways co-exist. It is expected that future studies on the enzymes involved in the biosynthetic pathways, in various plants and specific plant tissues, will add important information to this subject. We should remember that the identification that enzymes have a known chemical composition is more sensitive than evaluating minute levels of phytohormones and phytohormone precursors. The site of enzymes and their approximate levels can be evaluated by immunolocalization. For the identification of enzymes involved in BR biosynthesis, the isolation of mutants that interfere with the pathway at specific stages is a successful approach. For example, the mutant *det2* is deficient in several intermediates, such as campestanol. In *det2*, the respective enzyme DET2 does not function and conversions between campesterol to campestanol are inhibited. Interestingly, DET2 is rather similar to a mammalian steroid, $5\alpha$-reductase, that catalyzes the reduction of 3-oxo, $\Delta^{4,5}$ steroids as testosterone. The DET2 can rescue mammalian mutants that are deficient in this $5\alpha$-reductase, but it is also important to note that DET2 failed to rescue other mammalian mutants that are defective in the reduction of $3\beta$-hydroxy, $\Delta^{5,6}$ steroids, such as cholesterol. This suggests that at least parts of the BR synthesis pathways are phylogenetically very "old" and were developed before the branching of animals from plants.

The general strategy to reveal the biosynthetic pathway of BR is to induce mutation (up to recently it was frequently done with Arabidopsis) and then to apply several suspected intermediates until one of the latter restores the wild-type phenotype. Indeed, such studies revealed numerous mutants that are defective in specific enzymes. Moreover, several of these enzymes are very similar to mammalian enzymes involved in steroid metabolism: for example, CPD of Arabidopsis that converts cathasterone to teasterone (Fig. 28)

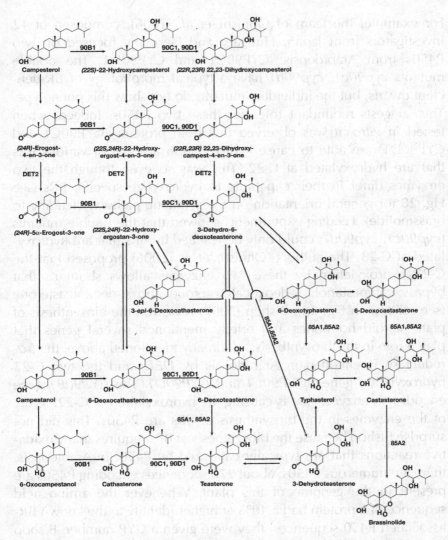

**Fig. 28.** Revised BR biosynthesis pathway. (From Bisop, 2007).

has a 50–90% identity with conserved-binding domains of microsomal cytochrome P450s.

More recent studies on the biosynthesis of BRs focused on several cytochrome P450 proteins that catalyze oxidative reactions.

For example, the team of Ohnishi *et al.* (2006), composed of 12 investigators from Japan, Hungary and Germany focused on two P450s from Arabidopsis: CYP90C1 and CYP90D1. The double mutants *cyp90c1, cyp90d1* have a typical morphology of BR-deficient dwarfs, but the individual mutants do not show this phenotype. This suggests redundant roles for these two P450s. Indeed, when tested *in vitro*, it was observed that both proteins (CYP90C1 and CYP90D1) are able to cause hydroxylation at C-23 of various BRs that are hydroxylated at C-22. This was so even though the two enzymes differ in their capability to hydroxylize specific BRs (see Fig. 28 for general orientation on the synthetic pathways leading to brassinolide). Feeding experiments showed that the double mutants (*cyp90c1, cyp90d1*) could only be rescued by BRs that are hydroxylated at C-23. The authors (Ohnishi *et al.*, 2006) proposed that the C-23 hydroxylation by these two enzymes allows shortcuts that bypass campestanol, 6-deoxocathasterone and 6-deoxoteasterone (see scheme in Fig. 28). Bishop (2007) reviewed the biosynthesis of plant steroid hormones and briefly mentioned several genes that play a role in this biosynthesis. As already mentioned above, the $5\alpha$-reductase is encoded in Arabidopsis by *DET2* and the two C-23 hydroxylation genes, *CYP90C1* and *CYP90D1*. The *CYP90B1* gene encodes an enzyme that is causing the hydroxylation of C-22. Many of the enzymes in the biosynthesis of BRs are P450s. This did not surprise Bishop because the biosynthesis of BRs requires many oxidative reactions that are typically catalyzed by P450. In Arabidopsis, there are numerous P450s: about 270 sequences encoding P450s are present in the genome of this plant. Whenever the amino acid sequence of a protein had a 40% or higher identity to the known BR-associated P450 sequences, they were given a CYP number. Bishop constructed a phylogenetic tree of 11 CYPs and found that some of them (such as CYP90C1 and CYP90D1) are closely related while others (such as CYP90A1 and CYP72C1) are rather distant.

Up to 1996, it was suggested that CPD (CYP90A1) is responsible for the C-23 hydroxylation. However, it was then found that the CYP90A1 is involved in an earlier step of the BR biosynthesis pathway: teasterone was found to rescue *cpd* mutants (see Figs. 26 and

28). It was then suggested that CYP90A1 catalyzes the oxidation at C-3 (e.g. of campesterol, hydroxycampesterol and dihydroxycampesterol). The DET2, that is a 5α-reductase, was shown to use as substrate compounds that are already hydroxylated at C-22 (Fig. 28), while CYP90B1 has a preference to hydroxylate campesterol rather than campestanol.

The biosynthetic pathways are still far from being clarified. Studies on specific tissues, even those that are part of the very same plant, may reveal differences in these pathways. A clue for such a difference came from the analyses of BRs in tomatoes: tomato fruits contain large amounts of brassinolide while castasterone predominates in vegetative tissues of the tomato plant.

The process of controlling the levels of BR biosynthesis causing BR homeostasis is an emerging subject of research. The details are still far from being fully discovered. This subject was studied by seven investigators from the Department of Plant Biology of the Carnegie Institution and the Department of Biological Sciences of Stanford University: He *et al.* (2005). (The present author does not like the name "Biological Science" because "Biology" is life science, thus "Biological Science" is "Life Science Science.") Contrary to some phytohormones (e.g. auxin), the biosynthesis of BRs takes place very near to the site that controls cell division and growth (in some cases, the very same cell). Consequently, the impact of BRs is tightly connected to their biosynthesis (in auxin, the transfer plays a major role). The perception and signal transduction of BRs will be discussed in the next section. Here, it should only be noted that BR receptors are at the cell surface and the signal transduction reaches the nucleus. BRs are involved in the accumulation of BZR1 and BZR2/BES1 in the nucleus. It was not known how BZR1 and BZR2/BES1 mediate the regulation of expression of BR synthesis genes. He *et al.* (2005) found that BZR1 is a transcriptional repressor that has a DNA-binding domain and binds directly to promoters of feedback regulation of BR biosynthesis genes. It was thus suggested that BZR1 coordinates BR homeostasis and by that affects growth responses.

A similar study was conducted at the Salk Institute for Biological Studies by Nemhauser and Chory (2004). They studied the direct targets of BIN2, BZR1, and BES1, and found that they are linked to the BR perception at the cell surface and the regulation of BR synthesis genes in the nucleus. They thus revealed details of the role of several factors that act in the negative feedback regulation of BR levels. The study of Nemhauser and Chory (2004) was an extension of previous studies (e.g. Yin *et al.*, 2002), which found that BES1 accumulates in the nucleus in response to BRs, regulating the gene expression and promoting Arabidopsis stem elongation. In fact, there are two kinds of BR-regulated genes in Arabidopsis: those that increased their gene expression and those that reduced it. The CPD gene — that encodes a cytochrome P450 enzyme — was found to be strongly repressed. Typically, the increase and repression of gene expression was "modest": there was only a 2–5 fold change while other phytohormones commonly change gene expression by 10–100 fold.

Part II of this book will examine the interactions of several other phytohormones with BRs and how they affect the development of plant organs. Here we should only note that in several cases, BRs interact with auxin. Early auxin responses are strongly up-regulated by BRs. Nemhauser and Chory (2004) used a clever approach to reveal the interaction of auxin and BRs; their idea was based on previous knowledge that promoters of most auxin-responsive genes contain an auxin-responsive element (T/AGTCTC). The investigators synthesized a construct containing five repeats of the element (termed DR5) and produced transgenic seedlings with DR5: GUS. They found that the GUS staining was enhanced in these seedlings when exposed to external BRs.

While numerous studies were devoted to investigating the biosynthesis of BRs, there is very little information on the degradation of these compounds.

## Perception and Signal Transduction

Over the past decade, several research teams have actively investigated the signal transduction of BR. These studies were reviewed by several authors, such as Clouse and Sasse (1998), Clouse (2002),

Peng and Li (2003), Nemhauser and Chory (2004), Wang and He (2004), Torii (2004), Vert *et al.* (2005), Li and Jin (2007), Gendron and Wang (2007), and Wang *et al.* (2008a).

A report on a putative leucine-rich repeat receptor kinase involved in BR signal transduction was published by Li and Chory (1997). This report already supplied the basic information about the BR receptor. The latter investigators based their study on an Arabidopsis mutant, *bri1*, which was found to be specifically insensitive to BR, but still sensitive to other phytohormones such as auxin, cytokinin, ethylene, abscisic acid and gibberellins. The *BRI1* gene was revealed by Clouse *et al.* (1996), and the gene that encodes the BRI1 protein was localized to the bottom of chromosome IV. Li and Chory found 18 BR-insensitive mutants that are alleles of *BRI1*. The gene was cloned and its expression pattern was examined, and a ubiquitously-expressed putative BR receptor kinase was thus revealed. It was located at the cell membrane and its extracellular region contained 25 tandem leucine-rich repeats (LRRs). Later analyses found that the number of LRRs is actually 24. The LRRs of BRI1 have considerable similarity to the LRRs of the CLV1 protein that is involved in the regulation of the shoot apex in plants (see Galun, 2007 for details). The BRI1 appeared to be constitutively and ubiquitously expressed throughout the development of Arabidopsis plants and in response to different light conditions.

A few years after the publication of the role and sequence of BRI1, Clouse (2002) provided in his review a much more detailed picture of BR signal transduction. Several additional components involved in this signal transduction were revealed at the surface of cells as well as in the cytoplasm and in the nucleus. It became evident that the plant system of BR signal transduction differs considerably from the signal transduction of steroid hormones in animal systems, although there is a considerable similarity in the biosynthesis of BRs and animal hormones from mevalonate to squalene-2-epoxide; the later biosynthetic steps of BRs and animal steroid hormones *do* differ. However, several plant enzymes involved in the BR biosynthesis can substitute faulty animal enzymes in steroid-hormone biosynthesis in animals. It was also found that the

signal transduction of steroid hormones is rather abbreviated in relation to the signal transduction of BRs. In the former systems, the
steroid binds to an intracellular receptor that consists of a variable
N-terminal domain that is commonly associated with transcriptional
activation. The receptor-ligand complex recognizes specific
sequences in the promoters of steroid-responsive animal genes
resulting in altered gene expression and steroid-mediated changes in
cell physiology. Animals do have a more elaborate signal transduction for some peptide ligands (such as the "transforming growth
factor $\beta$, TGF-$\beta$). It appears that plants adopted a somewhat similar
signal transduction process for the BR signaling. Nevertheless, plants
do not have intracellular receptors for BR and the receptors are at the
plasma membrane. Essentially, the BR receptors (e.g. BRI1) have
three main components: an extracellular ligand binding domain, a
single-pass transmembrane sequence and an intracellular kinase
domain (see Fig. 29). In his review, Clouse (2002) could already draw
the scheme of BRI1 as well as the scheme of the BR Associated
Kinase 1 (BAK1) and marked on these schemes the mutants that affect
the proteins' function.

The BAK1 was identified by the "yeast two hybrid" screen (a
method that reveals interaction between two proteins). BAK1 only
has 5 leucine-rich repeats in its extracellular domain and also lacks
the 70 amino acids island that exists in BRI1. The localization and
distribution of BAK1 in plant cells is similar to that of BRI1, and the
two are always associated with each other as was revealed when they
were marked by GFP. Mutants of *BAK1* have a less severe phenotype
than BRI mutants: knockout mutants of *BAK1* have a weak *bri1*-like
phenotype, while regular *bri1* mutants have a severe phenotype.
Several experiments indicated that there is normally a direct *in vivo*
interaction between BRI1 and BAK1 in the plant cell membrane. The
details of the interaction between BRI1 and BAK1 were only
provided recently (Wang *et al.*, 2008a) as shown in Fig. 30. According
to recent studies (e.g. Wang and Chory, 2006), BR first binds to the
island domains of the inactive BRI homodimers and the protein BKI1,
which in the absence of BR is associated with the intracellular region
of BRI1 (suppressing the phosphorylation of this region), is removed

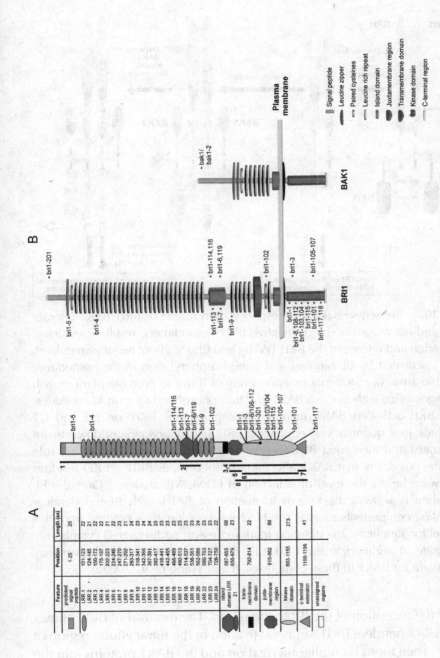

**Fig. 29.** Perception of brassinosteroids. (a) The BRI1 receptor (from Vert *et al.*, 2005). (b) BRI and BAK1 (from Peng and Li, 2003).

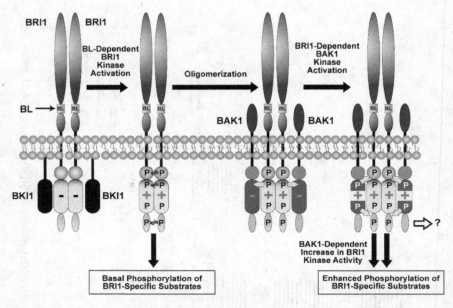

**Fig. 30.** The sequential transphosphorylation model of BRI1/BAK1 interaction and activation. BL binds inactive BRI1 homodimers, resulting in kinase activation and release of the BKI1 (Wang and Chory, 2006) negative regulator. BRI1, activated by BL binding and transphosphorylation in the homodimer (double arrows), can signal independently of BAK1 to promote plant growth or oligomerize with inactive BAK1 and transphosphorylate it on KD residues. This BRI1-activated BAK1 then transphosphorylates BRI1 on JM and CT residues, thus quantitatively enhancing BRI1 phosphorylation of downstream substrates and increasing BL-dependent signaling. The model does not rule out the possibility that BAK1 also transphosphorylates BRI1 on KD residues that were below the level of detection in LC/MS/MS analyses. Only the BL signaling is shown; the known interaction of BAK1 with BL-independent signaling components such as the FLS2 receptor kinase is omitted from this model for simplicity. The question mark represents undiscovered cytoplasmic substrate in addition to the BRI1 CD. BKK1 and SERK1 (not shown) may substitute for BAK1 in these complexes. (From Wang *et al.*, 2008a).

after the reception of the BR to the island. The removal of BKI1 allows the BR-dependent BRI1 kinase activation in the intracellular region of BRI1. Then there is an oligomerization and the BAK1 proteins join the complex and increase the kinase activity.

By this stage, we are already engaged with numerous proteins in the signal transduction of BR, hence the reader is referred to the following list.

## Proteins involved in BR signaling

**BRI**, the BR receptor kinase
**Components of BRI1** (see Fig. 29)
— LRRs, leucine-rich repeats
— Island domain
— Transmembrane domain
— Juxta-membrane region
— Kinase domain

**BAK1**, coreceptor to BRI1 (see Figs. 29 and 30)
**BIN2**, a negative regulator of BR signaling
**BES1**, a transcriptional modulator of BR-responsive genes; it binds to the CANNTG motif
**BZR1**, a transcriptional modulator of BR-responsive genes; it binds to the CTGT (T/C)G motif
**BSU1**, a nucleus localized in Ser-Thr protein phosphatase, preferentially found in elongating cells
**BSK1, BSK2 and BSK3**, homologous BR-signaling kinases that are phosphorylated by BRI1; they activate the BR signaling downstream of BRI1
**14-3-3 proteins**, a family of dimeric proteins that can modulate interactions between proteins; in rice, 14-3-3 proteins directly inhibit OsBZR1 function, reducing the latter's nuclear localization
**BKI-1**, inhibitor of phosphorylation at the intracellular region of BRI1.

The sequence of perception and transphosphorylation of the receptor was updated by Wang *et al.* (2008a). Their suggested scheme (Fig. 30) is a simplification of the actual process and several processes in this scheme are not yet fully understood. For example, BAK1 may also transphosphorylate BRI1 on kinase domain residues that were below the level of detection in the analyses. BAK1 may also interact with brassinolide-independent signaling components. Moreover, BAK1 could have yet undiscovered cytoplasm substrates in addition to the BRI1 cytoplasmic domain. Still, the main claim by Wang *et al.* (2008a) — that the binding of brassinolide to BRI1 is taking place before the association of BAK1 with BRI1 — seems to be valid. It should be noted that the clarification of BR perception was studied mainly in Arabidopsis and rice. In other angiosperms, the signal transduction may differ. For example, in tomatoes, BRI1 perceive not only BRs but also the peptide hormone system.

While the main steps of brassinolide (and possibly other BRs) perception is nearly clarified, the more downstream steps of its signal transduction are still part of very active research studies (e.g. Mora-Garcia *et al.*, 2004; He *et al.*, 2005; Vert *et al.*, 2005; Wang and Chory, 2006; Vert and Chory, 2006; Li and Yin, 2007; Gendron and Wang, 2007; Bai *et al.*, 2007; Gampala *et al.*, 2007; Tang *et al.*, 2008).

Downstream of the perception system for BRs (meaning BRI1, BKI1 and BAI1), there is the first link between the reception and the regulation of the BR-responsive genes. This is the "glycogen synthase kinase-3 (GSK)-like kinase brassinosteroid-insensitive 2" protein, termed BIN2. When BIN2 is active, it suppresses all the signaling that is downstream of it. However, when BR is bound to the extracellular regions of the receptors (BRI1 and BAK1), the interphosphorylated kinase region of these receptors inhibits the activity of BIN2. Active BIN2 negatively regulates the BR signaling by phosphorylating and inhibiting BZR and its homologue BZR/BES1 (BZR). The inhibition is apparent by repressing the binding to the promoters of BR-responsive genes. Is the phosphorylation of BES1 and BZR1 by BIN2, targeting these two proteins to degradation (e.g. by ubiquitination)? And is the change in location of BES1 and BZR1 from the cytoplasmic to the nucleus an obligatory component of the BR signaling? These possibilities were suggested by Li (2005), while Vert and Chory (2006) assumed that the degradation of BES1 and BZR1 play a minor role in the signal transduction of BR. The latter investigators provided evidence for the inhibition of binding of BES1 to the promoters of BR-responsive genes and thus stopping the BR signaling.

We should note that with the BR signaling, we are faced with a complicated system in which the expression of BR-responsive genes is regulated by transcription factors. A survey about the increase and decrease of proteins upon the addition of BR showed that hundreds of proteins are affected, probably by the same general system of BR signaling. The changes occurring at the level of specific proteins, in specific cells, causing direct changes in growth and cell division, as well as changes in the biosyntheses and metabolism of phytohormones, are probably all affected by the signal transduction of BR. How can the same transcription factors (BES1 and BZR1) cause such very specific changes in very specific plant cells and in specific developmental stages of the plant? We do not know the answer to this question yet, but it may well be that for regulating the expression additional factors are bound to BES1 and BZR1, and only after these factors bind to the transcription factors, the two transcription factors attain their specificity. The yet unknown factors may be unique to the specific cells.

Another open question is how do BR-activated receptors cause the inactivation of BIN2? In other words, it is not yet clear how the BRI1 kinase and the BAK1 kinase, located at the plasma membrane, transduce the signal to the cytoplasmic BIN2: no direct interaction has been observed between BRI1 (and BAK1) and BIN2 (see Gendron and Wang, 2007; Tang *et al.*, 2008). Candidates that may mediate between BRI1 and BIN2 are the BSK protein kinases. The latter possibility emerged from a recent study by Tang *et al.* (2008). Using proteomic studies and several other analytical procedures with plasma membrane proteins, they identified homologues of BR-signaling kinases (e.g. BSK1, BSK2 and BSK3). It was shown that red or green spots resulted from plasma membrane proteins of 7-day old *det2* Arabidopsis seedlings, that were treated either with 100 nM BR (and labeled red with Cy5) or mock-treated (and labeled green with Cy3) and separated by a two-dimensional electrophoresis procedure. BR treatment resulted in red spots and BR-repressed seedlings provided green spots. The proteins of the spots were identified and found to contain putative N-terminal myristylation sites that could mediate the localization of the spot proteins (BSKs) with the cell membrane. The localization was assumed to be adjacent to the intracellular kinase region of BRI1. In the absence of BR, there is no phosphorylation of the kinase region and the BSKs retain their close association with the kinase region. The BR treatment causes phosphorylation of the kinase region and the latter prompts the phosphorylation of BSKs, probably causing the separation of BSKs from the intracellular region of BRI1. In the model of Tang *et al.* (2008), the "free" and phosphorylated BSKs cause the inhibition of the effect of BIN2 and thus establish the link between the receptor and a downstream regulator. The inhibited BIN2 will then not be involved in the regulatory effects of the BZR1/BES1 transcription factors. Are the BSKs affecting BIN2 directly or are there additional factors that mediate between the phosphory-lated BSKs and BIN2? There is no answer yet to this question but it is quite definite that BSKs function downstream of BRI1 and upstream of BIN2.

The 14-3-3 is a family of dimeric proteins that can modulate the interaction between proteins (Aitken, 2006). These proteins, that are abundant in the brain, were first revealed in 1968, and purified and characterized in 1980. Since then, they have been the subject of numerous studies, mainly in animals (human brain and *Drosophila*), but some studies were also performed with plants. The unique name 14-3-3 stems from the initial isolation of these proteins by two-dimensional DEAE-cellulose chromatography, followed by starch gel electrophoresis: the 14-3-3 proteins eluted in the 14th fraction of bovine brain homogenate when these homogenates were applied to a DEAE cellulose column, and fractions 3.3 in the latter step included the 14-3-3 proteins. It was recently found that 14-3-3 proteins are involved in the signal transduction of BR of Arabidopsis (Gampala *et al.*, 2007) and of rice (Bai *et al.*, 2007). The study regarding Arabidopsis was a combined effort of an impressive number of 17 investigators in six different laboratories in the USA and China. These investigators found that BIN2-catalyzed phosphorylation of BZR1/BZR2 not only inhibits DNA binding, but also promotes binding to 14-3-3 proteins. Mutations of the BIN2-phosphorylation site in BZR1 abolish 14-3-3 binding, lead to an increased nuclear localization of BZR1 and also enhance BR responses in transgenic plants. Furthermore, BR defi-ciency increases cytoplasmic localization while BR treatment induces rapid nuclear localization of BZR1/BZR2. The investigators concluded that 14-3-3 binding is required for efficient inhibition of phosphory-lated BR transcription factors, largely through cytoplasmic retention. It should be noted that in the scheme of Gampala *et al.* (2007), the BIN2 is located in the cytoplasm while according to J. Chory and associates (e.g. Mora-Garcia *et al.*, 2004; Vert and Chory, 2006), BIN2 is consti-tutively in the nucleus.

Bai *et al.* aimed their study at the OsBZR1 transcription regulator that affects the expression of BR-responsive genes in rice. They actu-ally looked for proteins that interact with OsBZR1. By using various means, such as the suppression of OsBZR1 by RNAi, they found that OsBZR1 plays an essential role in BR response in rice. Plants with suppressed OsBZR1 were dwarfed, had erect leaves and a reduced BR sensitivity. By a yeast two-hybrid screen, the investigators found that 14-3-3 proteins interact with OsBZR1. Mutation of a putative 14-3-3 binding site of OsBZR1 abolished its interaction with 14-3-3 proteins, both in yeast as well as *in vivo*. Such mutants of OsBZR1 proteins suppressed the phenotypes of the Arabidopsis *bri1-5* mutant and showed an increased nuclear distribution when compared with wild-type protein, suggesting that 14-3-3 proteins directly inhibit OsBZR1 function, at least in part by reducing its nuclear localization. Based on all of the above, Bai *et al.* (2007) suggested a conserved function of OsBZR1 and an important role of 14-3-3 proteins in BR signal transduction in rice.

Another nuclear protein that probably has a role in the signal transduction of BRs is BSU1. This is a nuclear Ser/Thr phosphatase that was found to be involved in the dephosphorylation of the tran-scription factor BES1, consequently controlling the expression of BR-responsive genes. The details of the function of BSU1 may emerge out of future investigations.

# CHAPTER 5

# Cytokinins

## History and Tales with Moral Lessons

The early history of cytokinins (CKs) is not merely interesting but also demonstrative of the old Hebrew maxim that says: "The envy of scholars increases wisdom (or knowledge)." This was stated in the *Baba Batra* (pp. 20–1) of the Bavli Talmud. This history was reviewed, with humor, by Amasino (2005). It is a true tale about two prominent scholars: F.C. Steward, a British scientist who became a professor at Cornell University, and Folke Skoog, a Swede who became a professor at the University of Wisconsin in Madison. Both were pioneers of plant tissue culture studies. Steward used slices of carrot roots in his tissue cultures, and Skoog used slices of tobacco stems. Steward improved the cultures with fractions of coconuts. Skoog encountered variability in the reactions of his tobacco cultures. He tried to use coconut milk but was not satisfied with the reaction of the callus that was derived from greenhouse grown tobacco plants. He then intended to add coconut fractions to his cultures. Skoog thus approached Steward asking him to provide details on the best coconut fractions. Steward wrote to Skoog that he would not provide such details and told Skoog to take his hands off coconut fractions. This sparked off the fighting spirit in Skoog, who had been a competing athlete in the past and who had represented Sweden in the 1500 m race in the Olympic Games of 1932. Within minutes of receiving Steward's letter, Skoog approached Professor Frank Strong of the Department of Biochemistry at the University of Wisconsin, and established a collaboration. The bioassays were done in Skoog's lab and the biochemistry and fractions of the coconut meat were performed in Strong's lab. All of the above happened between 1950 and 1951 but no active fraction with consistent effects on the bioassays was obtained. In 1951, a very talented post-doctoral fellow from Ohio State University joined Skoog's laboratory and initiated another approach to improve the tobacco tissue culture. This was Carlos O. Miller. Miller chose yeast extract and found that it caused a massive amount of undifferentiated callus growth and increased cell division. First, Miller used yeast extract from an old bottle but when the extract of this bottle dwindled, a fresh bottle of yeast extract was ordered, but this yeast extract had no effect on the tissue cultures. After many fractionations and bioassays, Miller found that the active fraction of the old yeast extract was precipitated by silver nitrate. This led to the suggestion that the active fraction may include purines or pyrinidines. A source for these is DNA, and Miller chose herring sperm DNA. Here the story of the yeast extract was repeated: an old sample of herring sperm DNA was active in the bioassay while a new batch was not active. It was finally discovered that if the new DNA was autoclaved for 30 minutes, the activity was re-established. The purified active fraction was crystallized and identified as 6-furfurylaminopurine. Within days, the Strong laboratory synthesized 6-furfurylaminopurine and determined that the latter was identical to the crystals derived from the autoclaved DNA. This compound was termed kinetin (Fig. 31). This was the first CK compound but it is probably not a

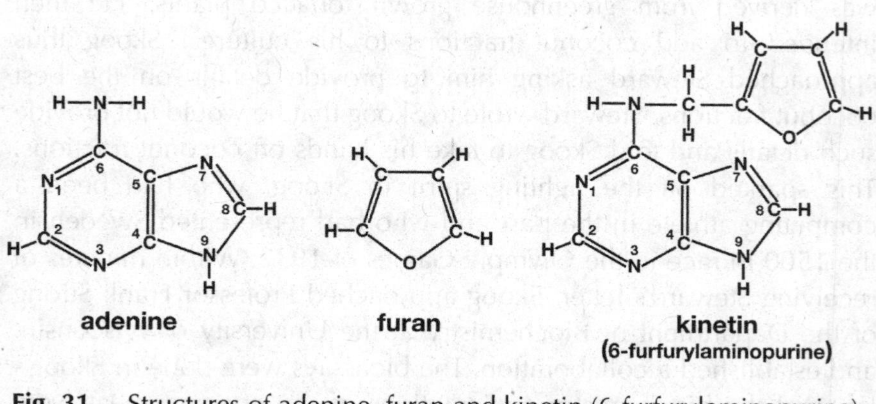

**adenine**　　　　　　**furan**　　　　　　**kinetin**
　　　　　　　　　　　　　　　　　　**(6-furfurylaminopurine)**

**Fig. 31.** Structures of adenine, furan and kinetin (6-furfurylaminopurine).

natural plant ingredient. The results of this study were published with the co-authorship of both the laboratories of Skoog and Strong (Miller *et al.*, 1955). However, the work in the two laboratories continued and additional purine-containing compounds were found to have a CK effect, such as 6-benzyladenine. The kinetin was quickly utilized by other laboratories. For example, Amos Richmond (an Israeli) and Anton Lang (then at UCLA) (Richmond and Lang, 1957) found that kinetin delayed leaf senescence.

At the same time, great efforts were being made by F.C. Steward to identify the active fraction from coconuts. He utilized a great amount of coconuts that became available after a storm in Florida. The extraction was by a big industrial device that the laboratory obtained from DuPont. A compound was isolated and identified as diphenylurea (a herbicide) and the results were also published in 1955 (Shantz and Steward, 1955). However, it turned out that this was an artifact: this industrial device was previously used by DuPont for the manufacture of diphenylurea, and the residual diphenylurea was extracted as if it came from the coconuts. Unfortunately, rivalry between scholars can also lead to very tragic results. An example for this is the rivalry between Nicolei Vavilov and Trofim Lysenko. Vavilov, a botanist and geneticist, became internationally famous due to his various studies, especially on geobotany and the origins of crop plants. He first worked in Saratov and then headed an institute in Leningrad. However, a few years before World War II, Vavilov fell out of favor with the Kremlin, which was then represented by Lysenko. Lysenko claimed (in the footsteps of the Russian investigator Michurin) that plants could be trained to resist stress and this resistance would then be inherited. Vavilov refused to accept this dogma, and as a result, Vavilov, together with two other scientists, was put to trial and the verdict was the death penalty. The two other scientists were killed but the verdict on Vavilov was changed to life in prison (in Saratov), but Vavilov died in prison after only about one year. A detailed biography of Nicolei Vavilov was given in a recent book (Peter Pringle, 2008). Lysenko became prominent and directed the agricultural research in the USSR. In 1948, he organized an "All Union" meeting to which all the geneticists of the USSR were invited. There, Lysenko detailed his dogma and blamed the USSR geneticists, claiming that while the USSR patriots sacrificed their lives in the defense of Stalingrad, the geneticists played with flies... An article was published in *Pravda* (the leading newspaper of the Soviet Union) during the "All Union" meeting, recommending to send all those who did not agree with Lysenko to Siberia. On the last day of the meeting, most of the geneticists stood up, one-by-one, and said that not because of the *Pravda* article, but due to their own convictions, they now agreed with Lysenko who was supported by the great friend of the Soviet scientists: Joseph Stalin. Those who did not accept Lysenko's dogma were indeed sent to Siberia. In spite of many years of great efforts, Lysenko was not able to "educate" wheat so that no wheat could be grown in the far north of the USSR. As a consequence of this, the Kremlin then sent Lysenko to Siberia. After several more years, in 1991, the whole USSR regime collapsed. The saying attributed to Hillel the Elder (who lived just before Christ, in Jerusalem) is: "He saw a scalp floating on the water and told the scalp that he was killed because he killed others and those who killed him will also be killed" (Mishna Pirkei Avot). It is as if Hillel had foreseen Lysenko's scalp...

## Occurrence, Summary of Roles in Plant Architecture and Utilization of CKs

A hunt for CKs in plant tissues started a few years after the chemical identification of kinetin. Already in 1960, Carlos O. Miller submitted a report on the discovery of a kinetin-like compound from maize grains (Miller, 1961) and two years later, in 1963, Letham undertook a detailed analysis of the crystals of this compound from 70 kg of immature maize grains. The compound was coined: "Zeatin," and the full chemical name of zeatin is: 6-(4-hydroxy-3-methyl-*trans*-2-butenylamino) purine. A shower of publications followed on the occurrence of CKs in several tissues of many plants. It became evident that natural CKs exist in all plants, although the levels of these CKs vary between plants and their tissues. Moreover, in addition to zeatin, other CKs were found in plants. These naturally occurring

CKs could be put into two major groups: isoprenoid CKs and aromatic CKs (Fig. 32). More details on these CKs were supplied in the review by Sakakibara (2006). The literature resulting from the intensive search of natural CKs in plants during 1961–1970 was updated by Skoog and Armstrong (1970). The latter review also handled the issue of CKs that have been found by several investigators in tRNA hydrolysates. However, it was not shown that CKs are involved in tRNA biosynthesis, neither that tRNAs are direct sources of CKs. Such information was to be published much later (Yevdakova *et al.*, 2008).

Information on the impact of CKs on the architecture of plant organs will be provided in Part II of this book, where the interactions between CKs and other phytohormones will also be discussed. Hence, here only a few cases of CK effects on plant tissues and organs will be mentioned. One of these CK effects takes place in the apical meristem where new leaves are initiated. There the KNOX protein promotes CK biosynthesis (Yonai *et al.*, 2005; Shani *et al.*, 2006), while GA inhibits the response to CK. Studies in rice indicated that a specific gene (*LONELY GUY, LOG*) is involved in the activation of CK, by converting inactive CK into the active free-base forms (Kurakawa *et al.*, 2007). The CK activity is thus very much focused on cells that require this activity for shaping the shoot apex. Other literature provided the notion that most of the active CK is produced in the roots and transported from there to other locations, such as the shoot apex meristem. However, a recent investigation revealed that localized CK biosynthesis also plays an important role in organ growth and patterning (Zhao, 2008). As described below, there are several routes in plants for the biosynthesis of CKs. Earlier studies presumed that the active t-zeatin (tZ) is the predominant CK in plants and that the c-zeatin (cZ) detected in plants is the result of isomerization. Later, it was discovered that plants also contain genes that are related to the bacterial IPT (isopentenyl transferase). Several such genes were found in the Arabidopsis genome. Moreover, plants also have tRNA-IPT genes. Hence, there are multiple *IPT* genes in plants, which suggests that there may be a tissue-specificity in the synthesis pathway that is typical for a given tissue or organ. Howell *et al.* (2003)

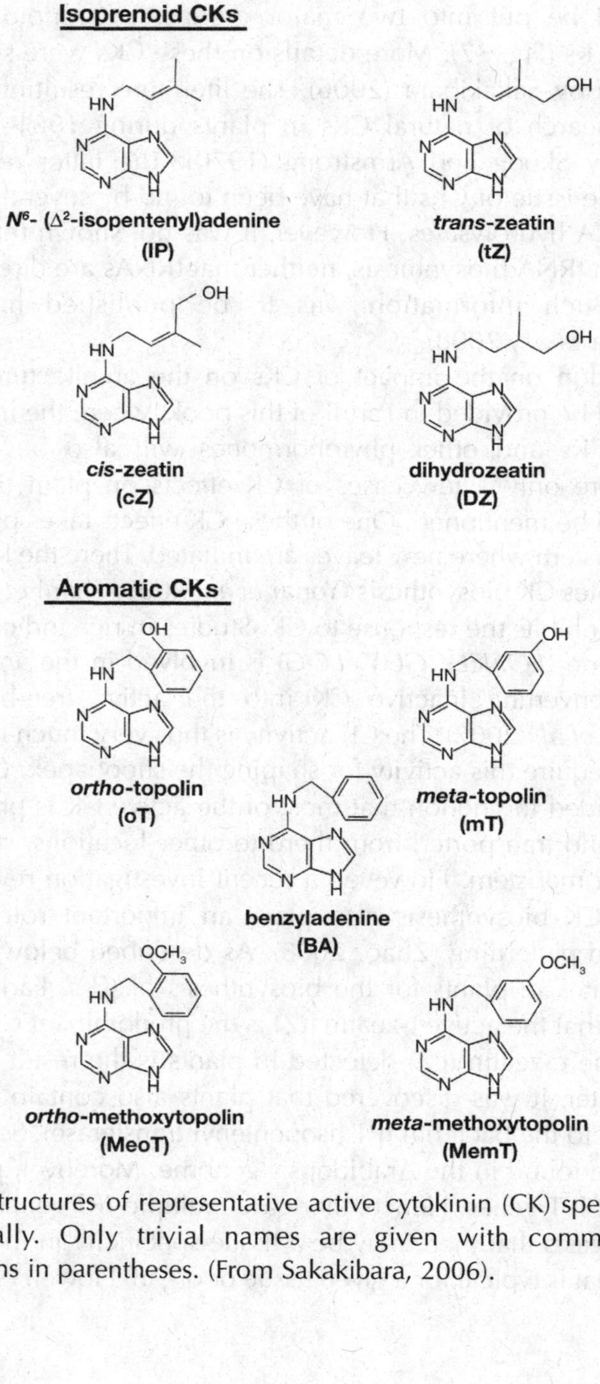

*Phytohormones & Patterning*

## Isoprenoid CKs

$N^6$- ($\Delta^2$-isopentenyl)adenine
(IP)

*trans*-zeatin
(tZ)

*cis*-zeatin
(cZ)

dihydrozeatin
(DZ)

## Aromatic CKs

*ortho*-topolin
(oT)

*meta*-topolin
(mT)

benzyladenine
(BA)

*ortho*-methoxytopolin
(MeoT)

*meta*-methoxytopolin
(MemT)

**Fig. 32.** Structures of representative active cytokinin (CK) species occurring naturally. Only trivial names are given with commonly used abbreviations in parentheses. (From Sakakibara, 2006).

who reviewed CK and shoot development took the petunia gene *SHOOTING* (*Sho*) as an example. Unlike the overexpression of the bacterial *IPT* gene, the *Sho* expression enhances the accumulation of IP-type CKs. The overexpression of some Arabidopsis *IPT* genes also results in the accumulation of a different spectrum of CKs than the overexpression of bacterial IPT. When a gene that encodes a CK oxidase, an enzyme that causes catabolism of CK, is expressed in transgenic tobacco plants, it leads to plants with stunted shoots, smaller apical meristems, and much reduced leaf production. On the other hand, transgenic plants that express the CK oxidase have fast-growing and more enlarged roots. These *in planta* experiments confirm earlier observations made in tissue culture, in which various levels of CK and auxin were added to the culture medium. For example, adding 5 μM of iP ($N^6$-($\Delta^2$-isopentenyl)adenine) and 0.9 μM IAA to the culture medium of Arabidopsis root explants, caused formation of shoots, while when only 0.9 μm IAA (but no iP) was added to the culture, medium roots were formed, but no shoots. Clearly, adding a phytohormone to cultured tissue does not simulate normal *in planta* conditions, because such a treatment does not represent *localized* changes in CK.

CKs are rarely used to improve crop productivity but they are used intensively by investigators who study plant patterning, and by those who intend to regenerate shoots from cultured explants or protoplasts (see Galun and Breiman, 1997). CKs are also used by the "talented" pathogen of plants: *Agrobacterium tumefaciens.* The latter introduces a gene (*Tmr*) into the host's genome that causes the biosynthesis of CK from a unique substrate: 1-hydroxy-2-methyl-2-(E)butanyl. By that and in combination with the induced biosynthesis of auxin, an active cell division is initiated in the infected root of the host, leading to the formation of crown galls. The CK biosynthesis mediated by the gene from *A. tumefaciens* is located in the plant's plastids of the infected cells and it differs from the biosynthesis of CKs that are mediated by the plant genes.

Experiments with two species of orchids (Blanchard and Runkle, 2008) indicated that spraying plants — that were shifted to inductive temperature (23°C) — with BA induced earlier flowering. Studies with maize (Young *et al.*, 2004) indicated that increasing the *in planta* degradation of CK caused male sterility, while the application of kinetin to such plants restored male fertility. Elevating the expression of a transgene, encoding the cytokinin-synthesizing IPT rescued the lower floret from abortion. In the spikelets of regular maize ears, the lower floret aborts.

## Biosynthesis, Glucosylation, and Degradation

The biosynthesis of CKs is a complicated issue because of several reasons and it is rather difficult for a novice to comprehend. One trivial reason is the nomenclature used by different authors. For example, what Crozier *et al.* (2000) termed $N^6$-($\Delta^2$-isopentenyl) adenoside 5' phosphate or in short: [9R-5-P]iP is termed by Sakakibara (2006) and by Mok and Mok (2001) as $i^6$AMP. Another complication arises because there are several routes to reach a final active CK as zeatin. The routes may vary in different tissues of the very same plant species.

### *List of acronyms: Cytokinin perception and signaling*

**AHK2**    Arabidopsis cytokinin receptor (<u>A</u>rabidopsis <u>h</u>istidine <u>k</u>inase 2)

**AHK3**    Arabidopsis cytokinin receptor (<u>A</u>rabidopsis <u>h</u>istidine <u>k</u>inase 3)

**AHP**      Authentic histidine containing phosphotransmitter including AHP1, AHP2, AHP3, AHP4 and AHP5

**AHP6**    Pseudo AHP, inhibits the activity of the authentic AHPs

**ARRs**    Arabidopsis response regulators, these include type A and type B; ARR1, a common type B response regulator;
*Type A response regulators*: relatively short, partly redundant, encoded by a family of 10 genes (e.g. ARR4, ARR8, ARR9, ARR5, ARR6, ARR7, ARR15) — all induced by cytokinins;
*Type B response regulators*: the C-terminal region has a DNA binding domain, a nuclear localization domain, and

a transcription activator (e.g. ARR1, ARR2, ARR11, ARR14, ARR18)

**BA**       $N^6$-benzyladenine

**CHASE**    <u>C</u>yclases/<u>h</u>istidine <u>k</u>inase <u>a</u>ssociated <u>s</u>ensory <u>e</u>xtracellular

**HPt**      Histidine-containing phosphotransfer

**CREs**     <u>C</u>ytokinin <u>R</u>esponse <u>F</u>actors (three of the six factors belong to the APETALA2-like class of transcription factors; the genes encoding three others rapidly form transcripts that are induced by CK in a type B ARR-dependent manner)

**IPT**      Adenosine phosphate, <u>i</u>sopentenyl<u>t</u>ransferase

***Tmr***    An *A. tumefaciens* gene encoding IPT that operates in the host-plant plastides

The routes may be at least partially redundant so that by suppressing one specific gene (i.e. by mutation) the biosynthesis of CK is not fully suppressed.

For a general orientation on the biosynthesis and metabolism, the reviews of Crozier *et al.* (2000), Mok and Mok (2001), and Sakakibara (2006) are recommended. A more recent review by the team of Sakakibara (Hirose *et al.*, 2008) is a useful update on this subject. Several research teams contributed to a better understanding of the biosynthesis and metabolism of CKs in plants (Kakimoto, 2001; Golovko *et al.*, 2002; Takei *et al.*, 2004a,b; Miyawaki *et al.*, 2006 and Zhao, 2008).

The active CKs are divided into two main groups according to the side chain attached to $N^6$: (1) isoprenoid CKs and (2) aromatic CKs. Most studies on CK biosynthesis were performed on the isoprenoid CKs: (1) $N^6$-(isopentenyl) adenine — *iP*; (2) *trans*-zeatin — *tZ*; (3) *cis*-zeatin — *cZ*; (4) dihydrozeatin — *DZ* (Fig. 32).

Figure 33 provides a model for the synthesis of isoprenoid CKs. The model shows two main pathways for CK biosynthesis. One pathway is termed methylerythritol phosphate (MEP) pathway and it commonly leads to iP and tZ. The other pathway is termed the mevalonate pathway (MVA), and it leads to a large faction of the cZ.

As shown for angiosperm plants in Figs. 33 and 34, dimethylallyl diphosphate (DMAPP) is conjugated with ATP, ADP or AMP, by the

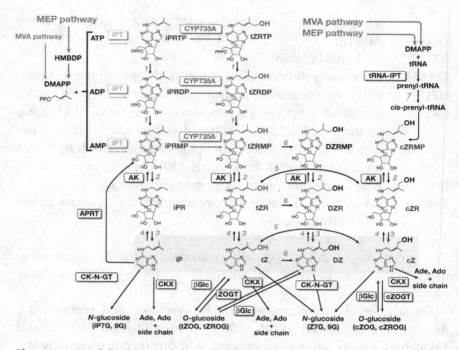

**Fig. 33.** Model of isoprenoid CK biosynthesis pathways in Arabidopsis. Isoprenoid side chains of $N^6$-($\Delta^2$-isopentenyl)adenine (iP) and *trans*-zeatin (tZ) predominantly originate from the methylerythritol phosphate (MEP) pathway, whereas a large fraction of the *cis*-zeatin (cZ) side chain is derived from the mevalonate (MVA) pathway (green arrows). Plant adenosine phosphate-isopentenyltransferases (IPTs) preferably utilize ATP or ADP as isoprenoid acceptors to form iPRTP and iPRDP, respectively (blue arrows). Dephosphorylation of iPRTP and iPRDP by phosphatase (I), phosphorylation of iPR by adenosine kinase (AK), and conjugation of phosphoribosyl moieties to iP by adenine phosphoribosyltransferase (APRT) create the metabolic pool of iPBMP and iPRDP. APRT utilizes not only iP but also other CK nucleobases. The CK nucleotides are converted into the corresponding tZ-nucleotides by CYP735A (red arrows). iP, tZ, and the nucleosides can be catabolized by CKX to adenine (Ade) or adenosine (Ado). cZ and tZ can be enzymatically interconverted by zeatin *cis-trans* isomerase. tZ can be reversibly converted to the O-glucoside by zeatin O-glucosyltransferase (ZOGT) and β-glucosidase (bGlc). CK nucleobases can also be converted to the N-lucoside by CK N-glucosyltransferase (CK-N-GT). The width of the arrowheads and lines in the green, blue and

IPT enzyme, to yield iPRTP, iPRDP or iPRMP, respectively. In Arabidopsis, iPRTP and iPRDP may then be converted to iPRMP before the hydroxylation of the side chain (by CYP735), or iPRTP and iPRDP are first hydroxylated by this enzyme (i.e. by CYP735) to produce tZRMP. The CYP735 is a cytochrome P450 mono-oxygenase enzyme. To become biologically active CKs, the nucleotides and the nucleosides are converted to nucleobase forms by deposphorylation and deribosylation, but the genes encoding the nucleotidase and nucleosidase have not yet been identified. Recently, a novel pathway was identified that directly releases the active cytokinin from the nucleotide, catalyzed by the cytokinin nucleoside 5′-monophosphate phosphoribohydrolase (LOG) that is shown by the scheme in Fig. 34. I noted above that the bulk of the cZ is synthesized by the MVA pathway but a conversion from tZR into DZR as well as from tZ to cZ is also possible, meaning that cZ can also originate from the MEP pathway. Moreover, a tRBA-isopentenyltransferase (tRNA-IPT or IPPT) that was earlier identified in microorganisms (e.g. *Escherichia coli*, *Saccharomyces cerevisiae*) was detected by Golovoko *et al.* (2002) in Arabidopsis, meaning that CKs — and especially cZ — may also originate from pre-tRNAs (by degradation of the *cis*-hydroxy-isopentenyl-tRNA). The Arabidopsis gene was expressed in *S. cerevisiae* and could replace the yeast gene. The quadrolyte mutants that suppressed the MEP pathway of Arabidopsis caused an elevated level of cZ, indicating the requirement of tRNA IPT for cZ (Miyawaki *et al.*, 2006).

As noted above, there are differences in the details of the biosynthetic pathways in different tissues, even of the same plant species. For example, there are several IPT genes in Arabidopsis; among these, *AtIPT3* is predominantly expressed in the phloem. This specific gene is up-regulated by nitrate. Also, other macronutrients were found to

---

**Fig. 33. (*Continued*)** red arrows indicate the strength of metabolic flows. Flows indicated by black arrows are not well characterized to date. tZEDP, tZR 5′-diphosphate; tZRTP, tZR 5′-triphospate; 2, 5′-ribonucleotide phosphohydrolase; 3, adenosine nucleosidase; 4, purine nucleoside phosphorylase; 6, zeatin reductase; 7, CK *cis*-hydroxylase. (From Sakakibara, 2006).

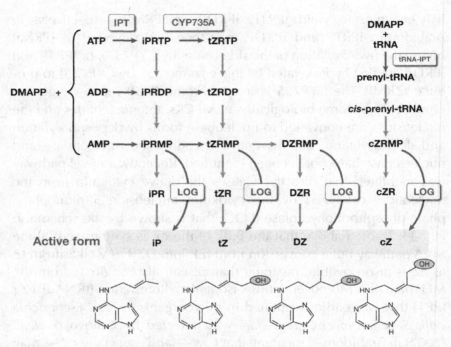

**Fig. 34.** A current model of cytokinin biosynthesis and the two known activation pathways. iPRMP, iP riboside 5'-monophosphate; tZRTP, tZ riboside 5'-triphosphate; tZRDP, tZ riboside 5'-diphosphate; tZRMP, tZ riboside 5'-monophosphate; DZRMP, DZ riboside 6'-monophosphate; cZRMP, cZ riboside 5'-monophosphate; DZR, DZ riboside; cZR, cZ riboside; LOG, a cytokinin nucleoside 5'-monophosphate phosphoribohydrolase. Blue arrows indicate reactions with known genes encoding the enzyme, and grey arrows indicate that the genes have not been identified. In this scheme, only biosynthesis and activation steps are drawn. Further details are shown in Sakakibara (2006). (From Hirose *et al.*, 2008).

increase the expression of *AtIPT3* (see details and references in Hirose *et al.*, 2008). The spatial distribution of CK-related gene expression in Arabidopsis was schematically summarized by Hirose *et al.* (2008). The scheme includes genes that are involved in the biosynthesis of CKs as well as genes involved in CK signaling (Fig. 35). The scheme of Fig. 36 shows that given genes that encode

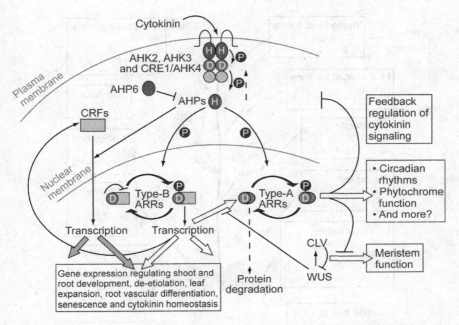

**Fig. 35.** Model of cytokinin signaling. (From To and Kieber, 2008).

enzymes in the biosynthesis pathways of CK were detected in specific plant organs. For example, among the *AtIPT* genes there is the following localization: *AtIPT1* — in the root procambium in axillary buds and in mature floral tissue; *AtIPT3* — in the phloem of leaf and root vasculatures; *AtIPT5* — in the abscission zone of fruits and in the root cap; *AtIPT7* — in trichomes and in the root elongation zone. Clearly, these are merely examples of what was found until the end of 2007. It is very possible that additional localizations may be found in future studies. What we should note here is that the different genes with different promoters enable the plant great flexibility in the synthesis of CK in specific tissues and in response to different internal and external affectors. As in Arabidopsis, there are also several *IPT* genes in rice.

Several studies indicated that CKs may be synthesized "locally" where they play their role in cell division; however, on the other

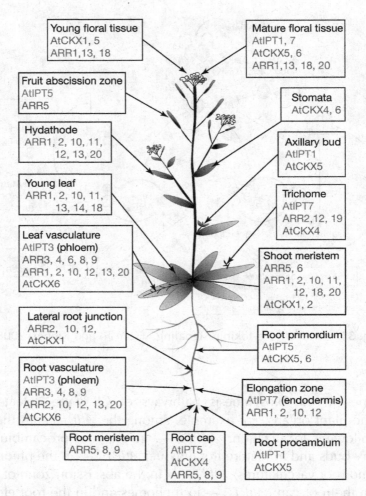

**Fig. 36.** Spatial distribution of cytokinin-related gene expression in Arabidopsis. Type A *ARRs* are in violet; Type B *ARRs* are in blue. (From Hirose *et al.*, 2008).

hand, CKs are also long-range signals: they are synthesized in the roots and move acropetally. The xylem sap of plants contains tZ-type CKs and roots are probably a major site of tZ production. As presented in Fig. 36, genes for CK synthesis were also detected in the phloem and in addition, CKs were detected in the phloem sap.

Phloem sap predominantly contains iP-type and cZ-type CKs. Hence phloem CKs may function as a basipetal or a systemic signal. The details of CK transport are not currently fully known. A number of CK transporters were identified (see Hirose *et al.*, 2008) but these can also transport purines and nucleosides. Such identified transporters are termed Equilibrative Nucleoside Transporters (ENT). Several genes that encode these transporters were revealed in Arabidopsis (*AtENT1–AtENT8*) and in rice (*OsENT1–OsENT4*), in different locations of the respective plants. In Arabidopsis, there are also genes (*AtPUP1, AtPUP2*) that encode purine permeases that facilitate CK nucleobase uptake; but again, these transporters have a broad substrate specificity: they mediate the uptake (by cells) of several adenine derivatives, including caffeine.

There is a regulation of CK that involves the levels of macronutrients supplied to plants. This issue was especially studied with nitrogen compounds and performed in Arabidopsis (Takei *et al.*, 2004b; also see the review of Sakakibara, 2006). Takei *et al.* (2004b) followed the changes in several IPTs' levels of expressions in Arabidopsis by using transgenic plants into which the promoters of specific IPT genes fused to a fluorescent label and were then analyzed. The investigators found that *AtIPT1* was predominantly expressed in the vascular stele of roots. *AtIPT3* was expressed in the phloem companion cells. *AtIPT5* was expressed in the lateral root primordia and in the pericycle. *AtIPT7* was expressed in both the vascular stele and in the phloem companion cells in roots. When grown in solutions containing different concentrations of $NO_3^-$ or $NH_4^+$, the levels of expression of *AtIPT5* were positively correlated with the concentrations of the nitrogen compounds. When the supply of nitrogen was limited, the *AtIPT3* expression was rapidly induced by $NO_3^-$ in the seedlings. This increase of *AtIPT3* levels was accompanied by an accumulation of CKs. The authors concluded that AtIPT3 is a key determinant of CK biosynthesis in response to changes in the availability of $NO_3^-$.

The levels of active CKs are regulated by several means, such as side chain removal and conjugations that are not reversible, leading to inactive conjugants.

There are several types of glucosylations in CKs. The glucosylation can be at the hydroxyl of the side chain (to form O-β-glucosylzeatin). Such side chain glucosylations are considered reversible and the glucosylated CK may serve as a "storage" CK. On the other hand, glucosylation of the purine moiety is probably not reversible and thus serves to lower the level of CKs that are too active. The glucosylation may take place at 3, 7 or 9 carbon of the purine moiety. These are the N-glucosylated CKs; they are not biologically active and are resistant to CK oxidases. There are also other conjugations, such as the attachment of alanine to the carbon 3 of the purine moiety. Such conjugants are not biologically active and seem to be performed by an enzyme that uses free base CKs (such as zeatin) for its conjugation. The $N^3$ amino acid conjugants are not hydrolyzed *in vivo* and hence they may also act to reduce the level of active CKs in the plant.

The CK oxidases in wheat and probably also in other plants are inhibited by diphenylurea. Therefore, the application of diphenylurea to plants interferes with the plant's capability to control CK levels. The application of diphenylurea may therefore elicit responses that are similar to the application of CKs to plants.

## Perception and Signal Transduction

The receptors of CKs were identified relatively recently, in 2001. They were reported almost simultaneously, but independently, by two Japanese research teams: Inoue *et al.* (2001) of Osaka, Tsukeba, and Chiba and Ueguchi *et al.* (2001) of Nagoya. The Inoue team found that there is a two-component system of CK perception in Arabidopsis that can be induced by cytokinins. They identified a gene, which they termed as *CRE1*, for cytokinin response 1. Mutants of this gene (*cre1*) showed reduced response to CKs. *CRE1* encodes a histidine kinase and has similarity to the two-component complex that regulates the osmosensing MAP kinase cascade in yeast. CKs can activate *CRE1* to initiate phosphorelay signaling. The other team, Ueguchi *et al.* (2001), identified three genes that are highly homologous in Arabidopsis and termed them AHKs (Arabidopsis his kinase): *AHK2, AHK3* and *AHK4*. These encode sensor histidine kinases located in the plasma membrane. These genes were found to be expressed in several organs but AHK4 was mainly expressed in roots. Another sensor, AHK1, has been previously implicated in osmotic response. Later, studies showed that AHK4 is identical to CRE1. The three partially redundant CK receptors are thus CRE1/AHK4, AHK2, and AHK3. The receptors of CKs were further studied in more recent investigations (e.g. Higuchi *et al.*, 2004; Mähönen *et al.*, 2006) and reviewed by several authors (Ferreira and Kieber, 2005; Müller and Sheen, 2007; To and Kieber, 2008). A scheme by To and Kieber summarizes the perception and signal transduction of CKs (Fig. 35). The main steps of perception and signaling of CKs are the following. CKs are first perceived by the CK receptors AHK2, AHK3, and CRE1/AHK4, at the plasma membrane. This is already a multistep process because it activates a multistep phosphorelay. The AHK2, AHK3, and CRE1/AHK4 proteins consist of an extracellular cytokin-binding CHASE (cytoplasmic his kinase) receiver domain. Cytokinin binding to the receptor activates autophosphorylation of a conserved histidine in the protein's kinase domain. The phosphoryl group on the histidine is transferred via a conserved aspartic acid residue on the receiver domain and further transferred to a conserved histidine on an "authentic" AHP (AHP1, AHP2, AHP3, AHP4, or AHP5) in the cytoplasm. A reverse phosphorylation may take place: CRE1/AHK4 that is not bound to CK can mediate a reverse phosphoryl transfer reaction to dephosphorylate an AHP.

The AHPs can be translocated into the nucleus to transfer the phosphoryl group to a type A ARR (see list of acronyms on page 118) or type B ARR received domain on a conserved aspartic acid. The phosphoryl transfer to and from the authentic AHPs can be inhibited by the "pseudo AHP" (AHP6) in the protoxylem. The type B ARR contains a C-terminal DNA binding domain, which is repressed by the unphosphorylated N-terminal receiving domain. The phosphorylation of type B ARR releases this repression to induce the transcription of a subset of CK-regulated targets including type A *ARR* genes and the CRFs (Cytokinin Response Factors). The latter factors (CRFs) are proteins that can be activated via AHPs to accumulate in the nucleus and initiate transcription. Type B ARRs and CRFs have partially over-lapping transcriptional targets. They mediate downstream CK-regulated processes, such as shoot and root development, de-etiolation, leaf expansion, root vascular differentiation, senescence and CK home-ostasis. Type A ARRs have "negative" roles. Phosphorylated type A ARRs are activated to mediate downstream processes, such as feed-back regulation of CK signaling, inhibition of meristem development, and phytochrome function. The type A *ARR* genes were detected in Arabidopsis and they differ in the specificity of their function (i.e. ARR4 is involved in phytochrome function). The protein turnover of several type A ARRs, such as ARR5, ARR6, and ARR7, is regulated by phosphorylation; unphosphorylated type A ARRs of this group are targeted for degradation, whereas phosphorylated type A of these ARRs are more resistant to degradation. Now let us turn our attention to the shoot apex differentiation (see Galun, 2007), because the transcription of *ARR5–ARR7* and *ARR15* (all type A *ARR* genes) is repressed by the transcription factor WUS, in the meristem. We should recall that an interaction between WUS and CLV maintains a feedback signaling "loop" that regulates the population of the apex meristem cells. Meristem function is in turn inhibited by type A ARRs, resulting in a feedback signaling "loop" between CK and the shoot-meristem-specific signaling pathway. It appears that specific members of gene families (such as *AtIPT* and *AtARR*) are expressed in specific Arabidopsis tissues and organs. For example, in the fruit abscission zone, it is the genes *AtIPT* and *AtARR5* that are expressed while in the endodermis of the root-elongation zone, the genes expressed are *AtIPT7* as well as *AtARR1, AtARR2, AtARR10,* and *AtARR12* (see Fig. 36 for detailed information). Complicated? Yes, it is. We should recall the words of Harry S. Truman (US President, 1945–1953) who addressed his fellow politi-cians and said: "Those who cannot endure the heat should not enter the kitchen."

While the description of perception and signal transduction summarized in the previous pages is based primarily on Arabidopsis studies, the perception and signal transduction of CK in other plants that are phylogenetically distant from Arabidopsis (such as maize and rice) seem to be similar to those revealed in Arabidopsis (e.g. Yonekura-Sakakibara *et al.*, 2004; Du *et al.*, 2007).

# CHAPTER 6

# Jasmonates

## Introduction, History, and Main Impacts on Development and Stress

Jasmonates are a group of compounds that are related to jasmonic acid (JA) and methyl jasmonate (MeJA). These, together with the common precursors of jasmonate 12-oxyphlytodienoic acid (OPDA) and dinor-OPDA (dnOPDA), are collectively termed octadecanoids. Figure 37 shows some of these compounds. Jasmonates are commonly included in phytohormones (Crozier *et al.*, 2000), but there is a problem with such an inclusion. The term "hormone" is usually defined as "a substance formed in some organs and carried as fluid to another organ or tissue, where it has a specific effect." This is the definition of the Webster's World College Dictionary. Is there a "flow" of jasmonates from the site of their biosynthesis to the site of their effect? There is yet no clear answer to this question. However, as it is commonly accustomed by those who handle phytohormones, jasmonates *will* be included in this book among plant hormones. The prostaglandins — that are closely related to jasmonates — are included among animal hormones. In a way, in regards to the transport of MeJA, we face a similar situation as we did with ethylene. As ethylene, MeJA is volatile and as such should be able to move in the plant tissue without any specific transporters. Moreover, exposing intact plants to MeJA is considered a systemic treatment of MeJA. It should also be noted that, as will be detailed below, the known receptor of jasmonates in plants is complexed not with JA itself but only with a conjugate between JA and an amino acid (e.g. JA-isoleucine and JA-valine). Such a complex leads to the derepression

131

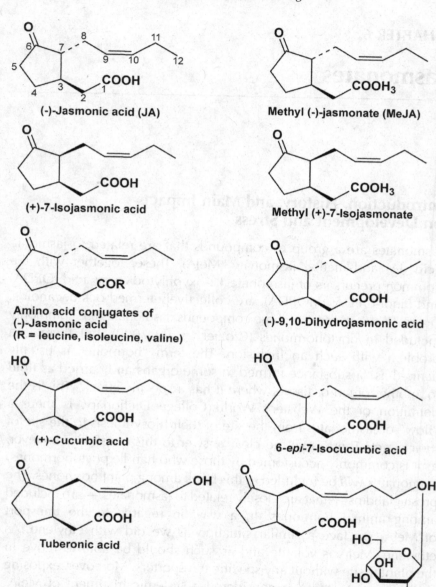

**Fig. 37.** Structures of (−)-jasmonic acid and related compounds. (From Crozier *et al.*, 2000).

of jasmonate-responsive genes that involves the ubiquitination of repressors that interact with the promoters of such genes. The strategy of derepression of responsive genes is also maintained in plants on the impact of other phytohormones (e.g. auxin and gibberellins), as detailed in previous chapters.

The scent of jasmonates was known since ancient times, but the identification of MeJA in the essential oil extracted from *Jasminum grandiflorum* and *Rosmarinus officinalis* was probably only published by Demole *et al.* in 1962. The free acid (JA) was identified from a fungus, *Lasiodiploida* (*Botryodiploidi*) *theobromae*, and recognized by them as a plant growth inhibitor in 1971. The identification of jasmonates in angiosperms (e.g. *Artemisia absinthium*) and the finding of their inhibitory effects on plants were carried out in the laboratories of Takahashi and Ueda and Kato of Japan (e.g. Yamane *et al.*, 1980, and Ueda and Kato, 1980, respectively). At about the same time, an intensive research on jasmonates started at the Institute of Plant Biochemistry of the University of Halle in Germany (then German Democratic Republic — DDR), where the investigators detected JA in *Vicia faba* (Dathe *et al.*, 1981) and revealed the difference between the inhibitory effects of JA and abscisic acid (JA does not delay the development of wheat seedlings). A detailed investigation of JA was then conducted by the German team of Halle (e.g. Meyer *et al.*, 1984; Miersch *et al.*, 1986). These investigators used various techniques, such as TCL, GC, GC-MS, HPLC, radio immunoassay, and bioassay to detect JA and MeJA in various tissues of a great number of bicot and monocot plants. They found a great variability in the levels of JA from over 500 ng/g (fresh weight) to only traces. Interestingly, Meyer *et al.* (1984) found only MeJA (but no JA) in the coniferous tree Douglas fir.

The intensive study of jasmonates was continued in Halle in the period when this university belonged to the DDR, as well as after the unification (around 1990) with the Republic of Germany. The studies before the unification, performed in Halle as well as other research centers, were reviewed by Benno Parthier (1990). Parthier reviewed the various effects of JA and MeJA on plants. The impact of these jasmonates on plants is quite variable and the enantiomer (i.e. the (+)

form of MeJA versus the (−) form of MeJA) can differ greatly. There were phenomena of inhibition of pollen germination by jasmonates but seed germination was not affected. In general, the jasmonates facilitated leaf senescence, and reduction of photosynthetic activity was reduced by MeJA treatment, while isolated chloroplasts were not affected by MeJA, suggesting that the effect of MeJA on photosynthesis is not a direct one. One possible explanation was that the treatment of leaves with MeJA suppressed the incorporation of amino acids into the two subunits of Rubisco (LS and SS): the transcription and/or translation of plastid protein could be suppressed by MeJA. While some proteins of leaves were reduced by jasmonate treatment (as was the case with barley leaves), there was a substantial production of "new" proteins in the treated leaves. Such a burst of new proteins was not detected in roots. The reprogramming of gene expression of isolated leaves was already apparent 3–5 hours after jasmonate treatment, and it is an early phenomenon under stress conditions.

In later publications on jasmonates and their precursors, the octadecanoids, Wasternack and Hause (2002) reviewed in detail the signals in plant stress responses and in development. These authors considered biotic factors as pathogens and herbivores; symbiotic relationships as in nitrogen-fixing nodules and mycorrhiza. They also considered numerous abiotic effectors, such as low temperature, frost, heat, high-light, ultraviolet light, oxidation, hypoxia, wind, touch, nutrient imbalance, and water deficit. They were dealing with genes that encoded a variety of proteins and the expression of which was either up-regulated or down-regulated in response to jasmonates. The effects updated for 2002 are summarized in Table 3. The list of proteins in this table should serve only to provide the awareness that there are many such proteins. Table 3 does not mention specific plants, neither specific tissues. Additional information on these up-regulations and down-regulations came in later years and is outside the scope of this book, but can be found in later reviews on jasmonates (e.g. Howe, 2004: Lorenzo and Solano, 2005; Schilmiller and Howe, 2005; Wasternack, 2004, 2007; Pauw and Memelink, 2004; Schaller *et al.*, 2004). An example of a detailed study on the changes in gene expression in which jasmonate signaling is involved,

## Table 3. Proteins Encoded by Genes Exhibiting Altered Expression in Response to Jasmonate

**Up-regulation**

Enzymes of JA biosynthesis
LOX
AOS
AOC
OPR3

Enzymes in secondary metabolism
Strictosidine synthase
Chalcone synthase
Phenylalanine-ammonia lyase
Berberin bridge enzyme
Myrosinase

Antinutritional proteins
PIN1, PIN2
Cystatin (cys proteinase inhibitor)
Polyphenoloxidase
Putrescine-N-methyltransferase
(nicotine formation)
Cysteine endopeptidase

Pathogenesis-related (PR) proteins
Chitinases
Glucanases
PR-10
Defensins
Thionins

Proteinases
LAP
Carboxypeptidase

Proteins of cell wall formation
Glycerin-rich proteins
Hydroxyproline-rich proteins

Stress-protective proteins
Osmotin
JIP23
Threonine deaminase
Glutathione S-transferase

Vegetative storage proteins
VspA
VspB
Soybean LOX

Seed storage proteins
Napin
Cruciferin

Proteins of signal transduction
Calmodulin
Prosystemin
Polygalacturonase
Phospholipase A1
ACC oxidase (ethylene)

Proteins with function in generation of ROS
NADPH oxidase
Peroxidases

**Down-regulation**

Chlorophyll a/b-binding protein
Rubisco
D1 of photosynthetic apparatus
Light-harvesting complex II
ELIPs

From Wasternack and Hause (2002).

is the work by McGrath *et al.* (2005) performed by seven Australian and two German investigators. These investigators used Arabidopsis in their study and screened 1534 transcription factors (TFs) following changes in transcripts after jasmonate application or after inoculation with a pathogenic fungus (*Alternaria*). They found that there were 134 TFs that significantly changed their expression after the treatment. Of these, 20 TF genes were induced after both MeJA and fungal infection. These included 10 members of the APETALA2/ethylene response factor genes. They also revealed that one of these TFs, AtERF4, acted as a "novel" negative regulator of JA-responsive defense gene expression, provided resistance to the fungal pathogen *Fusarium oxysporum*, and antagonized the JA inhibition of root elongation. On the other hand, AtERF2 is a positive regulator of JA-responsive defense genes and provided resistance to *F. oxysporum*. AtERF2 also enhanced the JA inhibition of root elongation. The study suggested that Arabidopsis is capable of coordinately expressing multiple repressor- and activator-type AP2/ERFs during pathogen challenge, to modulate defense gene expression, and affect disease resistance.

There are several effects of jasmonates on plant development, such as tendril coiling (Weiler *et al.*, 1993), trichome differentiation (Liu *et al.*, 2006), stamen development (Mandaokar *et al.*, 2006), and tuberization in potato (Abdala *et al.*, 1996). More details on such effects will be provided in Part II of this book where the role of JAs in the shaping of specific organs will be reviewed.

## Biosynthesis

The commonly studied pathway of jasmonate biosynthesis is the LOX system (as shall be detailed below). However, in saturated and unsaturated fatty acids, an oxygen can be inserted at the $C_2$ by $\alpha$-dioxygenases ($\alpha$-DOX). Such insertions were already reported to exist in leaves of tomato and Arabidopsis by Hamberg *et al.* (1999).

### List of acronyms and abbreviations for jasmonates

| | |
|---|---|
| **ACX** | Acyl-CoA oxidase |
| **$\alpha$-LeA** | $\alpha$-linolenic acid |
| **AOC** | Allene oxide cyclase |
| **AOS** | Allene oxide synthase |

| **ERF1** | Ethylene response factor 1 |
|---|---|
| **HPL** | Hydroperoxide lyase |
| **HPOD** | Hydroperoxyoctadecadienoic acid |
| **HPOT** | Hydroperoxyoctadecatrienoic acid |
| **IP$_3$** | Inositol triphosphate |
| **JA** | Jasmonic acid |
| **JAME** | Jasmonic acid methyl ester (also abbreviated as MeJA) |
| **JAR1** | JA conjugate synthase |
| **JIP** | Jasmonate-induced protein |
| **$\alpha$-LeA** | $\alpha$-linolenic acid (18:3) |
| **LOX** | Lipoxygenase |
| **LTP** | Lipid transfer protein |
| **MFP** | Multifunctional protein |
| **MGDG** | Monogalactosyl diacylglyceride |
| **OPDA** | *cis*(+)-12-oxophytodienoic acid |
| **OPR** | OPDA reductase |
| **OYE** | Warburg's old yellow enzyme |
| **PA** | Phosphatidic acid |
| **PIN** | Proteinase inhibitor |
| **PLD** | Phospholipase D |
| **PUFA** | Polyunsaturated fatty acid |
| **SAG** | Senescence-associated gene |
| **TD** | Threonine deaminase |

The stereospecific insertion of oxygen by $\alpha$-DOX yields unstable 2-hydroperoxide derivatives. The level of $\alpha$-DOX products is up-regulated by pathogens. Hence the $\alpha$-DOX system appears to play a role in the plant pathogen interactions. The $\alpha$-linolenic acid (i.e. from intra-plastidic lipid membranes) can also be metabolized by another system that involves phytoprostanes that are stress-related compounds. For more details on these $\alpha$-DOX and phytoprostane systems, the reader is referred to the review of Wasternack (2006). Hereafter, we will mostly deal with the LOX pathway. To avoid long chemical names and enzymes, acronyms and abbreviations will frequently be used instead, as shown in the list above.

# The LOX Pathway

In the late 1970s and early 1980s, two research areas converged. In one area, as detailed above, jasmonates were being discovered in many plants as growth regulators and involved in stress responses, and in another area, the study focused on the lipoxygenase-catalyzed oxygenation of polyunsaturated fatty acids that yield fatty acid hydroxy peroxides. The latter area was intensively studied by Zimmerman, Vick and collaborators in the North Dakota University, in Fargo. The match between the two areas was established in 1983 (see Vick and Zimmerman, 1984). The latter publication already provided information on enzymes that convert OPDA to jasmonates and a rough pathway was suggested that probably operates in all angiosperms (the authors found this metabolism in maize, eggplant, flax, oat, and wheat). Further work on the biosynthesis of jasmonates was reviewed by Creelman and Mullet (1997). A further examination on the pathway of jasmonate biosynthesis was provided by the review of Baker *et al.* (2006). In this review, the main steps of jasmonate biosynthesis were separated into those that take place in the chloroplasts (up to OPDA) and those steps that take place in the peroxisomes (that include cycles of $\beta$-oxidation), where JA is produced.

As mentioned above, the study of jasmonates had already started in the very early 1980s (e.g. Dathe *et al.*, 1981) at the Institute of Plant Biochemistry in Halle. After the unification of Germany and Claus Wasternack joining the jasmonate research team, the team succeeded in advancing the jasmonate research considerably and made major contributions to the understanding of jasmonate biosynthesis. Wasternack reviewed the work on jasmonate biosynthesis in his institute as well as in other laboratories (Wasternack, 2004, 2006, 2007; Delker *et al.*, 2007). Additional important publications have recently been written by authors of other laboratories (e.g. Schilmiller *et al.*, 2005, 2007; Schaller *et al.*, 2004). Incidentally, the publication by G.A. Howe of Michigan State University (Schilmiller *et al.*, 2007) almost completely ignored the work in Halle; out of over 20 existing publications from Halle, only a single publication on jasmonates

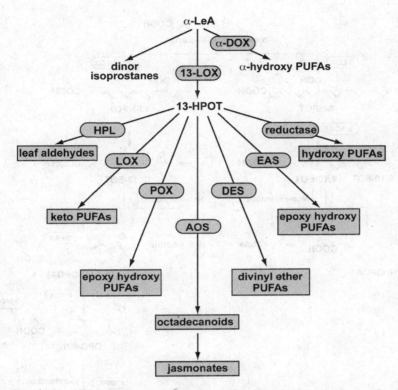

**Fig. 38.** The different branches of the LOX pathway. Metabolism of α-LeA by non-enzymatic formation of dinorisoprostanes, α-dioxygenase (α-DOX)-catalyzed formation of α-hydroxy-PUFAs and 13-lipoxygenase (13-LOX)-catalyzed formation of the 13-hydroperoxide of α-LeA (13-HPOT). 13-HPOT can be the substrate of a reductase, an EAS, a DES, an AOS, a POX, an LOX, and an HPL. (From Wasternack, 2006).

from the fungus *Fusarium oxysporum* (Miersch *et al.*, 1999) was mentioned.

Figure 38 shows the main "branches" of the metabolism of α-linolenic acid. The middle branch represents the LOX pathway that leads to octadecanoids and jasmonates. The main steps of the biosynthesis in the LOX pathway that lead to (+)-7-*iso*-JA and then to (−)-JA are shown in Fig. 39. The intracellular location of the JA biosynthesis is schematically provided in Fig. 40. The steps from α-linolenic acid to

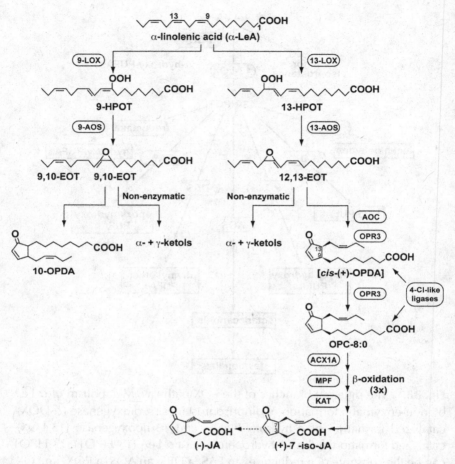

**Fig. 39.** The 9-LOX and 13-LOX pathways, and JA biosynthesis. (From Wasternack, 2007).

OPDA take place in the plastid (or the chloroplast in green tissue). The β-oxidation cycles that lead to (+)-7-*iso*-JA take place in the peroxisomes. A detailed description of all the metabolic steps that lead to JA, MeJA, and conjugates between JA and amino acids (as isoleucine), is beyond the scope of this book; however, a few remarks on these steps are provided below.

There are two main types of LOX enzymes that insert oxygen into PUFAs. In one type, 9-LOX, the insertion is at carbon 9 (C-9) of the fatty acids' backbone; the other type is 13-LOX which is prevalent in

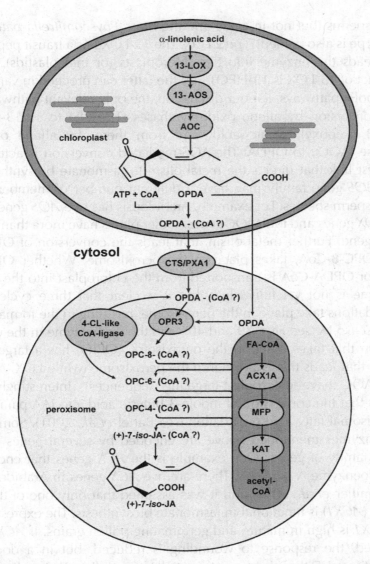

**Fig. 40.** Intracellular location of JA biosynthesis. The first half of JA biosynthesis is located in the chloroplast where OPDA is formed by the action of a 13-lipoxygenase (13-LOX), a 13-allene oxide synthase (13-AOS), and an allene oxide cyclase (AOC). Upon activation and transport into the peroxisomes, reduction by the OPDA reductase 3 (OPR3) and β-oxidation by fatty acid β-oxidation occur. OPC, oxopentenylcyclopentane. (From Wasternack, 2006).

angiosperms (but not in algae and the moss *Physcomitrella patens*). This type is also coined *Type 2* LOX. The 13-LOX has a transit peptide that leads the enzyme into the chloroplasts (or into plastids). The product of 13-LOX is 13-HPOT and the latter can proceed in various metabolic pathways. For our discussion, the only relevant pathway is the conversion by allene oxide synthases (13-AOS) to 12,13-EOT (12,13 S-epoxylinolenic acid) and from there, by allene oxide cyclase (AOC), to OPDA. The AOS-mediated conversion is actually the first link that directs the metabolism to jasmonate biosynthesis. The *AOS* gene family may have a different number of members in angiosperm species. For example, Arabidopsis has one *AOS* gene but six *LOX* genes and four *AOC* genes. Other plants have more than one *AOS* gene. Further metabolism, that leads the conversion of OPDA into OPC-8-CoA, takes place in the peroxisome. Whether OPDA itself or OPDA-CoA is transported from the chloroplast into the peroxisome is not yet fully clear, but it is clear that three cycles of β-oxidations take place in the peroxisome, resulting in the formation of (+)-7-*iso*-JA (see Figs. 39 and 40), and the first enzyme in the conversion that takes place in the peroxisome (OPR3) has a targeting signal that leads this enzyme into the peroxisome (while LOX, AOS, and AOC have chloroplast-targeting sequences). Interestingly, it seems that the conversion of indole-3-butyric acid into IAA parallels peroxisomal fatty acid β-oxidation (see Bartel *et al.*, 2001). Some of the enzymes mentioned above are encoded by several genes of a given family of genes. One example is the *ACX* genes that encode acyl-coenzyme A oxidases. There are five *ACX* genes in Arabidopsis (Schilmiller *et al.*, 2007) and it was assumed that only one of these genes (*ACX1*) is functional in jasmonate biosynthesis. The expression of *ACX1* is high in mature and germinating pollen grains. If *ACX1* is mutated, the response to wounding is reduced, but in a double mutant (*acx1,5*) there is no reaction at all to wounding. The application of JA restores the response to wounding in the mutants. Moreover, the response to the pathogenic fungus *Alternaria bassicola* is different: the double mutant (*acx1,5*) does respond to the infection with this fungus. It thus seems that different *ACX* genes are activated by different kinds of biotic stress.

It should be noted that the formation of all the enzymes in the JA biosynthesis pathway is increased by JA treatment, suggesting a positive feedback. The effect of JA was revealed only by the external application of JA but not as a result of the internal increase of JA. It is also worthwhile to note that some of the enzymes involved in JA biosynthesis are located in specific tissues: for example, AOC is confined to vascular bundles and even sieve elements include AOC protein.

The OPDA itself may activate certain genes. As OPDA can be esterified, the play between esterified and free OPDA may have a role in defense against wounding (i.e. by insects).

Many oxidated compounds generated by the LOX pathway show cell toxicity and antimicrobial activity. The question is how are they transported systemically in the plant? Clearly, plants contain lipid transfer proteins (LTPs); these could serve as carriers of toxic substances or carriers of signals for the defense compounds. In this connection one should recall the difference between jasmonate in biological stress resistance and other phytohormones, such as auxin. For the morphogenetic effect of auxin, this hormone has to be directly led to specific cells, while the jasmonates, when serving as antistress hormones, have no specific target. They have to move to all the plant's organs that may be exposed to a further biological stress.

JA can undergo several kinds of further metabolisms. MeJA is more active than JA in many plants. The conjugation of JA with amino acids, by the JA conjugate synthase (JAR1), is important for the signaling of JA. JA-Ile accumulates mainly in leaves, flowers and mycorrhizal roots. The *jar1* mutant does not produce JA-Ile but the application of JA-Ile can complement this mutant.

## Signal Transduction

Up to a few years ago, there was a great gap between the awareness that jasmonates are involved in stress resistance and development, and the understanding of the molecular mechanisms that lead to specific jasmonate responses. What was known resulted from numerous mutants in which the jasmonate responses were altered. Thus, when Arabidopsis became a major model plant, numerous genes were identified that modulated the jasmonate signaling in this plant. Among the genes identified were genes involved in the regulation of protein stability (probably via the ubiquitin-proteasome pathway), such as *COI1*, *AXR1*, and *SGT1b*. Another gene, *MPK4*, encodes a signaling protein while additional genes encode transcription factors (e.g. *AtMYC2, ERF1, NPR1,* and *WRKY70*). It was also discovered that in order to serve their tasks as stress responders and maintain their role in development, there should be an interaction between jasmonate signaling and other phytohormones, such as ethylene, abscisic acid, and salicylic acid signaling.

The finding that *COI1* (Coronatine Insensitive 1; coronatine is a fungal compound that has a jasmonate-like effect) and *AXR1* (auxin resistant1) are involved in the jasmonate signaling indicated the involvement of ubiquitination in the signaling of jasmonate: *COI1* was found to encode an F-box protein that is an important component of the ubiquitin-mediated protein degradation (see Devoto *et al.*, 2002).

This already suggested that as in other phytohormones (e.g. auxin, gibberellin), ubiquitination is a component of jasmonate signaling. Indeed, COI1 is present in a functional E3-type ubiquitin ligase complex, and plants that are deficient in a different component of this complex, the SCF, also show impaired jasmonate responses.

Interestingly, the very same protein that plays a role in the jasmonate signaling may also serve other phytohormones' signaling. One such protein is AtMYC2, which is a basic helix-loop-helix transcription factor for light, abscisic acid, and jasmonate signaling pathways in Arabidopsis (Yadav *et al.*, 2005). The MYC2 (also termed JIN1) probably binds to the G-box-element in the promoters of JA-responsive genes.

By the end of 2006, there were several players known to participate in the final steps of jasmonate signaling. There was the E3 ubiquitin ligase and a complex of SCF$^{COI1}$ ready to cause the degradation of a transcription repressor as an enhancer of transcription of responsive genes (e.g. MYC2), but what is the transcription suppressor? The answer was given in two consecutive publications of *Nature* on 9 August 2007 by Thines *et al.* (2007) and Chini *et al.* (2007) who independently revealed the JAZ proteins (mainly in Arabidopsis). The latest developments (until the summer of 2008) on the involvement of JAZ proteins in the signal transduction of jasmonates were summarized in several reviews (e.g. Katsir *et al.*, 2008a; Chico *et al.*, 2008; Kazan and Manners, 2008; Staswick, 2008). The fundamental nature of this involvement is presented in Fig. 41.

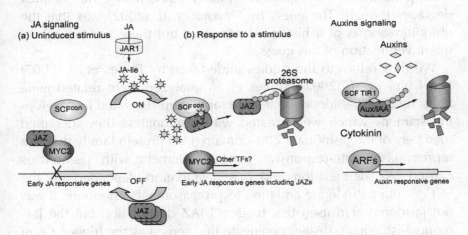

**Fig. 41.** Switching JA responses on and off. Model for the negative feedback loop regulation of JA responses and comparison to the auxin-signaling module in Arabidopsis. (a) In an uninduced situation, MYC2 activity is repressed by the direct binding of JAZ proteins. (b) Upon a stimulus, synthesis of JA-Ile by JAR1 promotes the interaction of JAZ repressors with SCF^COI1. Ubiquitination by SCF^COI1 promotes JAZ protein degradation by the 26S proteasome releasing MYC2 and probably other TFs. Activation of MYC2 induces transcription of early JA-responsive genes, including the *JAZs*. *De novo* synthesis of JAZ proteins restores MYC2 repression and turns the pathway off. JA-signaling and auxin-signaling modules are strikingly similar; both are composed of an F-box (COI1 or TIR1/AFBs) that in the presence of the hormone targets a repressor (JAZs or Aux/IAA) for proteasome degradation, releasing a transcriptional activator (MYC2 or ARFs). (From Chico *et al.*, 2008).

To better understand this issue, let us go back to the work of Devoto *et al.* (2002). These investigators studied the Arabidopsis protein COI1 and found that COI1 is involved in the signaling of jasmonates by affecting the transcription of jasmonate-responsive genes. It was also discovered that *in vivo* COI1 produces a complex that is typical to ubiquitination, namely SCF$^{COI1}$. However, what is it that connects this complex to a repressor of transcription in jasmonate-responsive genes? Devoto *et al.* (2002) rightfully expected that COI1$^{COI1}$ is complexed in the presence of jasmonates, and thus forms a ubiquitin-mediated degradation tool. What is it that the complex degrades though? The guess by Devoto *et al.* (2002) was that the ubiquitination is of a histone deacetylase but there was no subsequent verification of this guess.

We now return to the studies undertaken by Thines *et al.* (2007) and Chini *et al.* (2007). Thines *et al.* analyzed eight related gene transcripts of Arabidopsis that were rapidly up-regulated in developing stamens which were treated with jasmonates. They identified members of the jasmonate ZIM-domain (JAZ) protein family that can repress jasmonate-responsive genes. Treatment with jasmonates caused JAZ degradation and this degradation depended on the SCF$^{COI1}$ ubiquitin ligase and the 26S proteasome. Furthermore, it was not jasmonic acid itself that triggered JAZ degradation but the jasmonoyl-isoleucine (JA-Ile) conjugate that served as the trigger. Chini *et al.* (2007) started their research with the characterization of a mutant called *jasmonate-insensitive3-1* (*jai3-1*). Both Thines *et al.* and Chini *et al.* found that overexpressing the full length of their genes had little or no effect on jasmonate perception. However, overexpression of shortened forms, that encoded proteins that are truncated at the carboxyl terminus, reduced the plants' sensitivity to jasmonates (Yan *et al.*, 2007). In addition, full length proteins were unstable *in vivo* when treated with jasmonates but truncated proteins resisted degradation. The various experimental results fitted with the assumption that these proteins (full length) are inhibitors of the transcription of jasmonate-responsive genes, when not complexed with jasmonates (the formation of the complex required the carboxyl terminus). The proteins that can repress transcription were termed JAZ proteins.

In Arabidopsis, there are about 12 genes that encode JAZ proteins. It is not clear why Arabidopsis "needs" a dozen JAZ proteins; possibly this offers a great versatility of regulatory potential.

What is briefly and schematically described in Fig. 41 consists of several steps. (1) "Normally" the JAZ protein is complexed with MYC2 and the complex promotes DNA suppression of transcription in a jasmonate-responsive gene. (2) When jasmonate is elevated (by treatments causing a response to a stress), there is a conjugation between JA and an amino acid (e.g. isoleucine) forming JA-Ile. (3) The JA-Ile is complexed with both JAZ and the COI1 component of the SCF$^{COI1}$. The COI1 component can therefore be considered as the receptor of jasmonate hormones. (4) This triple-complex is ubiquitinated by the attachment of polyubiquitin. (5) After attachment of polyubiquitin to JAZ, the latter is led to a proteasome and degraded. (6) This degradation frees the promoter of the jasmonate-responsive gene from JAZ, and the transcription of the jasmonate-responsive gene is derepressed.

There is one correction to the above description: the JAZ is not sitting directly on the promoter's DNA; it sits on a transcription factor as MYC2 and prevents the latter's role of inducing the transcription. When JAZ is eliminated, by ubiquitination, the transcription factor can then cause the transcription of the jasmonate-responsive gene.

As indicated by Wasternack (2006), there is a positive feedback loop: an increased level of JA will activate a jasmonate-responsive gene that will lead to further synthesis of jasmonates.

Coronatine is a phytotoxic virulence factor produced by some pathovars of *Pseudomonas syringae*. It has a structural similarity to JA-Ile. Coronatine can trigger the COI1-JAZ interaction and thus interfere with the JA-related stress resistance.

In spite of the great progress achieved during recent years in the field of JA signal transduction, many important questions remain. One of these questions is how the intracellular level of bioactive JAs is controlled in response to developmental and environmental cues. Also, why is the level increased in some cells but not in others? Clearly, the latter question is part of a more general question: how does a small group of cells "know" where it is located? It is also not yet known whether JA-Ile promotes the degradation of all JAZs. Could other JA derivatives also fulfill the same role as JA-Ile? Apparently, most JA responses are controlled by the activity of only a single E3-ligase, $SCF^{COI1}$, so how is the specificity of response maintained? Is it possible that interaction between the jasmonates and other hormones, such as auxin, ethylene, and abscisic acid, contributes to the specificity? The interactions between several phytohormones (as those involving JA-auxin-salicylic acid) are discussed in the updates given on jasmonate signaling by Kazan and Manners (2008).

The abovementioned description points to JA-Ile (but not jasmonic acid or MeJA) as the key inducer of the jasmonate signaling, but a further modification (*in vivo*) of JA-Ile is not yet excluded. It should be noted that in auxin, the conjugation with certain amino acids serves as a "storage" for this hormone: the separation of the amino acid can then quickly elevate the level of free (and biologically active) auxin. The conjugation of JA with the amino acid isoleucine fulfills a very different purpose. In addition, while JAR1 is the enzyme that causes the conjugation between JA and isoleucine, there may be additional enzymes that cause conjugation and the conjugation is not only to isoleucine. Other amino acids such as tryptophan, valine, and leucine were also recently found to be conjugated to JA in tobacco (Wang *et al.*, 2008b).

Plants such as tomato, potato, and Arabidopsis evolved rather complex systems to provide resistance against abiotic and biotic stresses. This became apparent in the various means that plants use to resist biotic stress, such as herbivores, as well as fungal and bacterial pathogens. In spite of these evolved resistances, there are no cultivars in which these resistances caused complete immunity, although the loss of yields could be reduced. Moreover, while the plants evolved resistances mediated by the jasmonate-signal transduction, pathogens "fight back." This fighting back is exemplified by the bacterial pathogen *Pseudomonas syringae*. *P. syringae* produces a viral factor, coronatine (COR). In studies with tomatoes (*Solanum lycopersicum*), Katsir *et al.* (2008b) found that COR can replace JA-Ile in binding to the COI1 complex. Moreover, *in vitro* the binding of

COR is about 1000-fold more active than the JA-Ile binding. In other words, COR and JA-Ile are recognized by the same receptor and hence the pathogen's COR exerts its virulence by functioning as a potent agonist of the JA-Ile receptor system. It should not surprise us if additional cases are found in which a plant pathogen develops means to interfere with the plant's resistance that are mediated by the JA signal transduction.

# CHAPTER 7

# Abscisic Acid

## Historical Background

Already in the early 1950s, a fraction was separated from plant extracts by paper chromatography, that inhibited coleoptile elongation. The fraction was named "$\beta$-inhibitor complex." In subsequent studies, high levels of the $\beta$-inhibitor complex were correlated with the suppression of sprouting of potato tubers and other phenomena such as bud dormancy in deciduous trees, and terms such as "abscicin" and "dormin" were used. Addicott et al. (1968) reported the structure of "abscisinII," and this compound (Fig. 42) was finally termed *Abscisic Acid* (ABA). Four enantiomers of ABA are possible: (S)-ABA, (R)-ABA, (S)-2-*trans*-ABA, and (R)-2-*trans*-ABA. The (S)-ABA is the naturally occurring enantiomer. Contrary to what was suggested in the early years of ABA investigation, ABA does not induce abscission. This activity is attributed to ethylene: neither is ABA responsible for the dormancy of potato buds or buds on deciduous trees. However, increased ABA levels are correlated with seed maturation, development of desiccation tolerance, and suppression of vivipary (e.g. in certain maize mutants). Moreover, ABA is involved in a much-studied physiological phenomenon: water-stress induced closure of stomata.

While in the early years of ABA research, this hormone was identified in plant extracts by relatively "coarse" methods, such as thin layer chromatography (TLC) and high performance liquid chromatography (HPLC) combined with mass spectrometry (MS), in the 1980s, much more sensitive procedures of detecting ABA, based on immunoassay, were developed, that were sensitive to 0.05–2.5 pg

151

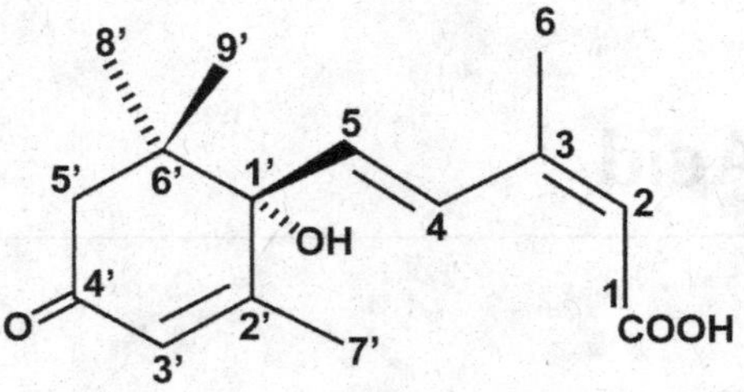

**Fig. 42.**   Numbering system of (+)-S-abscisic acid.

of ABA (see Zeevaart and Creelman, 1988, for references). Furthermore, the analytical methods included precaution to avoid artifacts of chemical conversion during the extraction and purification. Consequently, ABA was detected in various plants (although at very different levels) such as mosses, ferns, algae, and angiosperms. ABA was also detected in several pathogenic fungi. Moreover, there are reports of ABA in animals, but ABA is most likely introduced in animals via food, rather than being synthesized in animals itself.

## Biosynthesis and Catabolism

The early efforts to understand the biosynthesis of ABA were reviewed by Zeevaart and Creelman (1988) as well as by Crozier *et al.* (2000). This biosynthesis was further updated by Seo and Koshiba (2002), Schwartz *et al.* (2003), and Nambara and Marion-Poll (2005).

ABA belongs to a class of metabolites termed *isoprenoids* (or terpenoids). They are all derived from five-carbon ($C_5$) building units, isopentenyl diphospate (IPP). When the metabolite contains one isoprene unit, it is termed *hemiterpene*. More complex metabolites may result from a "head-to-tail" or "head-to-head" addition of units. Thus, monoterpenes are $C_{10}$ isoprenoids, sesquiterpenes are $C_{15}$, diterpenes are $C_{20}$, triterpenes are $C_{30}$, and tetraterpenes (that include carotenoids) are $C_{40}$ (see Rodriguez-Concepcion and Boronat, 2002, for updates and details). Up to the 1950s, it was assumed that IPP in all organisms is derived from mevalonic acid (MVA). However, further studies indicated that there is another precursor of IPP: methylerythriol 4-phosphate (MEP). Most organisms use either the MVA or the MEP pathway. The MEP pathway exists in most eubacteria and in the human malarial parasite *Plasmodium falciparum*, but it is absent in archaebacteria, fungi, and animals. Plants use both the MVA and the MEP pathways for isoprenoid biosynthesis, but the two pathways are localized in different compartments: the MVA pathway exists in the cytoplasm while the MEP pathway takes place in the plastids. In both pathways, IPP leads to geranylgeranyl diphospate (GGPP) and from there —in several steps — to ABA. Rohmer and collaborators were the pioneers in elucidating the MEP pathway (see Rohmer, 1999). Rodriguez-Concepcion and Boronat (2002) detailed the various studies that clarified the methylerythriol phosphate pathway in the biosynthesis of isoprenoids in bacteria and plastids. They also provided a rough scheme for the MEV pathway in the cytosol and the MEP pathway in the plastids of plant cells (Fig. 43). Fungi use the MEV pathway to synthesize ABA. Many of the studies in fungi were performed with phytoparasites of the genera *Botrytis* and *Cercospora*. Proposed ABA biosynthesis pathway by the "direct" cyclization of $C_{15}$ terpenoid (via the MEV pathway) in these two genera is schematized in Fig. 44. The pathway goes from a $C_{40}$ precursor (zeaxanthin), through violaxanthin to 9-*cis*-violaxanthin, and from there it is cut by the NCED to xanthoxin, finally resulting in ABA. It uses the MEP pathway and takes place in angiosperm plastids (Fig. 45). The great number of genes involved in the biosynthesis of ABA in plants (Figs. 43 and 45) suggests that the final level of ABA can be regulated at numerous steps through changes in the expression of the respective genes; such changes were indeed reported (see Schwartz *et al.*, 2003). However, it seems that the changes in the expression of the genes that encode NCED has a predominant effect on ABA formation.

During the early years of ABA studies, investigators assumed that all ABA is synthesized in the roots of angiosperms and then flown upwards to sites where it performed a physiological role, such as maintaining turgor in leaves and closing stomata. Such an exclusive biosynthesis does not exist, although ABA can move in

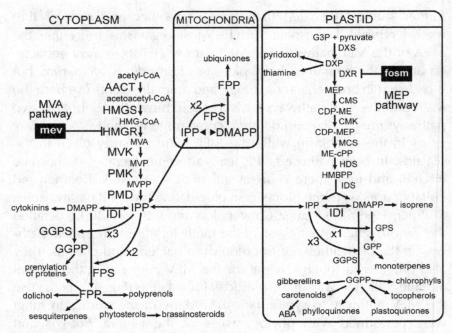

**Fig. 43.** Isoprenoid biosynthesis pathways in the plant cell. HMG CoA, hydroxymethylglutaryl CoA; MVP, 5′-phosphomevalonate; MVPP, 5′-diphosphomevalonate; HBMPP, hydroxymethylbutenyl 4-diphosphate; FPP, farnesyl diphosphate; ABA, abscisic acid. Enzymes are: AACT, acetoacetyl CoA thiolase (FC 2.3.1.9.); HMGS, HMG-CoA synthase (FC 4.1.3.5.); HMGR, HMG-CoA reductase (FC 1.1.1.88); MVK, MVA kinase (FC 2.7.1.36); PMK, MVP kinase (FC 2.7.4.2); PMD, MVPP decarboxylase (FC 4.1.1.33); IDI, IPP isomerase (FC 5.3.3.2); GPS, GPP synthase (FC 2.5.1.1); FPS, FPP synthase (FC 2.5.1.10); GGPS, GGPP synthase (FC 2.5.1.29); DXS (FC 4.1.3.37); DXR, DXP reductoisomerase (FC 1.1.1.267); CMS (FC 2.7.7.60); CMK (FC 2.7.1.148); MCS (FC 4.6.1.12); HDS; IDS, IPP/DMAPP synthase. The steps specifically inhibited by mevinolin (mev) and fosmidomycin (fosm) are indicated. (From Rodriguez-Conception and Boronat, 2002).

the vascular tissue. In this book, the roles of ABA in plant physiology are not discussed, but it is clear that ABA biosynthesis can take place in various plant sites such as in leaves and developing seeds.

As for the catabolism of ABA in plants, the information is far from being complete. There are two types of reactions that cause ABA

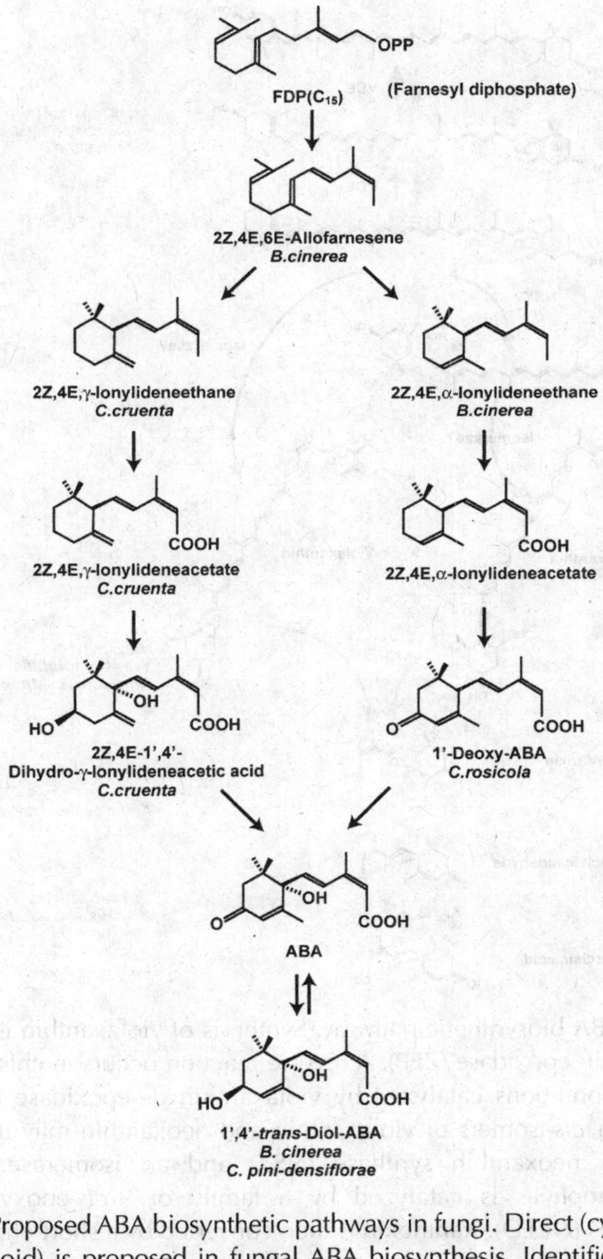

**Fig. 44.** Proposed ABA biosynthetic pathways in fungi. Direct (cyclization of C15 terpenoid) is proposed in fungal ABA biosynthesis. Identified potential intermediates with fungal species are shown. (From Nambara and Marion-Poll, 2005).

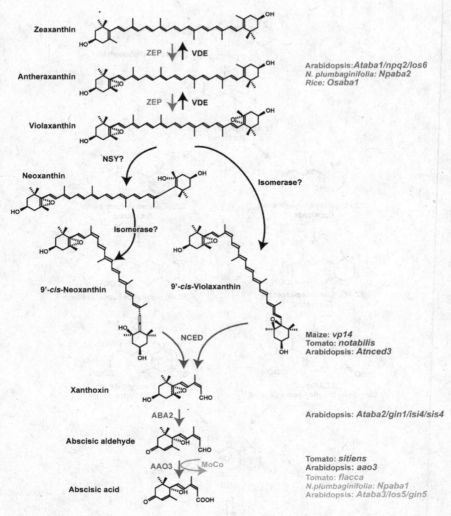

**Fig. 45.** ABA biosynthetic pathway. Synthesis of violaxanthin is catalyzed by zeaxanthin epoxidase (ZEP). A reverse reaction occurs in chloroplasts in high light conditions catalyzed by violaxanthin de-epoxidase (VDE). The formation of *cis*-isomers of violaxanthin and neoxanthin may require two enzymes: a neoxanthin synthase (NSY) and an isomerase. Cleavage of *cis*-xanthophylls is catalyzed by a family of 9-*cis*-epoxycarotenoid dioxygenases (NCED). Xanthoxin is then converted by a short-chain alcohol

catabolism. One of them is hydroxylation. Hydroxylation can take place in $C_7$, $C_8$, or $C_9$ (see Fig. 46). The hydroxylation at $C_8$ is thought to be the predominant ABA catabolic pathway. The other type of catabolism is conjugation. The carboxyl (at the $C_1$, see Fig. 46) and the hydroxyl groups of ABA, and its oxydative catabolites, are potential targets for conjugation with glucose. The most widespread conjugate is the ABA glucosyl ester (ABA-GE). In the past, the ABA conjugates were considered to be physiologically inactive and it was thought that they accumulated in the vacuoles during aging. However, more recent studies indicate that ABA-GE may serve as a long-range transport of ABA, and when reaching its destination, the $\beta$-D-glucosidase releases the ABA. The ABA-GE was also identified in *Citrus* as an allelopathic substance. The ABA-GE content of *Citrus junos* fruit peel reaches 132 mg.g$^{-1}$ dry weight. Such levels of ABA-GE are very inhibitory for the roots and hypocotyls of lettuce since concentrations of 0.3 $\mu$M or higher are detrimental to lettuce seedlings. Can ABA-GE serve as a kind of selective weed-killer? It is unlikely. The same authors (Kato-Noguchi and Tanaka, 2008) found that when ABA-GE was exogenously applied to Arabidopsis seedlings, the growth of the latter hypcotyl was reduced by 50% (when the seedlings were treated with 1.8 $\mu$mol/l). They suggested that ABA-GE penetrated the roots of the seedlings and was first hydrolyzed by the apoplastic ABA-$\beta$-D-glucosidase, causing the release of ABA and inhibiting hypocotyl growth. It thus seems that ABA-GE is not merely a "dead-end" of ABA metabolism; it may well be converted back to ABA and play a role in plant physiology. The liberated ABA can then reach the xylem and be transported to other parts of the plant.

---

**Fig. 45. (*Continued*)** dehydrogenase (ABA2) into abscisic aldehyde, which is oxidized into ABA by an abscisic aldehyde oxidase (AAO3). AAO3 protein contains a molydenum cofactor activated by a MoCo sulfurase. A list of defective mutants — which have been named separately depending on species or selective screens — is given on the right side of each enzymatic step. (From Nambara and Marion-Poll, 2005).

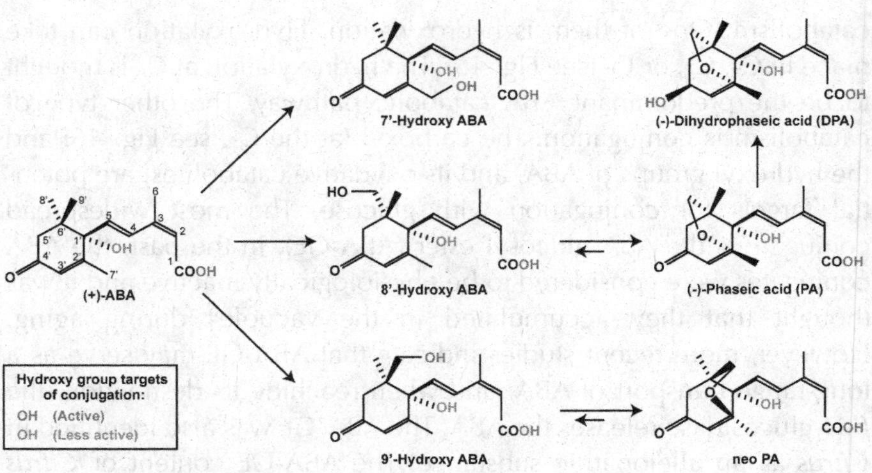

**Fig. 46.** ABA catabolic pathways. Three different hydroxylation pathways are shown. The 8'-hydroxylation is thought to be the predominant pathway for ABA catabolism. (From Nambara and Marion-Poll, 2005.)

## Perception and Signal Transduction

Until a few years ago, the receptor of ABA was enigmatic. Then, within the course of just a couple of years, information from several directions appeared in the literature. One such publication was from Winnipeg, Canada (Razem *et al.*, 2006). The authors of this report combined information from the autonomous pathway of floral initiation in Arabidopsis with the results of protein interactions obtained from large-scale screenings, as well as the knowledge that ABA can suppress the transition to flowering. The transition from vegetative growth to flowering in Arabidopsis involves several pathways (e.g. autonomous, vernalization, and photoperiod pathways), and the final known step is the interaction of *LEAFY* and *APETALA1* expressions that will cause the transition. However, upstream of these interacting two genes, there is a plethora of various proteins that are encoded by a variety of genes. A central gene is *FLC* (*FLOWERING LOCUS C*). More information on the genetics of the transition to flowering is provided by He and Amasino (2005) and in Fig. 53 by Galun (2007).

The FLC is a MADS box transcription factor and a potent repressor of transition to flowering. FLC expression is repressed by regulators of the transition to flowering of the autonomous pathway and the vernalization pathway of Arabidopsis. In the autonomous pathway, FCA — which is an RNA-binding protein that binds a floral factor *FY* (*FLOWERING LOCUS Y*) — is involved. The latter complex will derepress the pathway that is repressed by FLC. However, Razem *et al.* (2006) claimed that due to the strong binding between ABA and FCA, the FY is replaced by ABA in the complex of FCA. When complexed with ABA, this FCA/ABA complex will not repress FLC; consequently, the transition to flowering will be delayed. The investigators (Razem *et al.*, 2006) found that not only is FCA binding (+)-ABA with high affinity, but also the interaction is stereospecific and follows receptor kinetics. Hence, one receptor of ABA was identified. This publication initiated a lot of interest among the investigators of phytohormones. Within a short period, the Razem *et al.* (2006) paper was cited 120 times. It was the most cited study among 95 reports on ABA receptors between 2005 and 2008.

A second ABA receptor was revealed in Beijing, China, by Da-Peng Zhang and associates (a team of 13 investigators). This team (Shen *et al.* 2006), has had previous interest in ABA and its binding to specific proteins. One such ABA-specific binding protein, termed ABAR, was isolated from broad bean leaves (*Vicia faba*) and was potentially involved in ABA-induced stomatal signaling (Zhang *et al.*, 2002). Shen *et al.* (2006) then isolated a complementary DNA fragment from the broad bean leaves, that encoded the carboxy-terminal of about half of 770 amino acids of the putative H subunit (CHLH) of the magnesium-protoporphyrin-IX chelatase (Mg-chelatase). The CHLH was previously reported to have several functions in plant cells. The Chinese investigators found that the cDNA of CHLH binds ABA specifically. They therefore analyzed the functions of ABAR in ABA signaling in Arabidopsis. These analyses indicated that ABAR/CHLH is an ABA receptor that regulates seed development, post-germination growth, and stomatal aperture. The ABAR protein possesses one binding site for ABA. The binding (as was reported for FCA-ABA) is highly stereospecific: only the physiologically active ABA — (+)-ABA — binds, while the (–)-ABA and the *trans*-ABA do not bind to ABAR. Furthermore, Shen *et al.* (2006) found that by reducing the expression of ABAR (by RNAi), the plants had a significant number of ABA-insensitive phenotypes in post-germination growth arrest by ABA application. Such silenced ABAR plants did not show the typical ABA treatment effect on stomata closure. Arabidopsis plants that overexpressed ABAR were also established. Such plants were reported to be ABA-hypersensitive. Altering the level of the ABAR expression affected the concentration of ABA in the various transgenic plants (it was always in the range of 0.2 $\mu$g.g$^{-1}$ dry weight). On the basis of their experimental results, Shen *et al.* (2006) suggested that the ABAR is an ABA receptor that perceives the ABA signal in seed germination, post-germination growth, and stomatal movement. They claimed that this emerged from the following evidence. Firstly, ABAR binds specifically to ABA. Secondly, transgenic down-regulation of ABAR expression results to a decline in the number of ABA binding sites and leads to ABA-insensitive phenotypes. Thirdly, *ABAR*-overexpressing plants have ABA-hypersensitive phenotypes with an elevated number of ABA binding sites. Fourthly, a loss-of-function mutant in *ABAR* resulted in an immature embryo.

A third receptor for ABA was claimed by Liu *et al.* (2007). These authors reported that mutations in the putative GPCR protein — that is encoded by *GCR2* — conferred moderately reduced sensitivity to ABA in several standard assays, including seed germination, stomatal movements, ABA-regulated gene expression, and K$^+$ channel activity in guard cells.

The picture that emerged from the three abovementioned publications (Razem *et al.*, 2006; Shen *et al.*, 2006; Liu *et al.*, 2007) initiated several reviews that put forward a comprehensive description of ABA perception and signal transduction (see Hirayama and Shinozaki, 2007; Spartz and Gray, 2008; McCourt and Creeman, 2008; as well as a meeting report by McSteen and Zhao, 2008). Hirayama and Shinozaki (2007) even attempted to summarize this subject in an elaborate scheme (Fig. 47). Then came the "downfall." First, recent publications questioned the validity of the report by Liu *et al.* (2007).

It was again not accepted in a meeting reviewed by McSteen and Zhao (2008). Hence, one potential receptor was eliminated. However, the real blow to ABA receptors happened on 10 December 2008 when the journal *Nature* published an editorial feature that tells the story of the publication of Razem *et al.* (2006). The leading investigator of this Canadian team, Robert Hill, wrote to *Nature* how he wished to retract the Razem *et al.* (2006) publication: the results could not be repeated and the first author of the publication (the postdoc Fawzi Razem) had left Winnipeg laboratory and could not be located. Hill claimed that experiments with proteins that are extracted from plant tissues could be tricky. There are also technical problems involved with the interactions of such proteins. Moreover, due to the fact that Razem *et al.* (2006) had used similar methodologies to the ones used by Shen *et al.* (2006) and Liu *et al.* (2008) to claim that CHLH and GCR2, respectively, are ABA receptors, Risk *et al.* (2008) stated that as a result, none of the three claimed ABA receptors was likely to be a true receptor!

In several of the phytohormones discussed above, the transduction process includes a component of ubiquitination. This ubiquitination involves the attachment of a polyubiquitin chain to a short-lived intracellular protein, causing it to enter the 26S proteasome where the protein is degraded. The degradation changes a transcription factor's activity and thus also changes the expression of a hormone-responsive gene. This apparently also happens in respect to ABA-responsive genes. Smalle *et al.* (2003) identified a mutant in Arabidopsis that expresses an altered RPN10. The proteasome consists of two main components. One is the 20S core protease (CP) and the other is the 19S regulatory particle (RP). It is assumed that RP helps to correctly identify the ubiquitinated proteins that are destined for degradation in the CP. The mutant with the altered RPN10 had typical phenotypes such as reduced seed germination, reduced growth rate, and a reduced number of stamens. Most phenotypes could be explained by hypersensitivity to ABA. The authors suggested that RPN10 affects a number of regulatory processes in Arabidopsis, most likely by directing specific proteins to degradation in the proteasomes. Further verification of such an involvement of

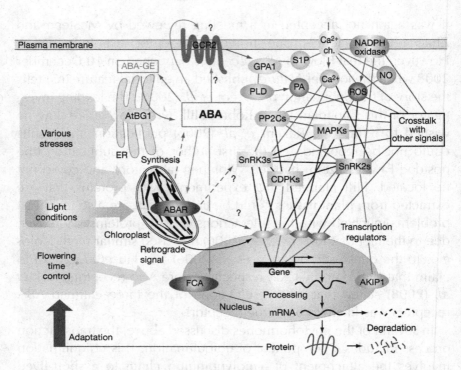

**Fig. 47.** Assembling components of the ABA response into a schematic of the cellular ABA response. ABA is produced in the cytoplasm from its precursor, which is synthesized in the plastid. ABA is inactivated as an ABA-glucose conjugate (ABA-GE) and is released by glucosidase activity. The ABA signal is received by (at least) three ABA receptors: GCR2, ABAR, and FCA. GCR2, a G protein-coupled receptor-like receptor, is localized at the plasma membrane, but its perception site has not been confirmed. GCR2 presumably transduces the signal through GPA1, the Arabidopsis α subunit of the trimeric G protein. In the ABA signaling pathway, second messengers, protein kinase, protein phosphatase 2Cs, and transcription factors form a web-like network. Cross-talk with other signaling pathways increase the complexity of the network. Connections between components are shown as lines. Red and black lines indicate demonstrated and predicted interactions, respectively; blue lines indicate predicted cross-talk. ABAR is Mg-chelatase, which is involved in the synthesis of chlorophyll and plastid-to-nuclear retrograde signaling, but the molecular basis for its action as a receptor is unknown. FCA is localized in the nucleus and is involved in the regulation

proteasomes with the transduction of ABA is needed. Such proof may potentially lead to the identification of the true receptor(s) of ABA.

Himmelbach *et al.* (2003) reviewed the relay and control of ABA signaling. They pointed out that ABA signaling involves a plethora of intracellular messengers. This probably reflects the function of ABA in integrating several stress responses and antagonizing pathways via "cross-talk," but it hampers the establishment of a unifying concept. For example, transcriptome analyses have unraveled more than 1000 genes that are differentially regulated by ABA. These ABA-mediated changes in gene expression translate to major changes in proteome expression.

This book focuses on patterning; therefore, studies on the signal transduction of ABA in response to tolerance of dehydration and cold-stresses will not be discussed. Those interested in these responses should turn to the review by Yamaguchi-Shinozaki and Shinozaki (2006). Readers interested in the signaling events in guard cells are referred to Pei and Kushitsu (2005).

A relatively recent publication by Teaster *et al.* (2007) handles the role of *N*-acylethanolamins (NAEs) in ABA signaling. In plants, NAEs are commonly elevated in desiccated seeds. NAEs, as well as ABA levels, were found to be reduced during seed germination. Moreover, both NAEs and ABA inhibited the growth of Arabidopsis seedlings, although the combined application of low levels of NAEs and ABA caused a more dramatic reduction in germination and growth than

---

**Fig. 47. (*Continued*)** of mRNA stability, which is required for flower induction. Chloroplast function and flowering-time control are crucial for survival. It is postulated that plants use ABAR and FCA as ABA receptors to integrate the ABA signal with other stress signals that affect chloroplast function and flowering-time control in order to adapt to various environmental conditions. $Ca^{2+}$ ch., $Ca^{2+}$ channel; CDPK, $Ca^{2+}$-dependent protein kinase; CP, chloroplast; ER, endoplasmic reticulum; MAPK, mitogen-activated protein kinase; NO, nitric oxide; PA, phosphatidic acid; PP2C, protein phosphatase 2C; ROS, reactive oxygen species; SnRK, SNF1-related kinase; S1P, sphingosin-1-phosphate. (From Hirayama and Shinozaki, 2007).

either component alone. After the NAE treatment, Arabidopsis seedlings had elevated transcripts of some ABA responsive genes. When an NAE degrading enzyme was overexpressed, it resulted in hypersensitivity to ABA. Some mutants that were insensitive to ABA exhibited reduced sensitivity to NAE. The authors suggested that NAE metabolism interacts with ABA in the negative regulation of seedling development and that normal seedling establishment depends on the reduction of the endogenous levels of ABA and NAE.

It cannot be excluded that additional metabolites interact with ABA. Hence, the riddle of ABA signaling is still far from being resolved.

# CHAPTER 8

# Peptide Phytohormones

## Background

There are several groups of peptide phytohormones that are actively investigated but the available information on these phytohormones is still limited. Hence, these peptide hormones will only be briefly mentioned here and toward the end of this chapter, a special section will focus on one group of these hormones: the CLAVATA3 hormones that play an important role in the shoot apex and the root apex. In the past, investigators focused on plant hormones that are mediated by small lipophilic compounds, such as auxins, cytokinins, gibberellins, abscisic acid, ethylene, brassinosteroids, and jasmonates. These were discussed in previous chapters. Peptide hormones were thought to occur in animals. This view has changed in recent years and a comprehensive review on peptide hormones in plants has been provided by Matsubayashi and Sakagami (2006) from the Nagoya University in Japan. A clue for the existence of the many groups of peptide phytohormones was the numerous receptors for peptides that were identified as being membrane-localized receptor kinases in plants. As it will be discussed, several of these presumed peptide hormones are involved in stress resistance or other phenomena that are not directly involved in the patterning of plants.

## Systemines

Systemines are involved in resisting plant pathogens such as insects. The systemines can be considered as defense proteins that include proteins that are toxic to the attacking insects. The proteinase inhibitors

I and II have been detected in some Solanaceae crops, such as tomato and potato. They prevent or reduce the capability of insects to use the plant tissue as food. It is typical for these defense proteins not only to be induced at the site of their initiation (an attacked leaf site) but also to be then produced in plant areas that are far away from the initial attack, indicating that a long distance signal transmission induces a *systemic* defense response. For example, a [$^{14}$C]-labeled 18 amino acid peptide that was formed in damaged leaves and then applied to a wounded leaf, was then detected in other leaves of the plant. This indicated that the respective peptide could move (or could induce another factor to move). This 18 amino acid peptide was termed *systemin* (tomato systemin, TomSys). It became feasible that the wound-induced TomSys that was released into the vascular system activated jasmonic acid biosynthesis in the surrounding vascular tissue, and then jasmonic acid acted as a long distance signal. The TomSys is derived from a proteolitic processing of a much longer precursor: the tomato prosystemin. Orthologs of prosystemin were detected in several Solanaceae species, such as potato, black nightshade, and pepper. Antisense inhibition of TomSys translation in tomato markedly reduced the systemic induction of proteinase inhibitor expression, while overexpression of prosystemin resulted in constitutive expression of the defense-response proteins, as if the plants were in a constant wounded state. A TomSys-insensitive mutant (*def-1*) was found to be deficient in the biosynthesis of jasmonic acid, indicating that jasmonic acid signaling is involved in the effects of TomSys. Studies with grafting and additional mutants suggested that indeed TomSys is not itself the long distance traveling signal but is instrumental in establishing this signal. The TomSys was found to mediate several processes in plant cells and is derived from linolenic acid of membrane lipids, thus inducing the jasmonic acid intermediate 12-*oxo*-phytodienoic acid, finally leading to jasmonic acid. The jasmonic acid may then further up-regulate the expression of the prosystemic gene via a positive feedback loop. While these studies on defense mechanisms were mainly conducted in tomatoes, similar mechanisms probably exist in other species of Solanaceae.

A receptor of TomSys was revealed and termed SR160. This receptor was found to be a member of the leucine-rich repeat receptor-like kinase (LRR-RLK) family and has similarities with the brassinolide receptors BRI1 mentioned in Chapter 4. Experimental results indicated that brassinolide and systemin signaling in tomatoes use the same cell surface receptor, although it seems that brassinolides do not compete with systemins for binding to BRI1/SR100, probably due to having distinct binding sites within the same receptor protein. The binding of each of the two ligands probably causes a different conformation change in the receptor.

## Phytosulfokines

Phytosulfokine peptides have a very different role to that of the systemines. The former are instrumental in determining when to continue cell division and when to stop this process. For example, *in vitro* cultured cells may readily form large calli or stop cell division. Hence, phytosulfokines were suggested as vital when sparsely plated plant cells (e.g. isolated protoplasts) were used for *in vitro* culture and callus formation. When plated at great dilution, no calli were formed, but when plated densely, such calli were readily formed. It was discovered in my laboratory (Raveh *et al.*, 1973) that cell division could be initiated in sparsely plated cells when "nursing cultures" were added. Protoplasts that were unable to divide due to their exposure to *γ*-rays, could serve as such nursing cultures. In the presence of a high density of *γ*-irradiated (none dividing) cells, the low-density plated (with a capability to divide) readily formed calli that could regenerate into plants. Hence we were able to regenerate functional *Nicotiana* plants from single protoplasts. The same method was also applicable to other plants; for example, low-density plated *Citrus* protoplasts could be regenerated into mature fruit-bearing plants (Vardi *et al.*, 1975). Subsequently, this "nurse culture" or conditioned medium was applied to many systems, indicating that cells can release a factor capable of inducing cell division in culture systems. The factors were not fully characterized, but experimental results hinted that these factors are sulfated peptides composed of only five amino acids. They were termed *phytosulfokines* (PSK). PSKs were then isolated from a variety of plants (asparagus, rice, maize, *Zinnia*, carrot, and Arabidopsis). All contain the five amino acids: Try, Ile, Try, Thr, Gln (or YIYTQ) in which the Try was sulfated. It was then found that PSK had additional effects on plant cells, such as somatic embryogenesis, adventitious bud formation, and adventitious root formation, although it did not become clear whether or not PSK plays a vital role in normal plant patterning.

The PSK is produced by the enzymatic processing of an about 80 amino acid precursor-peptide that has a secretion signal at its N-terminal. Several amino acids upstream of the five essential amino acids are found in all PSK precursors; other amino acids of the precursors are not conserved. In Arabidopsis, there are five genes for PSK, and PSK genes are expressed in various tissues of this plant. It also seems that dividing cells contain ample PSK but with maturation of the cells, the PSK level is drastically reduced.

Two proteins of 120 kD and 150 kD were found to bind PSK and were thus considered receptors of PSK. These were identified LRR-RLKs (Leu-rich repeat-receptor-like kinase) that were derived from the same gene. The receptor was named PSKR1. It is not yet clear what the function of PSK and PSKR1 is in the normal patterning of plants. Hopefully, future mutagenesis (formation of knockout mutants, especially in the specific LRR-RLKs) will supply such information.

## SCR/SP11

SCR/SP11 peptides have essential roles in the incompatibility system of several plants. One type of incompatibility in which the pollen of a given individual plant is unable to germinate on its own stigmata but will do so on other specific plants' stigmata, was thoroughly studied in *Brassica*. Since these peptides probably have no role in plant architecture, they will not be examined further.

## CLV3 (CLAVATA3)

The overall architecture of the shoot apical meristem in angiosperms is typically composed of the following components. There are three main layers of cells: the upper epidermal layer (L1), the subepidermal cell layer (L2), and the underlying layer (L3). Together they form a dome-like shape. There is some flexibility in the layers that become evident after surgical treatments of the shoot apical meristem (the SAM). The L1 layer cells will develop into leaves and flowers and shoot epidermis. The L2 cells will produce ground tissue and the internal tissue of leaves. The apical dome is commonly divided into three zones: the peripheral zone (PZ) that provides cells to the lateral organs; the rib zone (RZ) that will become the stem core, and the central zone (CZ), where cell division is slow and which furnishes cells to the PZ and the RZ. The SAM — especially of young angiosperms before flowering — uses mechanisms to maintain the relative size of the three zones. In particular, the SAM maintains a given volume of dividing cells in the CZ by a rather elaborate control mechanism. The *CLAVATA* genes play a central role in this control. When one of three *CLAVATA* genes is mutated (*clv1*, *clv2*, or *clv3*), the cell division volume of the SAM is greatly enlarged causing abnormal phenotypes. *CLV* genes also affect other plant organs; *clv* mutants have club-shaped carpels in their flowers — thus the name *CLV* was given to these genes. In spite of the similarity in name, these three genes encode quite different proteins. *CLV1* encodes a receptor-like serine/threonine kinase with an extracellular domain of leucine-rich repeats (LRR), a transmembrane domain, and an intracellular kinase domain. The latter domain is mostly missing from the protein encoded by *CLV2*. The two proteins, CLV1 and CLV2, form a het-erodimer that is stabilized by disulfide bonds. The CLV3 protein is very different from the two other CLV proteins. The *CLV3* gene encodes a 96 amino acid peptide from which a shorter peptide is processed and the mature polypeptide then serves as a ligand for the CLV1/CLV2 complex receptor (see Fig. 29 in Galun, 2007). There is also a difference in the localization of expression between these three genes. *CLV3* is expressed in the L1 and L2 layers of the CZ, whereas *CLV1* is expressed in the inner layer (L3) of the same zone. In order to activate the CLV1/CLV2 complex, the product of CLV3 should be transported from one site to another site of the SAM. Hence the CLV3 product can be regarded as a phytohormone. Once the CLV1 signal is activated, it causes the down-regulation of another gene: the *WUSCHEL* (*WUS*). The WUS product promotes stem cell division in the SAM. It was revealed that an overexpression of *WUS* leads to uncontrolled proliferation of stem cells, while mutations in *WUS* cause stem cell to differentiate (rather than divide). In mutants of *CLV1/CLV2*, there is an overproliferation of stem cells. On the other hand, there is a "loop": ectopic WUS can induce the activation of *CLV3*. In the absence of *WUS* expression, there is no CLV3 formation. This feedback "loop" is essential for maintaining an optimal balance of stem cells in the SAM. However, *CLV3* is also active in the root apical meristem (RAM), and there are indications that activation of a CLV-like signal may also control cell fate in roots and that CLV3 is also a regulatory component in RAM.

An embryo-surrounding region protein (ESR) was detected in maize. There the *ESR* gene encodes a secreted polypeptide. While the total sequence of *ESR* differs considerably from *CLV3*, both genes share (near their carboxy terminal region) a sequence that encodes a preserved 14 amino acid polypeptide. Further search in Arabidopsis revealed 25 genes that encode the conserved region, and the genes were named *CLAVATA3/ESR* related genes (*CLE*). The expression of these genes was not restricted to the SAM or the RAM, but were expressed in various tissues of the plant. One of these genes, *CLE40* is very similar to *CLV3*.

From experimental results, it is assumed that CLE and CLV3 peptides are first translated as precursor peptides and are then, after posttranslational processing, secreted as mature polypeptides. The *in situ* mature polypeptide of CLV3 is RTVP$^H$SGP$^H$DPLHH, in which P$^H$s are hydroxylated prolines. The latter functional peptide was isolated and sequenced as MCLV3 by Kondo *et al.* (2006). Ito *et al.* (2006) developed a bioassay with xylogenic cells to detect peptides that affect cell differentiation versus cell division. Mesophyll cells of *Zinnia* can differentiate tracheary elements when cultured in a certain medium. The investigators revealed a 20% methanol fraction that was named by them as Tracheary Element Differentiation Inhibitor (TDIF) that inhibited the development of tracheary elements in cultured mesophyll *Zinnia* cells. The TDIF was characterized and found to be a dodecapeptide with two hydroxyprolines. The mature

TDIF was derived from the carboxyl terminal of a long protein resulting in a polypeptide that differed from MCLV3 termed HEVHP$^H$SGHP$^H$NPISN, but which also contained two hydroxylated (P$^H$) prolines. It is still uncertain whether TDIF has a function in the patterning of plants.

## Rapid Alkalinization Factor (RALF)

Felix and Boller (1995) found that when mechanically wounded plant cells were added to suspension-cultured tomato cells, the culture medium quickly became alkalinized. This led to a screening of other wounded cells and a 49-amino-acid peptide was revealed to be associated with this phenomenon. The peptide apparently had no role in the defense of tomato cells, and the rapid alkalinization factor was termed RALF. Synthetic tomato RALF protein caused immediate arrest in root growth. RALF-like peptides were then found in several plants, but there is no information on how the mature 49 amino acid peptides are processed.

## ENOD40

This early nodulin, ENOD40, is a peptide that is induced by rhizobia bacteria in legume roots, in the process of producing nitrogen-fixing nodules in these roots. Peptides with similarity to ENOD40 were also detected in non-legume plants. There is not much information on ENOD40, and it does not seem to be involved in plant patterning.

## Polaris

The Polaris (PLS) peptide was revealed in embryonic roots from the heart-stage of plant embryos and in the lateral root tips. The peptide that has a predictable length of 36 amino acids is matured from a 4.6 kD protein. The role of PLS is not yet clear but mutated PLSs have a reduced primary root length. Such mutants also show hyper-responsiveness to exogenous cytokinin and reduced responsiveness to auxin. It was suggested that PLS is required for the correct auxin-cytokinin homeostasis, but there is not much evidence to support this suggestion.

## IDA

Inflorescence deficient in abscission (IDA) is another peptide for which there is very little information. The *IDA* gene encoding a 77-amino acid peptide was found in floral members at the abscission zone. The 77-amino acid peptide, that has an N-terminal secretion signal, is probably matured to a shorter peptide. It was suggested that the small peptide that is encoded by *IDA*, is a ligand of HAESA that is an Arabidopsis plasma-membrane-associated LRR-RLK, and that this LRR-RLK is involved in controlling floral organ abscission. HAESA is expressed at the base of petioles and the abscission zones of floral organs.

# CHAPTER 9

# Strigolactones and Branching

## Historical Background

The prevailing explanation about the regulation of branching in plants, up to about a decade ago, was "Apical Dominance." This explanation claimed that auxins move basipetally from the shoot apex, and very young leaves near the apex inhibit branching from leaf axils. The acropetal movement of cytokinins (SKs) was thought to promote such branching. Indeed, basipetal movement of auxin in the shoot was documented as early as 1933 when Thimann and Skoog reported that a compound — later to be identified as auxin — flows basipetally and could inhibit the growth of lateral buds. Moreover, cutting the uppermost part of the shoot initiated accelerated branching. However, when auxin was applied at the cut surface of the upper shoot, branching was inhibited. Although the experimental results of Thimann and Skoog (1933) were clear and reproducible, subsequent studies on branching of plant shoots that were conducted over a period of more than 60 years indicated that the regulation of bud outgrowth is controlled in a rather elaborated manner. The subject was thoroughly reviewed by Napoli *et al.* (1999). These authors already concluded that the shoot tip may not be the only suppressor of axillary bud outgrowth. One of the latter authors (Beveridge, 2000) from the University of Queensland in Australia, then reviewed the experimental results produced by herself and others that had been mainly conducted with pea plants (w.t. and *ramosus*, *rms*, mutants). The mutants *rms1* through *rms5* displayed increased branching at the basal and areal nodes. The idea was to detect long-distance signals involved in bud outgrowth. In summary, the various experimental approaches, that also included graftings between mutants and W.T. peas, resulted in a working model that included the activity of auxin and CK but also encompassed two unknown graft-transmissible signals. The ramosus mutants are a rather complicated issue. There are five genes of Rms but each has several alleles, and the various alleles cause some differences in the respective phenotypes. Beveridge (2000) suggested a working model shown in Fig. 48. Her experimental results indicated roles for two apparently "novel" graft-transmissible signals and provided only little direct evidence that auxin and CK are the main regulators of branching in peas. The work and the resulting model suggested that *Rms3* and *Rms4* control hormone perception, signal transduction, or other processes in the

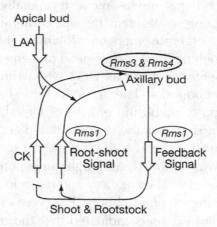

**Fig. 48.** Working model of long-distance signaling among the apical bud, axillary bud(s), and the remainder of the plant (shoot and root stock), together with the sites of action of the *Ramosus* genes *Rms1* through *Rms4* in pea. *Rms1* and *Rms2* act in shoot and root stock. *Rms3* and *Rms4* are shown associated directly with axillary buds based only on data indicating they act mostly in the shoot. Large unfilled arrows indicate direction of signal movement; line arrowheads indicate promotion; flat ended lines indicate inhibition. The root-shoot signal regulated by *Rms1* is shown as a branching inhibitor, but results to date are also consistent with *Rms1* down-regulating the level of a promoter. CK, cytokinin; IAA, indole-3-acetic acid. (From Beveridge, 2000).

system that occurs mostly in the shoot. Genes *Rms1* and *Rms2* act in root stock and shoot. *Rms1* probably controls the level or the transport of a signal — other than either auxin or CK — that moves in the direction: root-to-shoot. *Rms2* may control the level or the transport of a feedback signal — other than auxin — that moves in the direction: shoot-to-root. These suppositions are in line with a proposal by R. Snow made back in 1937! Beveridge's conclusion (2000) — if verified — is merely a "revolution" to the old and much simpler "Apical Dominance" hypothesis.

## Early Studies on a "New" Hormone Affecting Branching

Further information on the branching-inhibitory factors in pea was provided by Beveridge and her colleagues over the subsequent years (Foo *et al.*, 2005; Morris *et al.*, 2005). Foo *et al.* (2005) found evidence to characterize the protein encoded by *Rms1* as a carotenoid cleavage enzyme that acts with *Rms5* to control the level of an as yet unidentified mobile branching inhibitor, that is required for auxin-inhibition of branching. They concluded that the product of *Rms1* plays a central role in a shoot-to-root-to-shoot feedback system that regulates shoot branching in peas. Moreover, auxin positively regulates the *Rms1* transcript level so that auxin has an indirect effect on branching. But ... the transcript level of *Rms1* is also dramatically elevated in the mutants *rms3*, *rms4*, and *rms5* which do not contain elevated auxin levels. In *rms4* plants, grafting experiments suggested that an auxin-independent mobile feedback signal contributes to the higher level of the *Rms1* transcript. Hence, the branching and inhibition of branching in pea shoot constitute a rather elaborated system. Morris *et al.* (2005) inflicted another blow to the classical "Apical Dominance" hypothesis. These investigators used pea plants to test whether auxin inhibits the initial stage of bud release or subsequent stages of branching. They found that after decapitation the initial bud growth occurs *prior* to changes in auxin level or transport in the vicinity of the stem tissue of the emerging bud; this level is also not affected by the acropetal movement of exogenous supply of auxin.

The bud outgrowth is not stimulated in the time frame of changes in auxin. They therefore concluded that the decapitation triggers initial bud growth via an auxin-independent mechanism. They assumed that auxin operates *after* the initial stage of bud outgrowth, mediating the results of decapitation via autoregulation of buds that are already in transition toward sustained growth.

Peas are appropriate plants to study the control of branching, but when it comes to looking at this phenomenon by a molecular view, Arabidopsis has a clear advantage. Hence Van Norman *et al.* (2004) from the University of Utah in Salt Lake City turned to Arabidopsis. His team focused on *BYPASS1* (*BPS1*) that is an Arabidopsis gene that had previously no known function. They revealed that *BPS1* is required to prevent constitute production of a root-derived, graft-transmissible signal that is sufficient to inhibit leaf initiation, leaf expansion, and shoot apical meristem activity. These investigators also found that this root-derived signal is likely to be a novel carotenoid-derived molecule that can modulate both root and shoot architecture. Previous data indicated that ABA is also transported acropetally from the root to the shoot. However, experimental results indicated that the signal that is involved with the BPS1 gene is not ABA.

## More Evidence on the Impact of a Carotenoid-derived Compound on Branching

Working with Arabidopsis, the team of H.J. Klee (Auldridge *et al.*, 2006) added more important information on branching and the carotenoid cleavage by dioxygenases (CCDs). Five members of the CCD family in Arabidopsis are probably involved in the biosynthesis of ABA (see earlier in Chapter 7). The function of the other four enzymes is less clear. CCD1 is the only family member that is cyto-plasmic and since carotenoid cleavage is located in plastids, CCD1 is probably not involved in the cleavage that leads to the obscure hormone that controls branching. The genes *CC7/MAX3* and *CC8/MAX4* that encode intraplasmid cleavage enzymes seemed to have a vital role in branching. While Auldridge *et al.* (2006) did not yet come up with a chemical identification of a hormone that suppresses bud outgrowth, they did suggest the following model. Accordingly, the elusive apocarotenoid is generated at least in part by CCD7/MAX3 and/or CCD8/MAX4 that cleave a carotenoid molecule inside the plastid to make a mobile intermediate that is subsequently acted on by MAX1 to produce a signal perceived by MAX2, leading to the inhibition of branching.

Klee and associates (Snowden *et al.*, 2005) moved to petunia as an experimental plant. They revealed a similar branching control in petunia as the one in Arabidopsis. In petunia, there is a gene termed *Dad1/PhCC8* that encodes a hypothetical carotenoid-cleavage dioxynase (CCD). The acronym DAD stands for DECREASED APICAL DOMINANCE. The latter petunia gene is an ortholog of the *(MAX4)/AtCCD8* that was mentioned above. The *Dad1/PhCCD8* gene was found to be expressed in the roots and shoot tissues. The mutants, *dad1-1*, *dad2*, and *dad3* increased the branching of petunia. Overexpression of *Dad1/PhCCD8* complemented that mutant phenotype and RNA interference in the wild-type resulted in an increase in branching. However, there were also additional changes in the phenotype when *Dad1/PhCCD8* was lost. This suggested that the products of the Dad1/PhCCD8 enzyme are mobile signal molecules with diverse roles.

The study of bud outgrowth in petunia was continued by Simons *et al.* (2007), mainly through grafting experiments. Both *dad1-1* and *dad3* increased branching but could be reverted to new-wild-type by grafting on *dad2* root stocks; these root stocks contained the *DAD1* and *DAD3* genes. The *dad2* mutant cannot be reverted by grafting. The *DAD2* seemed to be expressed mainly in the shoot, and it probably acts in the same pathway as *DAD1* and *DAD3*, but not in a simple step-wise fashion. The *dad1-1, dad3* double mutant has additional phenotypes: decreased height, delayed flowering, and reduced germination.

## Tillering in Rice

A Chinese team (Zou *et al.*, 2006) studied the regulation of tillering in rice. The correct rate of tillering in rice is important for obtaining good yields. The rice genome contains a gene that is termed *HIGH-TILLERING DWARF1* (*HTD1*). This gene encodes a protein that is orthologous to Arabidopsis MAX3. A defect in *HTD1* leads to high tillering, and *htd1* mutants also have a dwarf phenotype. The Arabidopsis *max3* mutant could be rescued by overexpression of *HTD1* (by the introduction into Arabidopsis of $Pro_{35S}$:HTD1). This indicated that HTD1 is a carotenoid cleavage dioxynease that has similar functions as MAX3 (in Arabidopsis), namely the synthesis of a carotenoid-derived signal molecule. The *HTD1* gene is expressed in vascular bundle tissue throughout the rice plant. Auxin induces *HTD1* expression which suggests that auxin may regulate rice tillering partly through the up-regulation of the *HTD1* transcript. The removal of axillary buds restores the dwarfism to a more normal height, suggesting that the dwarfism is a consequence of too much tillering. The authors also suggested a feedback mechanism for the synthesis and perception of the carotenoid-derived signal in rice. The characterization of the *MAX* genes in Arabidopsis and identification of orthologs in petunia and rice indicates that there is a conserved mechanism for the control of shoot branching regulation in monocots and dicots.

## Studies by P. McSteen and O. Leyser

An update on the regulation of branching was provided in an important review by Paula McSteen (from Pennsylvania State University, US) and Ottoline Leyser (from the University of York, UK). These authors (McSteen and Leyser, 2005) combined their intimate knowledge in the patterning of monocots and dicots, respectively, to present a broad understanding of shoot branching in angiosperms. The concept of *phytomer* was instrumental for explaining the various types of branching. The phytomer is considered to be a link in the chain of the shoot that includes a stem segment, a node bearing, one or more leaves or leaf-like structures, and one or more auxillary meristems (AMs) in each leaf axil. Looking at the various phytomers of shoots provides the pattern of the shoot. The SAM can grow indeterminably, producing an "endless" succession of phytomers, the fate of which varies along the shoot axis. Alternatively, the shoot can undergo a determinate developmental program, producing a "fixed" number of phytomers that usually terminate in the production of reproductive structures (flowers). The Arabidopsis can serve as an example that maintains an indeterminate growth pattern. First, the rosette phytomers are produced which consist of a large leaf, a short stem segment, and a morphologically undetectable AM. At the transition to flowering, several phytomers are produced that consist of a small leaf, a greatly elongated stem segment, and a large indeterminate AM. Subsequent phytomers consist of a cryptic leaf, an elongated stem segment, and a determinate AM that forms a flower. The flower itself can be described as formed from phytomers with very short segments and "leaves" that are the floral organs (sepals, petals, stamens, and carpels). The phytomers of the tomato shoot are slightly different from the Arabidopsis phytomers but then lead to rather different shoot growth. The three components of the phytomer can differ substantially leading to very different shoots. It seems that the greatest variability is caused by the different AMs. The latter may be vestigeal (morphologically undetected) leading to unbranched shoots, or the AMs may be outgrowing; forming the branching or under some condition forming a floral bud. Moreover, the AM may stay dormant for an extended time and then continue to develop into a branch or a floral bud.

The authors noted that for many years the original idea of apical dominance resulted from the basipetal flow of auxin that inhibits branching (Thimann and Skoog, 1933). However, the authors claim that the inhibition of branching can also be explained in other ways: one of them is that basipetally, there is a reduction of auxin sensitivity. There were other issues that did not fit a simple branching inhibition model by the flow of auxin from the upper shoot. One of these is that by the use of radio-labeled auxin, the auxin fails to enter the auxillary bud. Also, directly applied auxin did not inhibit AM. Thus, auxin seemed to have a kind of "remote" inhibitory effect. One of the suggestions was that as cytokinin is promoting the outgrowth of buds, the presence of auxin inhibits the biosynthesis of cytokinin in the root so that less cytokinin is available to promote outgrowth. The other candidate for a second messenger is a signal that was not fully revealed at the time of this review: this unknown signal was found to be dependent on *MAX3* and *MAX4* in Arabidopsis, the *RAMOSUS* (*RMS1*) gene in pea and the *DAD1* gene in petunia. Loss-of-function in these three groups of genes caused an increase of branching. As noted above, much information on the branching effect of these genes resulted from grafting experiments that indicated an acropetal movement of the elusive signal. The *MAX3* and *MAX4* genes were found to encode divergent members of the carotenoid cleavage dioxygenase family. All the information that was available to the reviewers was consistent with the RMS/DAD/MAX signal(s) acting to relay the auxin message into the bud. Another gene of Arabidopsis, *MAX2*, that seems to play a role in the branching, encodes an F-box protein and may be involved in the signal transduction of the elusive signal; other suggestions also exist though. One of these alternative suggestions was that the active basipetal flow of auxin down the primary stem inhibits auxin outflow from the lateral buds into the primary stem polar transport stream, and that this export is required for bud outgrowth. This "ATA" model is not compatible with some experimental results and therefore was not further elaborated.

The control of branching in monocots has similarity with the branching of dicots. During vegetative growth of monocots, branches termed tillers, arise from the basal nodes and grow out in an acropetal sequence. The basal phytomers of monocots are characterized by short internodes, leaves with juvenile characteristics, and activate outgrowth of AM during early vegetative development. Tillers cause the bushy architecture of monocots, such as rice, wheat, and barley, and are an important issue in cereal yields. Maize* differs from the former cereals, as it was selected during domestication to only have a very few tillers, shifting the production of grains to the main shoot. Branches in monocots can also grow out from the upper part of the shoot and are termed "auxillary" or "secondary" branches. In some cases, the tillers produce such auxillary branches. In maize, the ear shoots are formed from auxillary branches that formed some nodes below the tassel. Grasses (i.e. members of the *Gramineae* family)

---

*The correct term should be *maize* but the USA literature commonly use the term *Maize*.

bear their flowers in spikelets that are considered to be short branches. In some genera of this family, the spikelets are formed in pairs while in other genera (such as in *Oryza*) they form single spikelets. In maize, the spikelets consist of two florets that are situated inside two, reduced leaf-like structures termed "glumes"; while in rice there are two sets of glumes, one of which is strongly reduced. These components can be considered as modified phytomers. Also, one can see conservation in the AM initiation between monocots and dicots, as there are genes in these two groups of plants that act in a similar manner. In addition, similar genetic mechanisms seem to regulate tillering and spikelet initiation. *TB1* in maize encodes a transcription factor. Changes in the expression of *TB1* cause the suppression of tillering, which probably took place during the evolution of maize from its ancestor teosinte. In maize, *TB1* is expressed in AM as soon as they become visible, while in teosinte, *TB1* is not expressed in outgrowing axillary buds. A gene orthologous to *TB1* was also revealed (*OsTB1*) and overexpression of *OsTB1* led to repression of tillering in rice.

The auxillary meristems of inflorescence are different to the vegetative shoot auxillary meristem. Here we shall not go into details but mention that the inflorescence of the maize ear has four types of AM. The first are the "branch meristems." There are two such meristems in maize that form the base of the tassel. Then there are the "spikelet pair meristems" and in addition, there are two floral meristems. In dicots such as Arabidopsis, there are only branch meristems and floral meristems.

The branch meristem of maize is a kind of a short indeterminate entity. Here I would like to note that in the axil of the upper shoot of cucumber there is also a kind of a short secondary shoot. If the lower flower formed on this secondary shoot is pistillate, then subsequent flowers (that are staminate) are suppressed, but if this first pistillate flower is removed by surgery or by hormone treatment (such as GA), the later staminate flowers will develop.

McSteen and Leyser (2005) concluded their review by stating that the phytomer concept provides a useful framework to analyze shoot structure in very diverse species. It appears that there is communality between all AMs, but there is also clear divergence between meristem types both within and between plant species. The divergence is not surprising due to gene duplication and mutation during evolution that changed the impact of various genes on AMs. The comparison between monocots and dicots is sometimes problematic. For example, decapitation and grafting experiments that were useful in dicots are not applicable in monocots. At the time this review was written, there were only hints of the elusive signal that could play a major role in branching (e.g. Sorefan *et al.*, 2003; Ward and Leyser, 2004; Brooker *et al.*, 2004). Genes were revealed that encode enzymes crucial for the processing of carotenoids into branching signal. On this basis, the team of Leyser as well as other teams were able to further study this elusive signal.

Additional studies in the laboratory of Leyser and by other teams vastly clarified the issue of the control of branching (Bainbridge *et al.*, 2005; Brooker *et al.*, 2004, 2005; Bennett *et al.*, 2006; Stirnberg *et al.*, 2007; Ongaro and Leyser, 2008; Gomez-Roldan *et al.*, 2008; Chen *et al.*, 2008; Umehara *et al.*, 2008).

Bainbridge *et al.* (2005) focused on the *MORE AXILLARY BRANCHING 4* (*MAX4*) gene in Arabidopsis and established its role. It encodes a long-rage, graft-transmissible signal that inhibits shoot branching. Moreover, buds of the mutant *max4* are resistant to auxin application, showing that *MAX4* is required for auxin mediated bud inhibition. The pea gene *RMS1* and petunia gene *DAD* are orthologous to *MAX4*. In spite of the similarity between these three genes, there are significant differences in the regulation of their expression. For example, *RMS1* is up-regulated by auxin in the shoot while *MAX4* is up-regulated by auxin only in the root and the hypocotyl. Both *RMS1* and *DAD1* are subjected to feedback regulation but there is no such feedback for *MAX4*. It seems that the most functionally significant point of interaction between auxin and *MAX4* is post-transcriptional.

In a further publication (Booker *et al.*, 2005), the authors characterized the four *MAX* genes of Arabidopsis. They found that branching is regulated by at least one carotenoid-derived hormone. All the four *MAX* genes seemed to act in a signal pathway. They proposed that

*MAX1* acts on a mobile substrate, downstream of *MAX3* and *MAX4*, which are immobile substrates. Furthermore, the authors identified *MAX1* as a member of the cytochrome P450 family. The Leyser team then looked at the regulation of auxin transport in Arabidopsis.

When the gene *AUXIN RESISTANT 1* (*AXR1*) of Arabidopsis is mutated (*axr1*), it confers a primary defect in auxin-regulated transcription. This will increase shoot branching and cause the resistance of bud to the suppression of outgrowth by apical auxin. Bennett *et al.* (2006) used the *AXR1* gene to further study the effects of *MAX1*, *MAX3*, and *MAX4* on branching that involves an elusive hormone. This "novel" hormone is a regulator of auxin transport. Such a regulation in the stem is sufficient to control bud outgrowth and is independent of AXR1-mediated auxin signaling. The authors suggested an additional mechanism for long-range signaling by auxin in which bud growth is regulated by competition between auxin sources, for auxin transport capacity, in the primary stem of Arabidopsis.

In an additional study by the Leyser team and a collaborator from the University of Cambridge (UK), the authors (Stirnberg *et al.*, 2007) focused on *MAX2*. They found that this gene encodes an F-box leucine-rich repeat protein. F-box proteins usually function as a substrate-recruiting subunit of the SCF-type ubiquitin E3 ligase protein ubiquitination. As *MAX2* is essential together with the products of *MAX1*, *MAX3*, and *MAX4* for the suppression of bud outgrowth, this finding strongly suggests that this suppression involves ubiquitination. Indeed, the *max2* mutants caused highly branched shoots. Transcript analyses and a transitional *MAX2-GUS* fusion indicated that *MAX2* is expressed throughout the plant, most highly in developing vasculature. *MAX2* acts locally (the product is not mobile) in the auxillary bud or in the adjacent stem or petiole tissue. The expression of a *MAX2* mutant that lacks the F-box domain does not complement *max2*. The authors concluded that auxillary shoot growth is controlled locally at the node by an $SCF^{MAX2}$, the action of which is enhanced by the mobile MAX signal. However, the signal itself has not yet been characterized by these investigators. Even in a review by Ongaro and Leyser (2008) the elusive hormone was not yet revealed chemically.

# Revealing the Branch-Suppressing Hormone

This hormone was revealed by two independent research teams: Gomez-Roldan *et al.* (2008) and Umehara *et al.* (2008). The respective studies were published sequentially in the same issue (September 2008) of *Nature*. The Gomez-Roldan (2008) work was performed by a group of 14 investigators from France, the Netherlands, and Australia. The Umehara (2008) work was performed by a team of 12 Japanese investigators.

The Gomez-Roldan *et al.* (2008) team briefly reviewed the inhibition of bud outgrowth in pea and petunia, noting that two respective mutants (*rms1* of pea, and *dad1* of petunia) are defective in the inhibition of bud outgrowth that occurs in the wild-type. Furthermore, they recorded that previous studies revealed a mobile signal which they termed SMS (shoot multiplication signal) and which moves acropetally in shoots and inhibits bud outgrowth. In the past, this led to a consideration of SMS as a hormone. It was also previously known (see above) that the shoot branching genes of Arabidopsis encode carotenoid cleavage dioxygenases (CCD7, CCD8, and also see Fig. 49) that are active in the synthesis of SMS. CCD7 and CCD8 were considered to be involved in sequential cleavage of $\beta$-carotene suggesting that SMS is carotenoid-derived. It should also be noted that CCD7 and CCD8 are conserved across the angiosperms (monocots as well as dicots). Also, grafting experiments indicated that RMS4 of pea and MAX2 of Arabidopsis (that encode F-box leucine-rich repeat proteins) confer response to SMS.

The carotenoid-derived SMS drew the investigators' attention to two very different phenomena other than branching controls: carotenoid-derived compounds termed *Strigolactones*. One of these phenomena is the parasitism between a number of plant species as host and other angiosperms — as *Striga* and *Orobanche* — as parasites. The parasitic plants have a "clever" strategy: their seeds will not germinate until they "sniff" the vicinity of a host. The parasite recognizes host roots by the release of strigolactones from these roots.

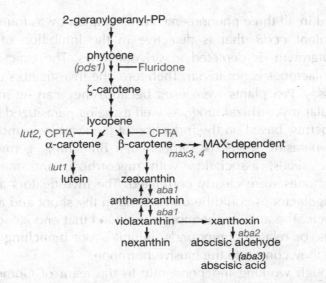

**Fig. 49.** This simplified scheme of carotenoid biosynthesis includes the sites of action for inhibitors and the position where mutations disrupt carotenoid biosynthesis (gray). *Pds1* and *aba3* are indicated in parentheses as they do not encode the carotenoid biosynthetic enzymes, but these mutations result in the disruption of carotenoid biosynthesis at the indicated positions. For simplicity, not all intermediates or enzymes are shown. (From Van Norman and Sieburth, 2007).

The parasite then reaches the roots of the host and forms "knots" in which the parasite utilizes the metabolites of the host. The other phenomenon concerns the symbiosis between arbuscular mycorrhizal fungi, such as Glomeromycota, and plants. This symbiosis is widespread and was revealed in almost all plants including monocots and dicots. Again, the mycorrhizal fungi sense the roots of the plants by the release of strigolactones from the roots. A correlation was found in the relations between the auxillary bud outgrowth and the formation of mycorrhizal association as well as parasitism between angiosperms. This already suggested that strigolactones are involved in all three phenomena. For example, it was found that the pea mutant *ccd8*, that is defective in the inhibition of auxillary bud outgrowth, is depleted in strigolactones. The exact detection of strigolactones is not trivial. Therefore, the investigators developed a bioassay. Pea plants were used because they can be infected by arbuscular mycorrhizal fungi, as well as being parasitized by certain angiosperms, based on the induction of the symbiosis and germination of parasitic seeds (of *Orobanche*). Branching, germination of parasitic seeds, association with mycorrhiza, and the level of strigolactones were clearly correlated. The investigators also found that strigolactones could be transported in the shoot and act at very low levels. The authors therefore concluded that endogenous strigolactones, or related compounds, inhibit shoot branching in plants. Hence, they constitute the elusive hormone.

Although working independently to the team of Gomez-Roldan *et al.* (2008), the Japanese team of Umehara *et al.* (2008) arrived at very similar conclusions. This team mentioned the known facts that shoot branching is a major determinant of plant architecture and that two known hormones are involved in the regulation of shoot branching, but that previous studies strongly suggested that an additional hormone is involved in this regulation. Several mutants in Arabidopsis, pea, and petunia that affect processing of carotenoids were previously found to affect shoot branching (e.g. mutants of *MAX3*, *RMS1*, *RMS3*, *MAX4*). Moreover, it was indicated that *CCD7* and *CCD8* encoded enzymes that catalyze sequential carotenoid cleavage. Also, *MAX1* encodes cytochrome P450 monooxygenase that is presumably involved in a later biosynthetic step of the elusive hormone. As the Gomez-Roldan team, the Japanese team also noted that *MAX2* and *RMS2* encode an F-box leucine-rich repeat, which probably acts as the substrate recognition subunit of an SCF ubiquitin E3 ligase for proteasome-mediated proteolysis. The latter team also recalled that strigolactones are involved in *Striga* and *Orobanche* parasitism as well as in the symbiotic interaction of plants with arbuscular mycorrhizal fungi. Thus, Umehara *et al.* (2008) also concluded that

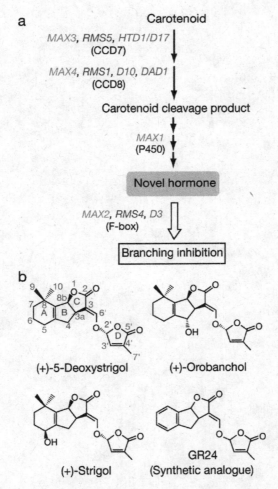

**Fig. 50.** (a) The novel branching inhibitor pathway and (b) chemical structures of representative strigolactones. (From Umehara *et al.*, 2008).

strigolactones — or a derivative of these compounds — are the elusive hormones that play a major role in the control of shoot branching. The pathway of the formation of the latter branching inhibitor and the chemical structures of representative strigolactones are shown in Fig. 50.

# PART II

# Plant Organs and Tissues

---

In Part I of this book, I focused on nine groups of phytohormones and presented the information on the history of their research, their distribution in the angiosperms, their biosynthesis, and their signal transduction and perception. In this second part of the book, my attitude will shift. Here, the focus is on the patterning of five major plant organs (and their respective tissues) and I will review the impact of phytohormones on these organs' patterning.

## The Phylogeny of Angiosperms

This book deals with angiosperms, although taxa of non-angiosperms will be mentioned occasionally. Scutt *et al.* (2006) reviewed the evolutionary perspective on the regulation of carpel development. Carpels are the significant "invention" of angiosperms and as such are considered one of the major evolutionary innovations of flowering plants. In order to place the angiosperms in the proper branches of the phylogeny of seed plants, we should consider the "tree" that was suggested by Scutt *et al.* (2006). It appears that the ancestors of all extant seed plants existed about 300 million years ago (MYA). The gymnosperms emerged about 200 MYA and the ancestors of all angiosperms appeared about 160 MYA (Fig. 51). It is also evident that angiosperms are a successful group, comprising of about 300,000 species, while there are presently only about 750 known species of gymnosperms. Among the angiosperms, the core-eudicots are the most successful group (with about 230,000 species), and this group was separated from other angiosperms about 115 MYA.

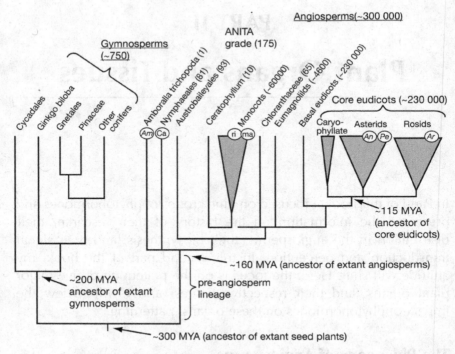

**Fig. 51.** The phylogeny of the seed plants, based on a consensus of molecular phylogenetic studies. The numbers of species in major clades are given in parentheses. Very large clades are represented by shaded triangles. The positions of certain species referred to in the text are indicated as follows: Am, *Amborella trichopoda*; An, *Antirrhinum majus*; Ar, *Arabidopsis thaliana*; Ca, *Cabomba aquatica*; ma, maize; Pe, *Petunia hybrida*; ri, rice. (From Scutt *et al.*, 2006.)

Some aspects of the control of the female reproductive development are conserved between the flowering plants and their "sister" group, the gymnosperms, suggesting their existence in the ancestors of all extant seed plants, already about 300 MYA. It is reasonable to assume that the evolution of the carpel in the angiosperms involved gene duplication, followed by changes in one of the duplicates and gradual elimination of the other duplicates. The difference between the gymnosperms and the angiosperms is that in the formers' ovules commonly occur as naked structures on leaf-like organs. In contrast, in the angiosperms, the ovules are enclosed and protected by a specialized female reproductive organ that is termed the carpel. The carpel probably provides several advantages to the angiosperms: it protects the ovules, the stigma tissues at the carpel's apex, appears to be efficient in capturing pollen, and also serves as a site for the selection of pollen in self-incompatibility (a mechanism that promotes out-breeding). Finally, the carpel develops into a fruit that protects the developing seeds and — in some cases — is helpful in the dissemination of the seeds. The genes involved in the specific development of the flower members (e.g. the Coen and Meyerowitz, 1991 model and subsequent models that were detailed in Chapter 12 of Galun, 2007) are not repeated here.

## The Shaping of Organs

The final architecture of a plant organ is shaped by the same overall strategy. The cells of evolving organisms undergo, in the various locations of these organs, one of four processes. There may be cell division, arrest of cell division, growth (extension) of cells, and arrest of cell growth. In all of these processes, phytohormones are involved. Before handling elaborated plant organs, we should start by looking at very simple organs — such as the hypocotyl of seedlings and trichomes — and see how phytohormones and the interactions between them shape such simple organs. Also, how do light/dark circles interact with the phytohormones? Referring to the update by Weiss and Ori (2007), we will take a look at five stages of plant development in Arabidopsis: (1) the germinating seed; (2) the emerging seedling; (3) the apex of the shoot; (4) the transition from a vegetative shoot to flowering; and (5) the development of the flower. Stages 3–5 will be discussed in detail in forthcoming chapters of this book, but it should be mentioned that several phytohormones were found to have a decisive effect on these stages. Thus, auxin (AUX), gibberellin (GA), and cytokinins (CK) have major roles in the shoot apex. During the transition to the flowering phase, AUX and GA promote this transition, while CK and ethylene (ET) retard this transition. During the

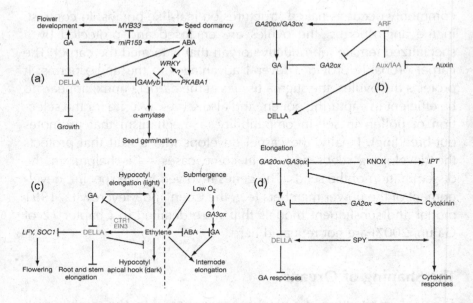

**Fig. 52.** Networks of interactions between GA and other phytohormones. (a) Network of interactions between GA and ABA. ABA suppresses GA responses through DELLA-dependent and -independent pathways. (b) Network of interactions between GA and auxin. Auxin promotes GA responses by destabilizing DELLA and by promoting the expression of GA biosynthetic genes. (c) Network of positive and negative interactions between GA and ethylene. Ethylene represses GA biosynthesis or suppresses GA responses via DELLA stabilization. GA promotes ethylene responses in dark- and light-grown seedling (apical hook formation in the dark and hypocotyl elongation in the light). Submergence promotes ethylene and GA synthesis in deepwater rice and *R. palustris*, and GA promotes ethylene-induced internode elongation. (d) Network of reciprocal interactions between GA and cytokinin. Two major players control the balance between GA and cytokinin. KNOX1 proteins control the balance between the two hormones in the SAM by inducing cytokinin production, directly inhibiting GA synthesis, and indirectly promoting GA deactivation. SPY regulates the balance between the response pathways of these two hormones via suppression of the GA signal and promotion of cytokinin responses. (From Weiss and Ori, 2007, where references to these schemes are provided.)

development of the flower and the germination of the seeds, GA promotes development while abscisic acid (ABA) retards this process. A demonstration of the cross-talk between GA and other phytohormones was provided in an update by Weiss and Ori (2007) who also illustrated this cross-talk in four schemes (Fig. 52). The antagonistic interactions between GA and ABA are outlined in Fig. 52a. GA and ABA play antagonistic roles in the regulation of numerous developmental processes. While GA is associated with the promotion of germination, growth, and flowering, ABA inhibits these processes. Antagonistic relationships between GA and ABA also regulate the transition from embryogenesis to seed germination. An old known phenomenon is that during cereal seed germination, the embryo releases GA to the aleurone cells, where they induce the transcription of genes encoding hydrolytic enzymes, supplying nutrients to the developing embryo. ABA suppresses the $\alpha$-amylase expression. It seems that the $\alpha$-amylase promoter contains a GA response element required for both its activation by GA and its suppression by ABA. Here comes an additional interaction: the formation of miRNA159 is enhanced by both ABA and GA. MiRNA159 suppresses *MYB33* while the latter seems to have a positive role in both the promotion of flowering and maintaining seed dormancy.

We learned in Chapter 3 that DELLA interferes with GA response. The scheme in Fig. 52b shows interactions between GA, auxin, and DELLA. Moreover, the level of biologically active GA is regulated by the promotion of its synthesis (mediated by the GA20ox and GA3ox genes) and its degradation (mediated by the gene GA2ox). In addition, auxins are probably destabilizing DELLA — opening the way for organ elongation.

There are additional interactions: GA-induced RGA degradation was inhibited in the *axr1* mutant, in which the auxin signaling is compromised. This suggests that auxin promotes the degradation of DELLA in root cells, in response to GA, allowing root elongation. When pea and tobacco were decapitated, not only the auxin source but also the supply of GA was reduced. Application of auxin reversed this effect. Auxin was shown to induce the expression of the enzyme for GA synthesis (GA20ox) in tobacco and Arabidopsis, whereas in pea it induced the expression of GA3ox that also causes active GA biosynthesis, while the GA-degrading enzyme GA20ox was suppressed by auxin in pea. The auxin signaling suppressors, such as AUX/IAA, were mentioned in Chapter 1; when the latter are degraded, the result is activation of the transcription factor ARF7. Accordingly, in the presence of the auxin receptor TIR1 (that is an F-box protein that mediates AUX/IAA degradation) there was ARF activation and suppressed auxin regulation of GA-biosynthesis gene expression. Hence, auxin may positively interact with GA at either the biosynthesis level or by promoting DELLA degradation.

The interaction between GA and the stress-related gaseous hormone ethylene (ET) is rather complex and depends on environmental conditions, as schematically shown in Fig. 52c. ET's central role is stressed in this scheme. ET represses GA responses via the stabilization of DELLA but is also antagonistic to ABA while ABA represses GA. ET is increased by submergence (low oxygen level) but the submergence also facilitates the synthesis of active GA. In fact, GA promotes ET responses in dark-grown and light-grown seedlings. GA also promotes ET-induced internode elongation. By inhibiting DELLA, GA has a role in facilitating the transition to flowering (via releasing the inhibition of *FLY* and *SOC1* by DELLA). By suppressing DELLA, GA is also increasing root elongation in seedlings, as observed in Arabidopsis. The scheme presented in Fig. 52d brings into the interaction two additional players: KNOTTED1-like homeobox (KNOX) and cytokinin (CK). Both KNOX and CK inhibit the activity of two enzymes that are involved in the biosynthesis of active GA: GA20ox and GA3ox. In addition, CK may have another effect: it causes the degradation of active GA by increasing the expression of the GA2ox gene. The KNOTTED1-like homeobox (KNOX) protein family also has an indirect effect: these proteins were shown to induce the expression of the CK biosynthesis gene ISOPENTENYL TRANS-FERASE7 (*IPT* gene in Fig. 52d). The increased CK may then degrade the active GA, and by that affect the DELLA level. The SPY protein was assigned a double role: it may regulate the balance between the response pathways of the two hormones (GA and CK) via suppression of the GA signal and promotion of the CK responses. As we will see when we deal with the emergence of new meristems in the stem apical meristem, the CK and GA may exercise a clear antagonistic effect. For example, the very early initiations of novel meristem in the apical dome require high CK and low GA signals; the later stage of the emerging of new meristems at the shoulder of this dome requires the opposite: low CK and high GA.

The interactions pointed out by the review of Weiss and Ori (2007) are far from describing all the interactions in which GA is involved. There are additional interactions in which three phytohormones are involved. For example, GA, AUX, and ET interact to promote the elongation of light-grown seedlings as well as in the differentiation at the stem apical meristem. Clearly, additional phytohormones (e.g. brassinosteroids, polypeptides, and strigolactones) can also take part in the interactions leading to a vast number of potential phyto-hormone interactions.

## Photomorphogenesis and Phytohormones

Plant tissues are constantly exposed to environmental, developmental, and metabolic cues. These cues are then connected with the elabo-rated network of phytohormones, as well as to chemical constituents such as enzymes and various controlling factors, which together shape the structure of plant tissues and organs. Advanced research tools enabled Jennifer Nemhauser (from the University of Washington in Seattle) to update our understanding of the interactions of phyto-hormones with photomorphogenesis, choosing to focus on a relatively simple organ: the shoot of a young seedling. Nemhauser (2008) focused on the Arabidopsis seedling. Light is a significant early environmental effector on seedling morphology: low light promotes "stem" elongation (i.e. the hypocotyl), while high light intensity causes cotyledon expansion. However, here already start the compli-cations. The different spectra of the light (i.e. blue light versus far-red light) have very different effects on the growth of the hypocotyl and the cotyledons. Moreover, phytohormones such as GA, AUX, BR, CK and ET, and their interactions have a profound role in the shaping of the young seedling. There is also an additional issue with the light signal: plants can sense changes in light/dark conditions and relate these changes to the operation of an "internal clock" (the "circadian clock").

From the brief description noted above, we can now start to deal with some details. First, there is an antagonism between light response and some phytohormones: AUX, GA, and BR. In Arabidopsis, light triggers a massive wave of transcriptional re-programming. This is happening primarily under the control of two sets of photoreceptors: the red/far-red sensing phytochromes (PHYA-E) and the UV-A/blue light sensing cryptochromes (CRY1 and CRY2). Light triggers the accumulation of the bZIP transcription factor ELONGATED HYPOCOTYL5 (HY5) that is required for hypocotyl inhibition in all light conditions. CONSTITUTIVE PHOTOMORPHOGENESIS1 (COP1) is a ubiquitin ligase that targets the HY5 for degradation in the dark. In the light, the COP1–HY5 interaction is disrupted allowing rapid accumulation of HY5 and reduced hypocotyl elongation. New, direct links were recently revealed between HY5 and hormone response. Many mutants having reduced auxin response are deteriorated in the dark, and both light as well as auxin responses require a common set of proteins involved in the ubiquitin-mediated protein degradation. The *hy5* mutant was one of the first ones to be identified with reduced light response. The wild-type HY5 may act through the repression of auxin-induced genes. By global transcriptome analysis of double mutants (*hy5, hyh*) that were exposed to six hours of white light, it was observed that there was an overexpression of genes encoding early response gene classes such as *Aux/IAA*, transcriptional co-repressors, and GH3 auxin-conjugating enzymes. Additionally, several genes were also mis-regulated in the double mutant. Furthermore, a genome-wide analysis of HY5-binding sites revealed evidence of direct binding of HY5 to the promoters of multiple genes required for auxin signaling (as AUX/IAAs and members of the auxin-response-factor family of transcription factors). HY5 was also found bound to the promoters of some ET-responsive factors (ERFs) and members of the DELLA gene family, as well as to transcriptional regulators from the ET and GA pathways. Clearly, the picture that already emerged points toward a multifactoral interaction in which several phytohormones take place: AUX, GA, and ET.

It was observed that defects in active GA biosynthesis lead to the phenomenon of the growth of dark-grown seedlings to appear as though they were grown in light. It appears that phytochrome-mediated light perception regulates levels of repressors of the GA response. As noted in Chapter 3, GA is perceived by the GA INSENSITIVE DWARF1 (GID1) receptors that promote the degradation of nuclear-localized growth repressors, named DELLA, via interaction with the F-box protein SLEEPY. It was revealed that a DELLA fusion protein is barely detectable in dark-grown rapidly elongating hypocotyl cells while this protein is rapidly accumulating in the same cells after exposure to light. Light inhibition was reduced in quadruple DELLA mutants. The effects involved in GA are downstream of the phytochrome-mediated changes in bioactive GA levels.

There is also a reciprocal effect: changes in hormone biosynthesis or response may influence the plant's response to light. This phenomenon was also covered in the review by Nemhauser (2008) and based on the CK perception that in Arabidopsis involved with "response regulators," termed ARRs. The phosphorylation on Type B ARRs relieves autorepression, allowing the expression of a number of early response genes — such as Type A ARRs — to be induced. The latter act as negative regulators of the pathway. Phytochrome B (PHYB) was shown to directly interact with a Type A response regulator, termed ARR4. It was then found that CK-induced phosphorylation of ARR4 modulates seedling sensitivity to red light through its direct interaction with PHYB. In short, we face interactions between light, phytochrome, CK, and growth characteristics.

A further interesting interaction is between the diurnal and circadian effects and the hormone network of plants. Plants (as most other organisms) possess an "internal clock" (the circadian clock) that enables them to sense environmental changes such as light and dark (short days vs long days). This ability has great advantages in respect to the adaptation of plants to a changing environment. For example, the transition to flowering will happen at the appropriate time with respect to the plant's sexual reproduction. Also, the growth of young seedlings is under the "clock" control and this includes hypocotyl elongation. Hence, when plants are grown under day/night cycles, the coincidence of specific light and time of the day cues cause an appropriate rhythmic growth. The two transcription factors PIF4 and PIF5 (that are involved in the regulation of DELLA levels) were identified as crucial integrators of the light and clock signal. Here, we face an interaction between "clock," light, and GA.

Another hormone network that is under "clock" levels concerns BR. The gene *CPD* that encodes a rate-limiting enzyme for BR biosynthesis was found to be under the dual control of phytochrome-mediated light response and circadian regulation. Indeed, the levels of endogenous BR were found to have clear diurnal fluctuations. Such clock-regulated fluctuations were also detected with additional hormones: ET, AUX, and GA. But... *de novo* changes in auxin biosynthesis did not change the regular circadian clock of plants.

The abovementioned review focused on a "simple" case of "patterning": the elongation of the hypocotyl of young seedlings. The interaction of phytohormones with the shaping of more complicated plant organs will be examined in Chapters 10 to 14. Here we are adding another "simple" case of "patterning": the root elongation of young seedlings. This case demonstrates the interaction of two phytohormones in growth. A gene, termed *BREVIS RADIX* (*BRX*), is strongly induced to be expressed by auxin. *BRX* is required for BR biosynthesis in the seedlings' roots. In the *brx* loss-of-function mutant, the roots of seedlings have a strongly reduced root meristem size and retarded root growth. These defects in root growth can be corrected by the application of BR or by constitutive expression of the *CPD* gene. In this case, it appears that a certain (low) level of BR is required for a normal growth promotion caused by auxin. When this BR is eliminated, the growth of the seedlings' roots is impaired.

In a further study on the interaction of BR and auxin on growth, Vert in collaboration with Walcher, Chory, and Nemhauser (Vert *et al.*, 2008) found a direct connection between the BR-regulated BIN2 kinase and ARF2 (a member of the auxin response factor family of transcriptional regulators). Phosphorylation by BIN2 results in a loss of ARF2 DNA binding and repression activities. The arf2 mutants are less sensitive to changes in endogenous BR levels, whereas a large proportion of the genes affected in an *arf2* background are returned to near-wild-type levels by changes in BR biosynthesis. Hence it was suggested that BIN2 increases expression of auxin-induced genes by directly inactivating repressor ARFs, leading to synergistic increases in transcription.

An additional facet on the molecular framework of light and GA interaction that concerns cell elongation, was recently studied by a team of 10 investigators from Spain and Switzerland (de Lucas *et al.*, 2008). There is a well-known antagonism between light and GA effect on the growth of hypocotyls in young seedlings: light leads to the inhibition of growth, whereas GA promotes etiolated growth that is manifested by increased hypocotyl elongation. De Lucas *et al.* (2008) attributed a central role in this process to the Arabidopsis PIF4 (PHYTOCHROME INTERACTING FACTOR4) that has a positive control on genes that mediate cell elongation. The investigators found that PIF4 is negatively regulated by the light photoreceptor PHYB and by DELLA proteins that have a key repressor function in GA signaling. The investigators detected a destabilization of PIF4 by PHYB in the light and found that DELLA block PIF4 transcriptional activity by binding to the DNA-recognition domain of this factor. GA negates such a repression by promoting the destabilization of DELLA, causing an accumulation of PIF4 in the nucleus. The degradation of PIF4 and of DELLA is probably processed by ubiquitination: the attachment of GA to DELLA facilitates the polyubiquitin binding to DELLA and thus transfers it to the 26S proteasome. The ubiquitination of PIF4 is facilitated in light, via PHYB.

## Control of Size

Plants, as well as other multicellular organisms, must have efficient means to control size. Although the size of an organ, such as a leaf or a petal, can be affected to some extent by environmental effectors, these differences are generally small, and there is a range of sizes in the various organs that is typical to the species. The situation in leaves is somewhat complicated because the same plant species may have very different types of leaves, such as Rosetta leaves in Arabidopsis that are very different from the leaves of a shoot that is in transition to flowering. The petals of a given plant species attain a certain size that is typical for this angiosperm species. The question then is how does the organ (e.g. the petal) "know" that it has reached its required size and consequently stops growing? While the petal constitutes a simple case, the same principal question can be asked for a variety of plant organs (e.g. fruit, seed, embryo, root hair, trichome). In order to stop growing once the organ reaches a required size, the organ should have a way of determining its size during the process of growth. Obviously, the same types of questions are of relevance in animal systems too, where investigators have already started to provide the relevant answers (e.g. Leevers and McNeill, 2005). The interest in mechanisms that control size in plants became evident in recent years (e.g. Sugimoto-Shirash and Roberts, 2003; Horiguchi *et al.*, 2006; Deprost *et al.*, 2007; Anastasiou and Lenhard, 2007; Anastasiou *et al.*, 2007; and Krizek, 2009).

Anastasiou *et al.* (2007) made an interesting contribution to the question of how plant organ size is controlled. These seven investigators from Germany and England focused on the Arabidopsis cytochrome P450 *KLUH (KLU)/CYP78A*, a gene that acts as a plant organ growth stimulator. In brief, the investigators found that *klu* loss-of-function mutants form smaller organs because of premature arrest of cell proliferation. *KLU* overexpression leads to larger organs with more cells. Hence, the Arabidopsis organs handled by these investigators (leaves and petals) attained different sizes primarily determined by the final number of the organ's cells, rather than the size of the cells themselves. They concluded that *KLU* promotes organ growth in a non-cell-autonomous manner. Yet it does not appear to modulate the level of any known phytohormone. These investigators' suggestion was that *KLU* is involved in generating a *mobile growth signal* that is different from the one in the classical phytohormone. The expression dynamics of *KLU* suggested a model that explains how the arrest of cell proliferation is coupled to the attainment of a certain primordium size. Moreover, they suggested that there is a common principle of size measurement in plants *and* animals. Could a common strategy go even deeper in the evolution? It was reported that some bacteria possess a mechanism that regulates the size of bacterial colonies: the dividing bacteria in the middle of the colony cause the release of a signal and this signal is sensed by the bacteria in the periphery of the colony; according to the *dilution* of this signal, the peripheral bacteria may stop their division.

Anastasiou *et al.* (2007) describe the growth of a plant organ in two successive and linked phases. The first phase is cell proliferation. Once sufficient cells have been produced, proliferation is arrested and cell expansion takes place. Such a division, though, may not be so clear-cut in all plant organs. However, several plant genes were identified that control the proliferation of cells. One of them is *AIN-TEGUMENTA* (*ANT*). In some cases, the lack of a specific factor will stop proliferation (e.g. *NUB, JAG*). Other genes have a self-division-arrest effect on very specific cells and the *duration* of the cell proliferation activity may also be controlled by certain genes (e.g. the E ubiquitin ligase *BIG BROTHER, BB*). The control of proliferation is probably determined by the *length of time* the proliferation takes place rather than by the control of the *rate* of proliferation. On the other hand, there are genes that control the expansion of cells, such as *ARGOS-LIKE1,* which acts downstream of BRs and by the cytochrome P450 *ROTUNDIFOLIA3* (*ROT3*) that is involved in the biosynthesis of BRs.

The basic questions that have to be asked are what is measured by the cells in a developing plant organ in order to enable them to make a "decision" about further growth, and how is this decision coordinated across the primordium of the novel organ? Anastasiou *et al.* (2007) came up with a suggestion that could answer both questions. Accordingly, there is a signaling via a mobile growth factor. If such a factor is produced from a limited source, as the very center of the primordium, the growth of the organ will change the factor's concentration (and/or the relative distribution) of the factor in the primordium. Growth will dilute the highly mobile factor with a largely homogenous distribution throughout the organ. However, a factor with a more restricted mobility the growth would push cells out of the factor's reach. In either case, this change in factor distribution could be used to measure the size of the premordium at any specific stage. The investigators first turned to the main plant phytohormones as the signaling factor for measured organ size and controlling the final size of organs (e.g. BR, AUX, CK). They looked for genes whose loss- and gain-of-function produce opposite effects in organ size. This search indicated that none of the common phytohormones fitted the role of the signaling factor. When the investigators searched for a candidate that would play a role in size determination in plant organs, they found the *KLU* gene from Arabidopsis. This gene encodes a cytochrome P450 CYP78A5. The KLU protein contained the hallmark features of functional cytochrome P450 enzymes, including a membrane anchor and a conserved heme-binding region that is used in catalysis. The KLU is specific to plants as other A class members of cytochrome P450 enzymes that were previously reported to fulfill plant-specific functions. In plants that were engineered to overexpress *KLU*, there was clearly overgrowth of leaves, sepals, and petals due to the increase in the number of cells. In the absence of KLU (in *klu* mutants) the reduced number of cells in the analyzed organs was due to earlier cessation of cell proliferation, while an increase of cells was due to an extended period of cell proliferation.

Interestingly, the expression of *KLU* is typical to the organ where it is expressed. For example, in petals, the expression of *KLU* is initially uniform in the very young primordia. Then the expression is limited to the periphery of the petal and later it ceases. Typically, the expression of *KLU* is *outside* the region of proliferation, although *KLU* expression may be detected also *after* the proliferation of cells ceased, suggesting that it is not the down-regulation of *KLU* that eventually causes the arrest of proliferation. There is another interesting feature of the KLU protein: it is localized to the endoplasmic reticulum (ER) and does not move in a similar manner to the movement of the classical phytohormones. Furthermore, experimental evidence suggested that KLU is not driving the movement of one of the classical phytohormones (GA, AUX, BR, CK). The assumption was thus made that KLU is involved in the generation of a "novel" (not yet identified) growth-stimulating compound that is mobile. Could the latter (yet unknown)

compound be "sensed" by the organ's cells and consequently regulate the timing of the arrest of their division? There is no answer yet
to this question but Anastasiou *et al.* (2007) assumed that with the
growth of an organ (such as a petal), the "novel" compound is
diluted. The cells of the organ sense this dilution as a signal to stop
division because the required size has been achieved.

In a recent review, Krizek (2009) reconsiders the problem of how
the final size of a plant organ is determined. Clearly, the number of
cells cannot be the only determinant of the final size, because there
are known cases in which cell division is arrested and the organ does
attain a normal final size. This phenomenon is termed "compensation"
as it involved the extension of the cells of a given organ beyond their
regular size. Krizek (2009) thus claims that there are commonly two
phases: the first phase is cell division, in a developing organ; then
comes the further growth of the organ by cell expansion. Krizek lists
nine key regulators of organ size in Arabidopsis. These genes seem to
affect size but have little effect on the form of the organs. Generally
speaking, auxin and BRs are important regulators of plant growth,
stimulating both cell division and cell elongation. One means by
which auxins control organ size is via the transcription factor *ARGOS*
(auxin-regulated gene involved in organ size) that acts upstream of
*AINTEGUMENTA* (*ANT*). Mutations in either *ARGOS* or *ANT* result
in plants with smaller lateral organs while the constitutive expression
of these genes under the 35S promoter will produce plants with
larger laterals. The expression of *ANT* may be negatively regulated
in maturing organs, by the auxin-response factor ARF2 that is a
repressor of organ growth. Hence, *arf2* plants have thick stems and
produce large seeds, embryos, sepals, and petals. The mature leaves
of this mutant show persistent *ANT* expression. There is a yet unclear
synergism between auxin and BR. However, it is noteworthy that the
auxin-response element (TGTCTC) is enriched in the promoters of
BR-responsive genes. Moreover, ARF2 mediates the cross-talk between
auxin and BR signaling (see Fig. 53 for additional interactions). Identi-
fying genes (e.g. in Arabidopsis) that have roles in the growth of
organs is far away from understanding how plants are able to control
the size of specific organs. The latter question was the subject of the

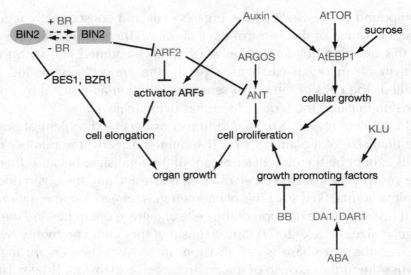

**Fig. 53.** Pathways controlling final organ size. Plant hormones are shown in blue, proteins promoting growth are shown in green and proteins restricting growth are shown in red. Auxin promotes growth through ARGOS and ANT while ARF2 negatively regulates *ANT* expression. ARF2 is negatively regulated by BIN2 allowing integration of auxin-signaling and BR-signaling pathways. The target specificity of BIN2 may be regulated by BRs. KLU promotes growth through the production of a novel growth-promoting signal while BB, DA1, and DAR1 restrict growth through negative regulation of unidentified growth-promoting factors. AtTOR and AtEBP1 likely regulate macromolecular synthesis and cellular growth to promote organ growth. *AtEBP1* expression is induced by sucrose and AtEBP protein is stabilized by auxin. (From Krizek, 2009.)

abovementioned study of Anastasiou *et al.* (2007), in which it was suggested how cells in an organ could evaluate the size of this organ in order to "decide" when to stop cell division. Nevertheless, we are still without answers to fundamental questions that should be answered before we can reach a full comprehension of size determination in plant organs. A meristem should "know" where in the plant it is located and consequently into which organ it will develop. It could be an indeterminate organ, such as an axillary shoot, or a determinate

organ, such as a petal. When the meristem is still very small, composed of a very small number of cells, many of the answers will be
furnished by the cells that are outside of the meristem; but with the
differentiation of this meristem, it will become gradually more
autonomous.

In mammals, insects, and yeast, there is a conserved signaling
network that regulates cellular growth in response to environmental
cues, such as nutrient statutes and stress. TOR is a Ser/Thr kinase of
the phosphatidylinositol-3-kinase-related kinase (PIKK) family. It is
present in two distinct multi-protein complexes (TORC1 and TORC2)
that differ in their downstream targets. TORC1 phosphorylates factors
such as p70 ribosomal S6 kinase (S6K) and eIF4E-binding proteins
(4E-BPs) to promote translation and ribosomal biogenesis, while
TORC2 regulates targets that include the actin cytoskeleton. In mammals, TOR activity is regulated by growth factors (insulin/insulin-like
growth factors), nutrients (amino acids), energy (AMP-activated protein
kinase), and stress (hypoxia).

Potential orthologs of TOR and other components in this signaling
pathway are present in the Arabidopsis genome, but functional
analysis of this pathway had previously been limited due to the fact
that mutations in the single Arabidopsis *TOR* gene (*AtTOR*) are embryo
lethal and Arabidopsis is insensitive to paramyxin. However, a recent
report describing RNAi lines in which *AtTOR* expression is reduced
but not eliminated, demonstrate the key role of this signal transduction
pathway of organ growth (Deprost *et al.*, 2007). In this case, AtTOR
RNAi lines produce morphologically normal but shorter leaves and
normal but shorter roots than wild-type plants, while Arabidopsis
plants that overexpress AtTOR have larger leaves and longer roots
than wild-type ones. The altered size results from the changes in cell
size, not cell number as seen above in the *KLU* system (where the
regulation of size was mainly by timing the duration of cell division).
A process similar to the *AtTOR* RNAi/AtTOR is the effect of another
regulator of cell growth in plants: *EBP1*.

There are additional repressors of growth in plants, such as *BIG
BROTHER* (*BB*) that is an E3 ubiquitin ligase. Thus, *bb* mutants produce larger sepals and petals, while overexpression of *BB* produces

smaller sepals and petals. *BB* seems to act by decreasing the number of cells, probably by mediating a proteasome-dependent degradation of growth-stimulating factors. Finally, there is an additional growth-restricting factor: *DA1*. In mutations of *DA1*, there are larger embryos, seeds, leaves, flowers, and thicker stems. The difference in size is probably due to the different *duration* of growth. There is a synergism between *da1* and *bb*: the double mutant produces flowers that are larger than in either *da1* or *bb* mutants.

The interactions of several of the factors, phytohormones, and genes that were mentioned above in dealing with the control of organ size, are schematically described in Fig. 53. It should be noted that the team of Anastasiou *et al.* (2007) — under the leadership of Michael Lenhard — was quite unique in attempting to also answer the question of how an organ's cells sense the size of its organ during its growth, in order to set the time when the further increase of the size should be stopped.

# CHAPTER 10

# Patterning of the Embryo

The process of embryogenesis was detailed in Chapter 7 of Galun (2007) and the reader is referred to this chapter for a general understanding of this process. In the present book I shall stress the involvement of phytohormones in embryogenesis. Some recent reviews on embryogenesis would also be helpful for understanding this process: first there is a recent book titled *Plant Embryogenesis*, edited by M.F. Suarez and P.V. Bozhkov (2008) that includes three chapters on plant embryogenesis: (a) "Arabidopsis Embryogenesis" by Park and Harada (2008); (b) "Maize Embryogenesis" by Fontanet and Vicient (2008); and (c) "Spruce Embryogenesis" by von Arnold and Claphan (2008). Published in the same year are articles by Lau *et al.* (2008), Brybrook and Harada (2008), Kaparakis and Alderson (2008), Muller and Sheen (2007), as well as publications on methodologies for embryo research, such as by Sauer and Friml (2008) and Kim and Zambryski (2008).

## Formation of the Zygote and the Endosperm

During the last stages of the angiosperm flower formation, the male and female gametophytes are differentiated. The male gametophyte is a simple entity composed of the pollen grain with its characteristic decorated cell wall and two nuclei: the vegetative nucleus that controls the metabolism of the male gametophyte during the "voyage" from the stigma toward the female gametophyte: the embryo sac (see Frankel and Galun, 1977, Section 3.1.1). The other nucleus may divide right in the pollen grain into two sperm cells, or this further division is delayed until the germination of the pollen tube. The female gametophyte is derived from the megaspore mother cell that usually undergoes two meiotic divisions, resulting in a "tetrade" of four haploid nuclei. In rare cases, such as *Allium*, there is only one meiotic division. Consequently, four (or rarely two) megaspores are produced. Further division can lead to different numbers of cells in the embryo sac. In Arabidopsis and many other genera the embryo sac finally contains eight haploid cells, all derived from one of the megaspores. This is the *Polygonum* type of the embryo sac, in which the mature embryo sac contains eight haploid cells: at one end of the embryo sac there are two synergids and one egg cell, in the middle there are two polar cells (that may ultimately fuse), and at the other end there are three antipodals. The role of the latter three cells is not clear.

The detailed description of the journey of the pollen tube (that protrudes out of the pollen grain through the germination pores) from the stigma to the embryo sac may take place in either one of two main ways. In plants with an "open style," the pollen tube grows down from the stigma on a papilae surface to a mucilage-filled canal, in the center of the style. The canal has a "transfer-type" wall comprising of the transmitting tissue that probably plays a role in the nutrition of the pollen tube. In the "solid type" plants, the pollen tube penetrates the intercellular spaces in the pistil tissue, finds its way to the transmitting tissue and from there reaches the ovary. The inhibition of the pollen tube at the stigma or in the pistil, in case of self-incompatibility, is outside the scope of this book.

In general terms, the growth of the pollen tube is from the stigma through the style and up to the microphyle of the ovule. After the entrance through the microphyle, the pollen tube penetrates one of the synergid cells and explodes there. From the exploded synergid cell, one of the two sperm nuclei enters the egg cell, while the other fuses with the two polar cells to produce a triploid endosperm tissue. The other sperm nucleus that enters the egg cell produces the zygote.

It is evident that the synergids release a chemical attractant that leads the pollen tube toward the embryo sac. The synergid cell that is penetrated by the pollen tube is termed "degenerate synergid." The latter synergid is destroyed by the penetration of the pollen tube.

There is a difference between the two sperm cells. One of them (probably the one that reaches the two central cells to produce the triploid endosperm) may contain more organelles than the sperm cells that fuse with the egg cell. The species *Torenia fournieri* (Scrophulariaceae) has a special feature: the embryo sac protrudes from the microphyle and is thus accessible to direct observation as well as the ablation of specific cells (by laser UV irradiation). Hence that attractant from the synergid could be followed: when one synergid was ablated, the migration of the pollen tube was disturbed, but when both synergids were ablated, the pollen tube did not enter the embryo sac. In fact, it seems that two guidances are required: guidance through the funiculus (that attaches the ovule to the carpel) and guidance through the microphyle into the embryo sac. Hence, at least in Arabidopsis, there seem to be two diffusible chemoattractants. Can we term these chemoattractants "phytohormones"? This question may be answered after the chemical composition of these attractants is determined. There is an enigma: how is the attraction taking place in plants that have no synergids, such as in members of the Plumbaginaceae family? It should be noted that in analogy to animals, once a pollen tube enters the embryo sac, the penetration of further pollen tubes is barred from such an entrance.

Is there a difference between the two sperm cells so that one specific sperm nucleus will fuse with the nucleus of the egg cell, while the other will fuse with the central cell nuclei? It appears that at least in *Plumbago zeylanica* there is a difference; the nucleus of the sperm cell that has more mitochondria tends to fuse with the central cells' nuclei, while the nucleus from the sperm cells that is relatively richer in plastids, will fuse with the nucleus of the egg cell. However, there is not yet a clear answer in regards to other species. We should recall that *Plumbago* belongs to the minority of the angiosperm genera in which organelles (mitochondria and plastids) are also inherited paternally.

## The Embryo: Structure, Genes, and Phytohormones

Initiation: after one of the sperm nuclei fuses with the nucleus of the egg cell, the process of embryogenesis is initiated. The elongated fused cell divides to result in the more or less round "upper" apical cells and a "lower" elongated basal cell. The elongated basal cell will then divide several times to form a "chain" of cells that connects the apical cell to the carpel, forming the hypophysis and, below it, the suspensor. The first transfer division in the basal cell, in Arabidopsis, parallels the vertical cell division in the apical cell. This latter division heralds the formation of the proembryo. After two additional cell divisions, the proembryo has eight cells. In a further division, the embryo has an outer protoderm and is now the early globular embryo. Additional cell divisions lead to the late globular embryo and after a "transition" stage, the heart-shaped embryo is finally formed. At its lower side is the hypophysis that connects the embryo to the suspensor.

Several genes that we will come across whilst dealing with the shoot apical meristem are beginning to be already expressed in the heart-shaped embryo (see Figs. 54 and 55 where auxin flow is also indicated).

Before we dive into the details of the impact of genes and phytohormones on the embryo's patterning, let us look at some of the respective genes. We will begin with genes that are already expressed in the early stages of embryogenesis.

*ATML1* (*ARABIDOPSIS THALIANA MERISTEM LAYER1*), a class IV homeodomain leucine-zipper gene, normally active later in the development of the L1 layer of the shoot apex, expressed in the apical but not in the basal cell of the very young embryo.

*WOX* (*WUSCHEL-RELATED HOMEOBOX*) genes that are differentially expressed in the apical and basal cells, *WOX2* and *WOX8* are already expressed in the egg cell and then in the zygote, but their expression becomes restricted to the apical cell and the basal cell, respectively, after the first cell division.

*PIN7* (*PIN FORMED7*) encodes an auxin efflux facilitator, and the protein is located in the apical membrane of the basal cells, following the asymmetric division of the zygote. There are additional *PIN* genes, such as *PIN1*, *PIN3*, and *PIN4*. *PINs* are partially redundant but the quadruple mutation has severe defects in apical–basal polarity.

*GN* (*GNOM*) encodes a guanine nucleotide exchange factor for ADP ribosylation (a factor involved in vesicular transport that is required for the proper localization of *PIN1*). The *gn* mutant embryos appear as ball-shaped structures without basal–apical polarity, and the plane of the first division of these embryos is often disturbed.

*YODA* encodes a component of a MAP (mitogen activated protein) kinase-signaling pathway. A loss-of-function of *YODA* causes defects in the elongation of the basal cell derivatives, whereas

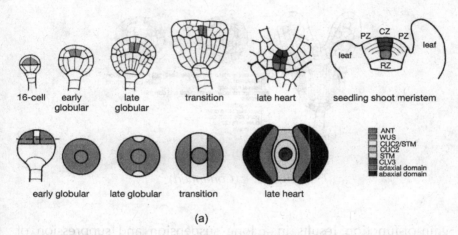

(a)

**Fig. 54.** Patterning of the Arabidopsis embryo. (a) The top row shows schemes of longitudinal median sections. The upper and lower thick lines represent clonal boundaries between the descendants of the apical and basal daughter cells of the zygote, and between the apical and central embryo domains, respectively. The bottom row shows cross-sections of the same stages as indicated by the dashed line on the left. CZ, central zone; PZ, peripheral zone; RZ, rib zone. The expression domains of early genes in the apical region are shown in color as indicated. (b) Development of the radial pattern. The top row and the illustration at the bottom left show schemes of longitudinal sections; the other illustrations in the bottom row show schemes of cross-sections through a root. The upper and lower thick lines represent clonal boundaries between the descendants of the apical and basal daughter cells of the zygote, and between the apical and central embryo domains, respectively. Cell types are shown in color as indicated. Vascular and pericycle cells are shown in lighter colors than stem cells. Gt, ground tissue; hy, hypophysis; lsc, lens-shaped cell; pc, pericycle; vp, vascular primordium. (From Laux *et al.*, 2004.)

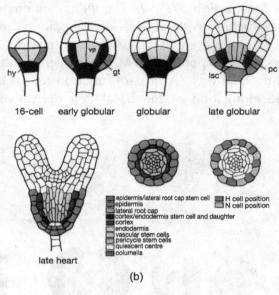

late heart

(b)

**Fig. 54.** (*Continued*)

gain-of-function results in a long suspension and suppression of normal embryo development.

## Genes Affecting the Apical Domain

*GURKE*, *gurke* mutant seedlings are defective in cotyledon and shoot apical meristem formation. They possess only a hypocotyl, a radicle, and a root apical meristem.

*TOPLESS*, *topless* mutants lack an apical domain; unlike *gurke* mutants, *topless* mutants may in some cases develop a root instead of a shoot, resulting in seedlings with roots at both ends. *TOPLESS* has been suggested to encode a transcriptional co-suppressor that inhibits basal region patterning in the apical domain.

*MERISTEMLESS* (*STM*), *WUSCHEL* (*WUS*) and *CUP-SHAPED* (*CUC*) are genes that affect postembryonic shoot apical meristem formation. Mutants of *STM*, *WUS*, and *CUC* result in seedlings with no — or reduced — apical meristems. While their impact is mainly

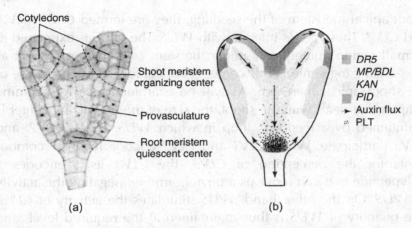

**Fig. 55.** Outline of the body plan at the heart stage. (a) Most aspects of seedling organization are apparent in the heart-stage embryo. Two inductive centers organizing the apical meristems (organizing center of the shoot meristem shown in green, and quiescent center of the root meristem in yellow) are connected by the vasculature of the primary body axis (pin). The vasculature is continuous with the first two lateral organs, the cotyledons. (b) Auxin is a primary source of positional information in the embryo and is channeled mainly through the epidermis and vasculature (green arrows). Auxin distribution is probably graded, and local accumulation maxima presumably trigger expression of the synthetic reporter DR5 (green). Auxin-dependent transcription in the embryo is partly mediated by the ARF gene *MONOPTEROS* (*MP*; pink) which is predominantly expressed in the provasculature. In the root, PLETHORA factors (PLT; red dots) accumulate in a gradient with a maximum at the QC and organize stem cell fates in a dose-dependent manner, presumably downstream of auxin. In the shoot, expression of the serine-threonine kinase *PINOID* (*PID*; blue) switches polar localization of PIN-FORMED auxin transporters, redirecting auxin flux toward the incipient cotyledons (green arrows). In addition, *KANADI* genes (yellow) directly or indirectly repress auxin responses along the hypocotyl flanks. (From Nawy *et al.*, 2008.)

on the shoot apical meristem of seedlings, they start to be already expressed in the embryo.

*CLV* genes are actually three different genes that start to be expressed in the developing embryo, but have their impact at the shoot apical meristem of the seedling; they are termed *CLV1*, *CLV2*, and *CLV3*. These genes interact with *WUS*. The *WUS* is expressed in a small region immediately under the stem cells that are known as the "organizing center"; this center is critical for the maintenance of the shoot apical meristem. As will be examined in a forthcoming chapter that deals with the shoot, the size of the organizing center is maintained by a feedback loop in which *WUS*, *CLV1*, *CLV2*, and *CLV3* participate. While *CLV1* and *CLV2* encode receptor components for the perception of *CLV3*, the *CLV3* itself encodes a polypeptide that can move as a phytohormone, negating the activity of *WUS*. On the other hand, *WUS* stimulates the activity of *CLV3*. The quantity of *WUS* is thus maintained at the required level, and consequently the volume of the organizing center is also kept at the required size.

Here we return to *STM*, which is another positive regulator of stem cell identity that is inhibiting the differentiation of stem cells (i.e. maintaining their undifferentiated cell division). This is achieved primarily by inhibiting *ASYMMETRIC LEAVES1* (*AS1*).

*CUP-SHAPED COTYLEDON1* (*CUC1*), as well as *CUC2* and *CUC3* that encode transcription factors, are primarily responsible for forming boundaries between cotyledons. These genes are expressed in a band across the apex of the embryo, coinciding with the expression of *STM*. After the heart-shaped stage of the embryo, *STM* expression becomes restricted to the shoot apical meristem, whereas *CUC* genes show complementary expression patterns within this band. Thus, *CUC* genes cause hurdles that permit two separate cotyledons and prevent excessive outgrowth of the incident shoot apical meristem.

*PINOID* and *ENHANCER OF PINOID* are two genes that have been implicated to play roles in generating cotyledon primordia: *PINOID* is a serine/threonine kinase that controls targeting of *PIN1* and *ENHANCER OF PINOID*. Dimeric mutations in these genes result in the precise deletion of cotyledons with no visible effect on hypocotyls and roots. The double mutation causes a reversal of *PIN1* localization in the epidermis. These results suggest that

*PINOID* and *ENHANCER OF PINOID* are required for the establishment of auxin maxima, which are essential for cotyledon formation.

## Genes Affecting the Patterning of the Central Domain

The central domain of the developing embryo contributes to the establishment of the hypocotyl, the radicle, and part of the cotyledons. The hypocotyl and the radicle make up the bulk of the embryonic axis, which can be viewed as cylinders with concentric layers of tissue (see Fig. 54). One of the earliest genes that were revealed as having a role in the embryo patterning in Arabidopsis was *FACKEL* (*FK*). In *fk* mutants, the resulting seedlings consist of a shoot apical meristem and cotyledons that are directly attached to the root. Then, additional genes — the mutants of which appeared similar to *fk* — were reported (as *HYDRA1* and *CEPHALOPOD*). The latter two genes encode enzymes involved in sterol biosynthesis. It was suggested that sterols act as signaling molecules for pattern formation along the shoot-root axis. Consistent with this hypothesis is the conclusion that *BRASSINOSTEROID-INSENSITIVE1* acts downstream of *FK*. However, it could also be that the *fk* mutant, as well as *hydra1* and *cephalopod*, cause a deficiency in cellulose and thus a defect in cell walls, resulting in an *fk*-like phenotype because a glucose-conjugated form of a sterol, sitosterol, is thought to be serving as a primer of cellulose synthesis.

Seedlings with a mutation in the *MONOPTEROS* (i.e. *mp*) have a shoot apical meristem as well as cotyledons, but are defective in both the central and the basal domains. The earliest change from normal to a new phenotype in *mp* mutant embryos was transverse, rather than longitudinal cell division in the apical cell that resulted from the zygote. This leads to an octant-stage embryo with four, rather than two layers of cells along the axis. At later stages these mutants had defects in the orientation of cell division so that the cells in the central domain failed to form characteristic cell files within the hypocotyl and the radicle. A similar mutant phenotype is also caused by mutation in *BUNELOS* (*BDL*) and *AUXIN RESISTANT6* (*AXR6*), all of which encode proteins involved in auxin signaling. *MP* encodes an auxin response factor (ARF5), a transcription factor that regulates auxin-responsive genes; *BDL* is an AUX/IAA protein (IAA12) that is thought to bind with *MP* and inhibit its transcriptional activation function until *BDL* is degraded in the auxin response pathway. *AXR6* encodes a cullin protein of the E3 ubiquitin ligase required for auxin responses. Auxin is probably transported in the embryo from the shoot apex through the hypocotyl and the radicle to the root apical meristem; therefore the products of *MP*, *BDL*, and *AXR6* are required to respond to auxin polar transport in order to establish the embryogenic axis. It is clear that we witness here another vital role of auxin in the normal embryo differentiation.

## The Root Apical Meristem

The root apical meristem is a tissue that is manifested early in the embryogenesis. It maintains a role in the development, at a later stage, of all "below-ground" plant organs, throughout the plant's life. This apical meristem consists of a group of slowly dividing cells, termed "quiescent center," that is surrounded by stem cells. The latter produce cell files that constitute the bulk of the root apical meristem. The "quiescent center" is thus analogous to the "organizing center" of the shoot apical meristem. The root apical meristem is derived from both the central and the basal domains. Initials that form the vascular cylinder and the ground tissue of the root come from derivatives of the central domain, while the quiescent center and the central root cap initials are derived from the hypophysis (see Fig. 54b).

At the globular stage, the hypophysis divides asymmetrically and gives rise to an apical "lens-shaped" cell that will lead to the quiescent center, and a basal cell from which the central root cap initial is derived. The timing of this division roughly coincides with the shift in auxin maxima, within the embryo, from the embryo proper to the hypophysis, that is probably mediated by PIN1, PIN4, and PIN7 auxin efflux facilitators. Also, *mp* and *bdl* mutants do not form lens-shaped cells of the hypophysis nor functional root apical meristems, suggesting a requirement for auxin signaling. All this information stresses the important role of auxin in normal embryo differentiation.

The response to auxin signals may be mediated, at least in part, by the *PLETHORA* genes: *PLT1* and *PLT2*. Digenic mutations in these genes (*plt1, plt2*) at the early heart-shaped stage of the embryo, cause misspecification of the quiescent center and the surrounding cells. The *PLT* genes encode a putative *APETALA2* domain transcription factor, the expression of which is independent of auxin and auxin-responsive transcription factors, including MP. At the globular stage of the embryo, these genes are expressed in the lens-shaped hypophysal cell and in the provescular cells. The expression domain though, later becomes restricted to the quiescent center and its surrounding cells. The *PLT* genes probably interact with two additional genes that are required for the specification of the quiescent center: *SCARECROW* (*SCR*) and *SHORTROOT* (*SHR*). When *PLT* genes are expressed ectopically, they specify a new quiescent center. Together with the *SCR* and *SHR* genes, the auxin-dependent *PLT* genes form a combinational signal that specifies stem cells. This suggests a role for auxin signal transduction already in the early embryonic specification of meristem domains.

The *HOBBIT* (*HBT*) is a gene that is required for cell division and cell type specification in the root apical meristem. It already has a role at the globular stage since *hbt* mutant embryos either do not form the lens-shaped cell of the hypophysis or undergo atypical divisions in this site. Furthermore, *hbt* mutant seedlings lack columella and lateral root-cup cells, as well as a quiescent center. There are also no cell divisions in the postembryonic root meristems of these mutants. Cell fates in the root apical meristem are probably also affected by the *hbt* mutants. HBT encodes a homolog of the CDC27 subunit of an anaphase-promoting complex that is required for cell cycle progression.

## Radial Pattern Establishment

When we look at the transverse section through the hypocotyl or the radicle of the mature (e.g. heart-shaped or torpedo-shaped) embryo, we can see the concentric arrangement of the protoderm, ground meristem, and procambium. However, the first manifestation of radial patterning already happens when the octant-stage embryo undergoes periclinal divisions, establishing an outer layer of protoderm cells and inner cells. Later in the development of the embryo, the globular stage is formed. Then the cells in the center of the central domain undergo periclinal divisions that lead up to the vascular cylinder.

Already in the upper cell, following the asymmetric division of the zygote, the *ATML1* gene is expressed. This overall expression is continued until the octant-stage. Thereafter, *ATML1* expression is restricted to the protoderm or the L1 cell layer. The *ATML1* is then not expressed in the inner cells of the embryo, but the expression continues in the epidermis of the shoot apex. A similar expression pattern was revealed for the gene *PROTODERMAL FACTOR2* (*PDF2*). The signal for this pattern of expression may derive from cells that are external to the embryo (maternal tissue); or the signals come from the cell wall of the zygote (the cell wall of the zygote is also a type of maternal tissue).

Mutations in two genes involved in cytokinesis — *KNOLLE* and *KEULE* — cause the expression of the epidermis-specific genes in subepidermal tissue and the formation of incomplete cell wall in embryos. Hence the signals that specify protoderm cell fate are not yet clear.

Auxin and a polypeptide phytohormone encoded by *CLV3* were mentioned above in the discussion of the patterning of the Arabidopsis embryo. Now we will take a look at the cytokinins. Mutation in the *WOODEN LEG* (*WOL*) gene is implicated with abnormal vesicular tissue that is already manifested after the heart-shaped embryo. In *wol* mutants the number of cell divisions in the hypocotyl and vascular cylinder are reduced, and the phloem is also reduced or even eliminated. *WOL* is allelic with *CYTOKININ RESPONSE1*. The latter gene encodes a two-component histidine kinase that serves as a cytokinin receptor. Hence CKs are also active in the patterning of the embryo.

The ground meristem initials in the root apical meristem undergo periclinal cell divisions that lead to the cortex and the endodermis layers. Mutations in *SHR* and *SCR* have defects in the cortex and the endodermis that can be observed already in the torpedo-stage of the embryo. Both *SHR* and *SCR* encode GRAS-type transcription factors that are required for asymmetric divisions of the ground tissue initials and endodermis specification. *SHR* is expressed in the vascular cylinder but the SHR protein moves to the adjacent ground tissue layer, where it activates *SCR*. *SCR* then promotes the asymmetric periclinal division.

## Initiation and Maturation of Embryos

Braybrook and Harada (2008) discussed the initiation and maturation of embryos and found similarities between zygotic embryogenesis and somatic embryogenesis. There are several ways of somatic embryogenesis. A common one is functioning in some genera, such as *Solanum* and *Daucus*. When tissue cultures of these genera are grown in a certain external phytohormone regimen, such as transfer from high auxin to low auxin, somatic embryos are initiated. There are also "natural" formations of somatic embryos, such as in the leaves of *Kalanchoe daigermontiana*, where plantlets are abundantly formed on the margin of leaves. The authors claim that the formation of somatic embryos is a manifestation of the totipotence of plant cells. They found that LEAFY COTYLEDON (LEC) transcription factors (TFs) establish environments that promote cellular processes characteristic of embryo maturation and somatic embryogenesis. They also found target genes that are activated by LEC TFs and they proposed that there is an effect of LEC TFs on the levels of ABA: ABA increases gradually during embryogenesis, from the heart-shaped embryo until after the torpedo-shaped stage. The ABA is functional in the arrest of growth and in the pre-mature germination of the mature embryo. The authors claimed that the arrest of embryo maturation is of great advantage and contributed to the evolutionary success of seed plants in which this arrest exists (e.g. gymnosperms and angiosperms).

On the other hand, GA levels are low throughout the maturation of the embryo, but increase during germination. During maturation there is accumulation of storage macromolecules, acquisition of desiccation tolerance, inhibition of precocious germination, and metabolic arrest resulting from the desiccation. This developmental arrest (combined with the ability to withstand further drying) enables the embryo to remain in a quiescent and/or dormant state until the seed encounters conditions that promote germination. However, the molecular mechanisms that cause an arrest in the cellular activity in the zygote, or in the cell during somatic embryogenesis are not yet known, but a close look at some genes may provide a clue. The Arabidopsis *LEC* genes, *LEC1*, *LEC2*, and *FUS3* encode two classes of TFs. *LEC1* is a HAP3 subunit of the CCAAT-binding TF, and *LEC2* and *FUS3* are closely related B3 domain TFs. Ectopic expression of all three *LEC* genes causes cells in vegetative and reproductive tissues to adopt characteristics of maturation phase embryos. Moreover, embryos with loss-of-function mutations in *LEC* genes are intolerant to desiccation and have defects in reserve accumulation. This, together with other observations, provides evidence for the assumption that the LEC TFs along with another B3 domain TF, ABI3, are master regulators of the maturation phase. It seems that *LEC* genes are also necessary for normal progression through the morphogenesis phase of zygotic embryogenesis. LEC2 and FUS3 bind specifically with the RY DNA motif that is present in the 5′ flanking regions of seed protein genes, which are rapidly up-regulated by the induction of these TFs.

LEC TFs have been implicated in repressing GA activity. The bioactive GA levels in the *lec2* and *fus3* mutant seeds are higher than in wild-type and the RNAs encoding the GA biosynthetic enzyme GA3 oxidase is elevated in *lec2* and *fus3* mutants but reduced in plants ectopically expressing *FUS3*. There seems to be a delicate balance between ABA and GA in the maturing of the embryos with ABA pushing toward the maturation process. Braybrook and Harada (2008) assumed that the LEC TFs might also affect the maturation through their involvement in controlling the balance between ABA and GA levels.

Much progress was achieved in recent years in understanding the complexity of the auxin pathway. A great part of the experimental work concerning this complexity was done with Arabidopsis as the model plant, but it can be assumed that essentially this pathway also operates in other angiosperms (see Lau *et al.*, 2008, for a recent review). The intracellular auxin is perceived by the TIR1/AFB-3 receptors (see Chapter 1 of this book). This triggers the degradation of the AUX/IAA proteins (auxin influx carrier). Under low auxin levels, AUX/IAA form dimers with AUXIN RESPONSE FACTOR (ARF) transcription factors thereby blocking the activity of at least the activated ARFs. Once freed from the AUX/IAAs, these ARFs can regulate the expression of auxin-responsive genes. The binding of AUX/IAA in the TIR1 receptor quickly triggers the binding of polyubiquitin to the AUX/IAA (in the presence of auxin), leading the AUX/IAA to be degraded in the 26S proteasome. The auxin signaling in plant cells — including cells of the embryo — may be divided into three "layers" that contribute to the complexity: (a) the site and timing of IAA biosynthesis; (b) the directional transport of the auxin (i.e. mediated by influx and efflux proteins, that can now be identified at the cellular level by immunohistological methods); and (c) cell- or tissue-responses to the level of auxin. It should be noted that in recent years, great progress has been made in the development of techniques to reveal spatial and temporal changes in the levels of auxins in the cells of Arabidopsis embryos (Sauer and Friml, 2008).

## Cytokinin in Early Embryogenesis

Biosynthesis of cytokinins (CKs) was reported to occur repeatedly in the root and transported toward the upper parts of the plant seedling (it was later found that CKs are also synthesized in certain sites in other parts of the plant). Muller and Sheen (2008) focused their attention on the early stages of Arabidopsis embryos and studied the formation of cytokinins in such embryos. These authors constructed a synthetic reporter to visualize universally cytokinin output *in vivo*. For that they synthesized a reporter in which a sequence that recognizes CK ("two component output sensor," TCS) was fused to luciferase — *TCS::LUC*. The TCS component included the concatenation of B-type Arabidopsis response regulator (ARR)-binding motif and a minimal 35S promoter. After carrying out several tests, the authors found that *TCS::LUC* is activated only by CKs but not by other phytohormones, such as AUX, ABA, and GA. In addition, the reporter was not activated in double CK receptor mutants. They also found that only B-type ARR family members promoted strong *TCS::LUC* induction, while A-type ARR family members inhibited CK-dependent *TCS::LUC* activity. The range of CK sensitivity was from 100 pM up to 1 $\mu$M, and the addition of a viral translational enhancer ($\Omega$) increased the LUC activity. The authors concluded that *TCS::LUC* could probably report even low levels of phosphorelay output triggered by any of the three endogenous CK receptors and relayed to any response regulator tested. Transgenic plants containing a reported GFP controlled by the TCS synthetic promoter (*TCS::GFP*) showed the same picture as was observed by *TCS::LUC*. Moreover, when seedlings were subjected to a short period of incubation with the CK synthesis inhibitor, lovastatin, the *TCS::GFP* expression was abolished. Adding CK restored the expression. They then detected the first distinct signal in the founder of the root stem cells: the hypophysis. This could already be observed in embryos at the 16-cell stage (very early globular embryos). By the transcription stage, the hypophysis had undergone asymmetrical division into a larger basal cell and an apical lens-shaped cell. The latter is the founder of the quiescent center. The large basal daughter cell and its descendents repressed *TCS::GFP* expression, whereas the apical lens-shaped cell retained its expression. By the heart-stage, a second phosphorelay output had appeared near the shoot stem cells' primordium. The expression of the A-type ARR genes was different as visualized by the ARR7::GFP activity. The activity of the latter construct was high, right after the division of the hypophysis; it was then increased in the basal daughter cell but was lower in the lens-shaped cell and its descendants. The authors concluded that additional input might control *ARR7* and *ARR15* expression.

A similar approach was used to detect auxin signaling: the use of another synthetic reporter *DR5::GFP*. With this reporter, the auxin signaling appeared to be highest in the hypophysis-derived basal cell — similar to *ARR7* and *ARR15* expression. The possibility was thus raised that auxin induces the CK-repressors *ARR7* and *ARR15*, which in turn prevented phosphorelay output in the basal cell lineage. Indeed the application of synthetic auxin analogs (2,4-D) supported this possibility. Furthermore, exogenous CK caused a broad expansion of *TCS::GFP* but left *DR5::GFP* expression unaffected. The experimental work supported the model that high endogenous auxin activity suppresses cytokinin output through stimulation of *ARR7* and *ARR15* transcription. It was strongly suggested that loss of *ARR7* and *ARR15* function, or ectopic CK signaling in the basal cell during early embryogenesis would result in a defective root stem cell system.

## Non-Dicot Embryos

Relative to the large amount of information that accumulated on embryogenesis in Arabidopsis, there is much less detailed information on other dicots. However, it can be assumed that the differences between the embryo morphogenesis of Arabidopsis (including the roles of phytohormones on this morphogenesis) are not substantial and that the Arabidopsis embryo can serve as a representative model for embryo architecture in dicots (see West and Harada, 1993, for review). When we shift our attention to monocots though, at least the final result of embryo morphogenesis is very different from the one in Arabidopsis. Among the various monocots the embryogenesis of maize probably received the greatest attention (see publications by Sheridan and Clark, 1987; Consonni *et al.*, 2003; Dolfini *et al.*, 2007; Fontanet and Vicient, 2008, where literature on embryogenesis of wild-type and mutants of maize is listed). The differences between the embryogenesis of maize and the embryogenesis of Arabidopsis are rather great (see stages of embryogenesis of maize in Fig. 56). At maturation, the maize embryo is already similar to a small plantlet with six or seven leaf initials. The duration of embryo formation from the establishment of the zygote up to the mature embryo in maize differs from the duration observed in Arabidopsis. In the latter plants,

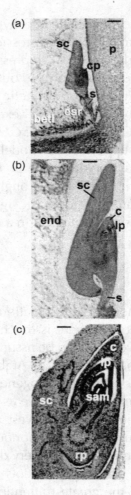

**Fig. 56.** Sections of maize embryos of different stages (days after pollination, DAP). (a) 12 DAP; (b) 16 DAP; (c) 30 DAP. Betl, basal endosperm transfer layer; c, coleoptile; cp, coleoptile primordium; end, endosperm; esr, embryo-surrounding region; lp, leaf primordium; s, suspensor; sam, shoot apical meristem. (From Consonni *et al.*, 2003.)

embryo formation only takes place within a few days while in maize the duration is almost two months. Until the late 1980s, most of the research on maize embryogenesis was involved in the isolation of the numerous mutants that showed phenotypes of abberrated embryos. The molecular understanding of embryogenesis in maize was meager, and very little progress in understanding the roles of phytohormones in the differentiation of maize embryos was achieved. It was even denied that a flow of auxin actually does have an impact on this morphogenesis. Thus, Sheridan and Clark (1987) suggested that "simple regulation by production of different plant hormones at different stages seems unlikely since the tissues appear to have the same limited repertoire of growth substances available as regulatory molecules at all stages of development." These authors tended to agree with Trewavas (1982) who claimed that the regulation of plant development is limited by phytohormones, and therefore the regulatory factor is sensitivity to growth substances rather than phytohormone levels. Hence, the morphogenetic changes of the maize embryo may at least in part be regulated by genes that control the appearance of specific receptor proteins in different regions of the developing embryo. We saw earlier that the level of phytohormones, at least in specific sites of the Arabidopsis embryo (e.g. at the dividing hypophysis), is tightly correlated with phytohormone presence. Also, Goldberg *et al.* (1994), in their review on embryo morphogenesis (based on embryos of Arabidopsis and *Capsella*), stressed the importance of phytohormones (mainly auxin) in embryogenesis. Since great progress has been made in recent years on the quantitation of phytohormones in plant cells and on the flow of phytohormones from cell to cell (especially with auxins), we should not be surprised when detailed information is published in the near future on the correlation of phytohormone levels and flow, and morphogenetic changes in the embryogenesis of maize and other Gramineae species.

# CHAPTER 11

# The Root Patterning and Phytohormones

In Chapter 9 of my previous book (Galun, 2007), I examined the patterning of angiosperm roots. Several aspects of root patterning have already been discussed in this book, such as: the root meristem in dicots (mainly based on Arabidopsis) and in monocots (mainly based on maize and rice); the initiation of root hairs; the initiation of lateral roots; the formation of adventitious roots; the impact of soil nutrients (mainly P and N), on root architecture, and the intercellular flow of macromolecules in the root (plasmodesmata). While a great number of genes that participate in the shaping of roots were discussed, only little information was provided on the roles of phytohormones in patterning the angiosperm roots. In the present chapter, available information regarding the impact of phytohormones on root architecture will be examined in detail. Among the various phytohormones, it seems that auxin has a major role in this architecture, but other hormones, such as cytokinins (CK), brassinosteroids (BR), ethylene (ET), and abscisic acid (ABA) also participate in certain aspects of root differentiation, and the latter phytohormones may interact with auxin in controlling the development of the angiosperm root.

The human mind "needs" drawers... in order to perceive an entity; this entity should be defined before it can be stored in our memory. There are fields of knowledge in which exact definitions are possible, as for example, in Euclidean geometry. Philosopher Baruch Spinoza

(1632–1677) in his work *Ethics* opened each of the five parts of the book with a set of definitions in the spirit of Euclidean geometry. Roots belong to those entities that have no simple definition. One may define a root according to its service to the plant (e.g. anchorage to the soil, furnishing the plant with soil minerals, symbiosis with nitrogen-fixing bacteria), or by its origin (from the lower cell of the divided hypophysis). However, there are always roots that do not fit any of these definitions. We shall therefore devote a whole chapter on roots without a definition of roots.

This chapter will include a short description of plant roots, handling mainly the root meristems. The Arabidopsis root shall serve as our main model of the dicot root. The description will include a brief comparison of dicot roots with monocot roots (e.g. maize and rice). Since such a description was detailed in Chapter 9 of Galun (2007), I will only present here a summary of this description. The differentiation of root hairs was also handled in some detail in Galun (2007) and not much additional information on this differentiation was obtained in recent years. Therefore, only a short section on root hairs will be provided in this chapter. On the other hand, because the formation of lateral roots and the role of phytohormones in the initiation and growth of lateral roots was recently the subject of many studies, this subject will be discussed below. There was also progress in our understanding of the impact of nutrients (e.g. phosphate and nitrogen) in the growth media, on the architecture of roots, as well as the involvement of phytohormones in this impact. Hence a special section will be dedicated to this subject.

## General Features of the Roots of Dicots and Monocots

We will begin with the dicot root. In our model plant, Arabidopsis, as well as in many other dicot roots, the seedling starts to produce the main (primary) root from which lateral roots branch out. Here comes the question of what is "apical" and what is "basal" in the main root. The confusion stems from the term "base" because seedlings have two "bases." One is the base of the above-soil part of the seedling, where the base of the shoot meets the upper-most part of the young root; but in reality the base of the roots is its uppermost part. When we go back to the embryonal stage, the hypophysis divides (asymmetrically) and then the uppermost cell that is derived from the hypophysis contributes cells to the shoot, while the lower cell that is derived from the hypophysis contributes cells for the root. The seedling then has two apices — one at the top of the shoot and one at the bottom of the root. There is a flow in the seedlings of IAA that starts from the shoot tip and the very young leaves below this tip, and it reaches the root tip. In the shoot, the flow is clearly "basipetal," but when it reaches the root and continues to flow in the same direction, it can then be viewed by the root as an "acropetal" flow. The polar flow of IAA in the young root is demonstrated in Fig. 57A (from Friml, 2003). The flows of IAA can now be followed by the immunolocalization of two types of proteins that either cause the in-flow of IAA into a cell (AUX1) or cause the out-flow of IAA from the cell (the PIN proteins). The same flow direction can be defined as basipetal in the shoot but acropetal in the root. As can be seen in Fig. 57A, IAA flows in the young root in two opposite directions. In the vasculature, it moves from the top of the shoot toward the root tip, and in the epidermis, IAA flows from the root tip inwards. The main root-cell files of roots are shown in Fig. 57B. The anatomical organization of the dicot root (Arabidopsis) and the monocot root (maize) differ considerably. There is also a great morphological difference between dicot roots and monocot roots. The dicots produce a major indeterminate root, from which lateral roots are derived, and the latter branch out again and again until an elaborate root system is established. In monocots, such as in members of the Gramineae (e.g. maize, rice), there is at the beginning a primary root, and lateral roots that branch out from this primary root, but there are additional types of roots, such as seminal roots

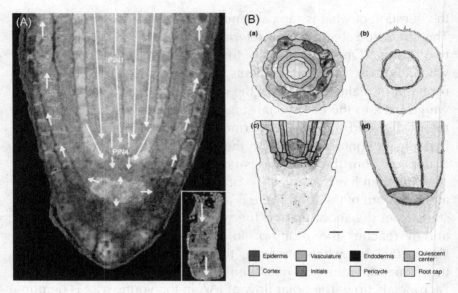

**Fig. 57.** Root tips of angiosperm seedlings. (A) Immunolocalization of the PIN3 protein in the Arabidopsis root apex and probable routes of polar auxin transport (white arrows). PIN3 (in green) is localized symmetrically in columella cells and apparently mediates lateral auxin distribution to all sides of root cap. After the root is turned by 90 degrees away from vertical (i.e. after a gravistimulus is applied), PIN3 rapidly relocates to the bottom side of columella cells (insert), and this proba-bly regulates flux to the lower side of the root. Auxin is further transported through lateral root cap and epidermis cells basipetally by a PIN2-dependent route (polar localization of PIN2 at the upper side of cells is depicted in blue). This basipetal transport also requires AUX1-dependent auxin influx. AUX1 is present in the same cells as PIN3 and PIN2. Auxin is supplied to the root cap by the PIN1- and PIN4-dependent acropetal route (which is depicted in white) (Friml, 2003). (B) Anatomical organization of the Arabidopsis (a,c) and maize (b,d) primary root in median longitudinal (c,d) and transverse (a,b) sections, showing the dif-ferent cell types and their relative positions. Notice that the upper margins of (c) and (d) represent the longitudinal positions of the root from where the transverse sections were obtained. The images are light micro-scopic (b,d) and electron microscopic (a,c) photographs that have been colored. Scale bars: (a,c) 10 μm; (b,d) 200 μm. (From Hochholdinger *et al.*, 2004.)

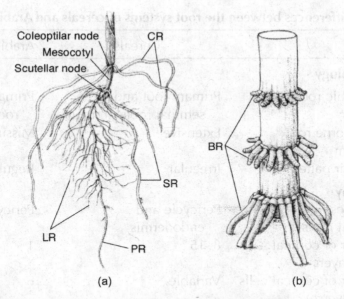

**Fig. 58.** Maize root system at different developmental stages. (a) 14-day-old maize seedling displaying primary (PR), seminal (SR), shoot-borne crown (CR), and lateral (LR) roots. Morphological seedling structures of cereals described in the text are labeled. (b) Above ground root stock of an adult maize plant composed of shoot-borne brace roots (BR). (From Hochholdinger *et al.*, 2004.)

and shoot-borne roots. In maize, there are also several "rings" of brace roots that grow out from the lower shoot of adult plants (Fig. 58).

We should notice that the Arabidopsis root displays many features that are common in all the dicots of angiosperms, but nevertheless, this root does not "represent" the dicot root. The Arabidopsis root is "minimalistic" with a small number of cells in their root apical meristem, such as a single layer of epidermis, a single layer of cortex, a single layer of endodermis, and a single layer of pericycle. The number of cells in the quiescent center is also very small (Fig. 59).

There are several major differences between dicot roots (such as those of Arabidopsis) and monocot roots (such as those of cereals — Gramineae) as indicated in the following list.

## Major differences between the root systems of cereals and Arabidopsis

|  | Cereals | Arabidopsis |
|---|---|---|
| **Morphology** | | |
| Embryonic root system | Primary root and seminal roots (in maize) | Primary root |
| Shoot-borne root system | Extensive | Missing |
| Root hair pattern | Irregular | Regular |
| **Anatomy** | | |
| Cell types forming lateral roots | Pericycle and endodermis | Pericycle |
| Number of cortical cell layers | 8–15 | 1 |
| Number of cortical cells (transverse) | Variable | 8 |
| Number of quiescent center cells | 800–1200 | 4 |
| Number of root initial cells | Few | Several hundred |

## The Apical Meristem of the Root

Figures 57 and 59 supply the structure of the apical meristem of dicots (Arabidopsis). However, as for its normal patterning, there was a surprise. A team of nine investigators from the Netherlands and Germany (Fiers *et al.*, 2004) used a mild heat shock treatment (32°C) of *Brassica napus* that triggers the development of embryos from these microspores, in order to identify genes that are expressed in the globular to heart-shaped stage embryos. They then identified one gene that encodes a putative extracellular protein that had great similarity to a CLV3/ESR-related protein (encoded by the *AtCLE19* of Arabidopsis). The *B. napus* gene (BnCLE19) encodes a protein that has a typical CLE box of about 14 amino acids. This box exists in CLE19, BnCLE19, and CLV3. The CLV3 was found to have a vital role

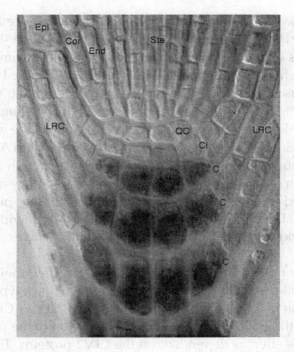

**Fig. 59.** Pattern of different cell types in a lugol-stained Arabidopsis root meristem. The quiescent center (QC) in the middle is surrounded by undifferentiated initial cells, such as columella initials (Ci) that give rise to differentiated lugol-stained layers of columella (C). Differentiated cell types such as epidermis (Epi), cortex (Cor), endodermis (End), stele (Ste), and lateral root cap (LRC) are indicated. The regular and invariant pattern of cell fates correlates with the auxin gradient, which reaches its maximum in the columella initials. (From Friml, 2003.)

in the maintenance of a proper volume of dividing cells in the central zone of the shoot apical meristem. It is coding for a polypeptide (in Arabidopsis) that can move in the intercellular spaces and then meet two proteins: CLV1 and CLV2 that are receptor-like serine/threonine kinases. The complexed, disulphide-linked of CLV1 and CLV2 to the ligand CLV3 then antagonizes the effect of WUS; the latter protein is leading to the maintenance of great volume of dividing cells in the central zone of the shoot apex. Thus, mutations in the

*CLV* genes will vastly increase the volume of dividing cells, while a mutation in *WUS* will eliminate the dividing cells in the shoot apex. More details about the control of the dividing cell volume are provided in the next chapter that deals with the shoot. Now back to the study of Fiers *et al.* (2004). These investigators found the expression of *BnCLE19* in various sites as the primordia of cotyledons in sepals, the cauline leaves in some pericycle cells, and in the maturation zone of roots. Mis-expression of *BnCLE19* or AtCLE19 in Arabidopsis (under the control of the CaMV35S promoter) resulted in a dramatic consumption of the root meristem. Also, the morphology of the pistil was affected by the mis-expression of CLE19: the pistils were pin-shaped. The results suggested that genes that encode proteins with a CLE box affect cell division in both the shoot apex and the tip of the root.

In a further publication, Fiers *et al.* (2005) verified their previous experimental results and found that even the short polypeptides of 14 amino acids, the CLE box that exists in CLV3, CLE19, and CLE40 is sufficient in triggering the consumption of the root meristem, but this effect is dependent on the CLV2 proteins. The authors thus proposed that these 14 amino acid peptides represent a major active domain in the corresponding CLE proteins, which interacts with — or saturates — an unknown cell entity, maintaining CLV2 receptor complex in roots and leading to the consumption of the root meristem.

The quiescent center (QC) of the root tip acts as a source of stem cells. The latter may further divide — or contribute to — differentiated cells in the root tip. Ortega-Martinez *et al.* (2007) found that ethylene (ET) modulates cell division in the cells of the QC. The division-induced cells may repress the differentiation of the surrounding cells. Following experimental conditions that either caused an increase of ET or the reduction of its level (e.g. by 2-aminoethoxyvinyl glycine — AVG), the authors obtained results that indicated that ET is part of a signaling pathway that modulates cell division in the QC, the stem cell niche during the postembryonic development of the root system.

# The Involvement of Brassinosteroids and Other Phytohormones in Root Development

The interaction of brassinosteroids (BR) and auxin was revealed soon after the discovery that BRs are phytohormones (e.g. Yopp *et al.*, 1981; Takeno and Pharis, 1982; Cohen and Meudt, 1983; Arteca *et al.*, 1983). An update on these interactions was provided by Bao *et al.* (2004) whom we shall discuss below in the section covering lateral root development. More recently, Li *et al.* (2005) studied the stimulation of plant tropisms by BRs and reported that BRs stimulated polar auxin transport in *Brassica* and Arabidopsis. These investigators mainly used roots in their experiments. Hence, BRs promoted the accumulation of the PIN2 (an auxin-efflux protein) from the root tip to the elongation zone. Brassinolide (BL) increased the expression of another gene, *ROP2*, during tropistic responses. Constitutive overexpression of *ROP2* resulted in an enhanced polar accumulation of PIN2 protein in the root elongation region and increased gravitropism. The authors suggested that *ROP2* modulates the functional localization of PIN2, through the regulation of assembly/reassembly of F-actins, and by that mediating the BR effects on polar auxin transport and tropistic responses. It should be noted that PIN2 is specifically involved in the intropism of roots and that other *PIN* genes (e.g. *PIN1*, *PIN3*) react to BL in a different manner than *PIN2*. The authors thus suggested a scheme that is presented in Fig. 60.

In a later study, Mouchel *et al.* (2006) investigated the involvement of the gene *BREVIS RADIX* (*BRX*) in the interaction between BRs and auxin. They used Arabidopsis roots in their study. The expression of *BRX* is required for optimal root growth. In the mutant *brx* there is a root-specific deficiency of BRs. In the mutant, there is a reduction of a rate-limiting enzyme in the BR biosynthesis pathway. The deficiency in *BRX* expression caused a reduction of about 15% in the transcription of root genes. The reduction could be restored to wild-type level by BR treatment. This means that a certain level of BR is required for the normal expression of many genes. Furthermore, auxin responsive gene expression was globally impaired in the *brx*

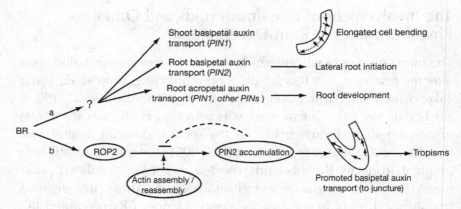

**Fig. 60.** Hypothetical model of BR-mediated regulation of PAT activities during plant tropisms and multiple developmental processes. (a) The regulation of diverse *PIN* genes by BRs suggests potential modulatory effects of BR on plant developmental processes, including hypocotyl elongation and bending, lateral root initiation, and primary root growth, through their effects on PAT activities. (b) In addition, ROP2 and actin assembly/reassembly may mediate BR RAT modulation of plant tropisms through the regulation of PIN2 accumulation and distribution. (From Li *et al.*, 2005.)

mutant. On the other hand, there is a "loop": *BRX* expression was strongly induced by auxin and mildly repressed by BL. This loop probably maintained a threshold of BR levels to permit optimal auxin action.

A mutation termed *iba response5* (*ibr5*) has a clear phenotype: the *ibr5* mutant exhibits, in response to exogenous auxin, a long primary root in Arabidopsis seedlings, and a reduction in lateral roots. In addition, *ibr5* plants are resistant to ABA. The team of Bonnie Bartel and associates from the Rice University in Houston, Texas, used the elongation of roots and hypocotyls to study the interaction of *ibr5* with various phytohormones. While the roots of wild-type seedlings were sensitive to several phytohormones and related compounds (e.g. IBA, IAA, 2,4-D, TIBA, ABA, and ACC), the *ibr5* mutants were insensitive to these compounds. The investigators revealed two classes of Arabidopsis mutants that suppressed the *ibr5* resistance to indole-3-butyric acid (IBA): those with restored responses to both the auxin precursor IBA and to IAA, and those with restored responses to IBA but not to IAA. Those mutants that were restored to IAA sensitivity were also restored to ABA responsiveness, while mutants that remained IAA-resistant also remained ABA-resistant. Furthermore, the investigators found that some suppressors restored sensitivity to both natural and synthetic auxins; others restored responsiveness only to auxin precursors.

Recently, Ubeda-Tomas *et al.* (2008) focused their attention on the role of GA/DELLA signaling in root growth. This role was mentioned in Chapter 3 of this book, which deals with GA, where it was indicated that GAs promote growth by targeting the degradation of DELLA repressor proteins. However, the site of action of GAs in the root was not clarified. Ubeda-Tomas *et al.* mapped the sites of GA action in roots. For that they expressed *gai*, a non-degradable mutant DELLA protein, and found that root growth was retarded specifically when *gai* was expressed in the endodermal cells of the roots. Furthermore, the experimental results of these investigators suggested that the endodermis represents the primary GA-responsive root tissue that regulated organ (root) growth and that endodermal cell expansion is rate-limiting for the elongation of other tissues and therefore of the root as a whole.

In an earlier part of this chapter (and in more detail in Galun, 2007), I reviewed the morphological differences between the root system of dicot seedlings (such as those of Arabidopsis) and the root system of young monocot plants (such as maize and rice). One of these differences is the development of crown roots that emerge from the base of the shoot. In rice, the crown roots establish themselves as the major root type. Zhao *et al.* (2009), a team of five investigators, of which four reside in Wuhan, China, and one resides at the Université Paris Sud, undertook the study of the regulatory mechanisms of crown root development. These investigators found that a WUSCHEL-related homeobox encoded by the *WOX11* gene is involved in the activation of crown root emergence and growth. The *WOX11* was expressed in the emerging crown roots and later in the cell division regions of these root meristems. The expression could be induced by the exogenous application of auxin and CK. Loss-of-function mutation of *WOX11* or down-regulation of this gene reduced the number of crown roots and their rate of growth, whereas overexpression of this gene caused precautionary crown root growth and vastly increased the total level of crown roots, which under these conditions also emerged from higher parts of the stem. It was also discovered that *WOX11* directly repressed the gene *RR2* that encodes a type-A CK-responsive regulator that was found to be expressed in crown root primordia, suggesting that *WOX11* may be an integrator of auxin and CK signaling that feeds into RR2 to regulate cell proliferation during crown root development. It should be noted that in addition to *WOX11*, there are several other *WOX* genes, such as *WOX3*, some of which were detected in Arabidopsis (e.g. *WOX8* and *WOX9*), hence the various WOX proteins may have different biochemical or transcriptional functions.

## Lateral Roots and Phytohormones

The branching of lateral roots from the main root and the formation of further branching from these lateral roots is an "old" subject of investigation. It has already been reviewed by McCully (1975) and Torrey (1986). After Arabidopsis became a favorite experimental model plant, it served to identify numerous mutations that affect out-branching of lateral roots. Two very good reviews by Malamy and Benfey (1997a,b) updated this subject; see also Chapter 9 of Galun (2007). Briefly, the formation of lateral roots is initiated from mature cells in the pericycle that are located in a special position with respect to the protoxylem poles. This radial positioning correlates with the xylem architecture. In Arabidopsis roots, lateral roots are initiated in the pericycle cells that are immediately adjacent to the two protoxylem poles. However, not all pericycle cells that are in this position will contribute to the new lateral root. Exogenous application of auxin, removal of the root tip, or the presence of additional nutrients can greatly increase lateral root initiation. This means that all properly positioned pericycle cells are a potential source of new lateral roots. The first "step" of lateral root formation is cell division of the pericycle cells in a given location. The pericycle cells are in the $G_2$ phase of the cell cycle and this initiation of cell division takes place at some distance away from the root tip, in the differentiation zone of the young root, where the pericycle cells are not actively dividing. The pericycle cells that are destined to become a lateral root start to form closely spaced cell walls in a perpendicular orientation to the root axis as a result of anticlinal divisions. The two layers of cells are formed by periclinal cell division. The outer layer of the divided cells then goes through a series of anticlinal divisions causing the formation of three cell layers, and subsequently four layers of cells are produced. A "bump" is then formed that reaches the epidermal layer of the main root. In further steps, the new lateral root emerges from the main root and further develops into a similar architecture as the tip of the very young main root. Detailed descriptions of the initiation and early development of lateral roots in Arabidopsis were provided by Malamy and Benfey (1997a, b) and were also summarized in Chapter 9 of my previous book (Galun, 2007).

There is evidence that auxin is involved in lateral root initiation. The application of auxin will result in an increase of initiation of lateral roots. Also, in Arabidopsis mutants that have an elevated auxin level (as in the "rooty" mutant), there is an increase of lateral root initiation. The auxin-resistant mutants axr1 and axr4 have a reduced number of lateral roots; but the exact way in which high auxin activate specific pericycle cells to develop lateral roots, is still unknown. Only a few mutants were identified in Arabidopsis that lacked lateral roots (alf4, "aberrant lateral root formation"). In total, the initiation and emergence of the lateral roots in Arabidopsis has great similarity with the development of the main root from the germinating embryo. However, there are probably some differences at the molecular level because the alf4 of Arabidopsis and the rtcs mutant of maize show defects only in lateral root formation but not in the initiation of the main root in the young seedlings.

Ivanchenko et al. (2006) noted that the initiation of lateral roots is an auxin-regulated process. The investigators chose to use tomato roots in their study and analyzed one of the dgt1 mutants that does not abolish the proliferative capacity of the xylem-adjacent pericycle cells but which did abolish later stages of lateral root development. The *Dgt* gene was found to be expressed (in wild-type plants) in lateral root primordia, starting from their initiation and upon auxin treatment, the expression was enhanced in the primary root meristem. In the roots of *dgt1* the auxin distribution was altered. The authors suggested that a type A cyclophilin is a product of *Dgt*, and that this cyclophilin is essential for the morphogenesis of lateral root primordia. The *dgt* mutations apparently uncouple patterned cell division in lateral root initiation from proliferative cell division in the pericycle.

A team of five investigators from Japan (Okushima *et al.*, 2007) studied the roles of two AUXIN RESPONSE FACTORs, ARF7 and ARF19, in the regulation of lateral root formation in Arabidopsis. These two FACTORs are transcriptional activators of early auxin responses. A double knockout mutation (*arf7 arf19*) is severely impaired in lateral root formation. They made a target gene analysis in *arf7 arf19* transgenic plants that harbored inducible forms of ARF7 and ARF19, and revealed that ARF7 and ARF19 directly regulate the auxin-mediated transcription of LATERAL ORGAN BOUNDARIES-DOMAIN16/ASYMMETRIC LEAVES-LIKE18 (LBD16/ASL18) and/or LBD29/ASL16 in roots. Overexpression of LBD16/ASL18 and LBD29/ASL16 induced lateral root formation in the absence of ARF7 and ARF19. These LBD/ASL proteins are located in the nucleus and dominant repression of LBD16/ASL18 activity inhibited lateral root formation and auxin-mediated gene expression, suggesting that these LBD/ASLs function downstream of ARF7- and ARF19-dependent auxin signaling in lateral root formation. Hence, the authors concluded that ARFs regulate lateral root formation via direct activation of LBD/ASLs in Arabidopsis.

De Smet *et al.* (2006) provided — by their experimental work — information on the control of initiation of lateral roots in Arabidopsis seedlings. They reported that lateral roots are shaped along the main axis in a regular left-right alternating pattern and correlate with a region of gravity-induced waving. This patterning depends on the influx carrier of auxin, the AUX1. They also suggested that the priming of pericycle cells for lateral root initiation takes place in the basal meristem (the region above the root meristem where the cells start to differentiate) and is correlated with elevated auxin sensitivity in this part of the root. This local elevated auxin responsiveness oscillates with peaks of expression at regular intervals of 15 hours. Each peak in the auxin-reporter maximum correlates with the formation of a consecutive lateral root. The auxin signaling in the basal meristem appeared to trigger pericycle cells for lateral root initiation.

A further study on the involvement of the local levels of auxin in root pericycle cells, in the initiation of lateral roots, was more recently carried out by Dubrovsky *et al.* (2008). These investigators found that local accumulation of auxin in the pericycle cells is a necessary and sufficient signal to specify these cells into lateral root founder cells. It was also suggested that specification of founder cells and the subsequent activation of cell division that leads to primordium formation represents two genetically separate events. The local activation of auxin response correlates absolutely with the acquisition of founder cell identity and this precedes the actual formation of a lateral root primordium through patterned cell division. The local production and subsequent accumulation of auxin in single pericycle cells induces the activation of auxin biosynthesis and converts the cells into founder cells.

## Formation of Root Hairs

Root hairs are formed from root epidermis cells. There are files of epidermal cells in Arabidopsis roots that do not produce hairs and files of cells that do produce such hairs. Epidermal cells that touch two files of cortex cells may produce hairs while epidermis cells that touch only one cortex cell will not produce hairs. It appears that the default situation of root hair formation is the initiation of such hairs. This means that if this initiation is not prevented, the file of epidermal cells can develop root hairs. This subject was elaborated in Chapter 9 of Galun (2007). No additional information on the role of phytohormones regarding the development of root hairs has been provided in recent years. Therefore, this subject will not be discussed in the present book.

## The Impact of Nutrients in the Medium on the Architecture of the Root

Arabidopsis seedlings are a convenient material to study the impact of root medium on the architecture of roots. Seeds can be planted in different media and then the growth of the primary root and the lateral roots may be visualized. Also, such seedlings can be moved from one medium (e.g. rich in phosphate) to another medium (e.g. poor in phosphate). The change in root patterning can then easily be followed. It was found that phosphates (Pi) and sulphates ($SO_4{}^{2-}$) have a great impact on root architecture. When Arabidopsis is planted in a Pi-rich medium, the roots maintain about the same architecture as in standard medium, whereas in media poor in Pi, the main root is much shorter but has a dense lateral root region, and secondary as well as tertiary lateral roots are formed. The effect of media either rich or poor in $SO_4^{2-}$ is less dramatic, but lateral roots are initiated closer to the root tip in poor $SO_4^{2-}$ medium than in $SO_4^{2-}$ rich medium. Changes in nitrogen (e.g. nitrate) levels in the root medium hardly change the root architecture. On the other hand, if nitric oxide (NO) is induced in the seedling of tomato plants (by sodium nitroprusside), the emergence of lateral roots is enhanced.

Svistoonoff *et al.* (2007) took the impact of Pi levels in the medium one step further. They asked: "How does the main root sense the level of Pi in the medium?" They found that it is through the tip of the primary root: when it is in contact with low Pi, the elongation of the main root is reduced and more lateral roots are produced. The sensing is involved with the expression of *LOW PHOSPHATE ROOT1* (*LPR1*) and *LPR2* that encode multicopper oxidases. These genes are expressed in the root cap. Hence, it is probably the root cap that senses the Pi level.

Ward *et al.* (2008) from the Purdue University in Indiana were asking if another element is involved in shortening the main root in a Pi-deficient medium. These investigators found that this shortening is actually the result of iron (Fe) toxicity. When the Fe level in a Pi-deficient medium was reduced, without increasing the Pi-level, the main root reached control length. It should be noted that interaction between Pi and Fe can occur in several locations. The presence of Fe in the soil can render Pi unavailable to the plant and Fe can also interact with Pi inside the plant, causing a shortage of Pi. The Fe toxicity is a major agricultural problem because it is estimated that about a third of agricultural soils are Pi-deficient due to the precipitation of Pi with Fe. It was noted that the changes in root architecture, due to changes in the level of Pi, vary in different plants. In some monocots, the changes are rather minute. Even in Arabidopsis there are differences between races.

As indicated above, low phosphate availability in Arabidopsis may cause a number of responses, such as inhibition of the elongation of the main root and greater formation of lateral roots (as well as increase in root hair elongation). Lopez-Bucio *et al.* (2005) intended to gain more insight into the regulatory mechanisms by which phosphorus (P) availability alters postembryonic root development. They thus performed a mutant screen to reveal genes involved in response to P deprivation. They revealed three low phosphate-resistant root lines: *lpt1-1*, *lpt1-2*, and *lpt1-3*, all of which had *reduced* lateral root formation in low P conditions (the wild-type develops increased lateral root formation in low P). It was then found that all these *lpt1* mutants were allelic to *BIG* that is required for normal auxin transport in Arabidopsis. Detailed characterization of lateral root primordia (LRP) development in wild-type and *lpr1* mutants revealed that *BIG* is required for pericycle cell activation to form LPR in both high (1 mM) and low (1 $\mu$M) conditions, but not for the low P-induced alteration (shortening) in the primary main root growth, lateral root emergence, and root hair elongation. Exogenously supplied auxin restored normal lateral root formation in *lpr1* mutants in the two P treatments. Treatment of wild-type Arabidopsis seedlings with brefeldin A — a fungal metabolite that blocks auxin transport — phenocopied the root development alterations observed in *lpr1* mutants in both high and low P conditions. This suggested that BIG participates in vesicular targeting of auxin transporters. The authors concluded that auxin transport and BIG function have fundamental roles in pericycle cell activation to form LRPs and to promote root hair elongation. The authors also claimed that the mechanism that activates root system architectural alteration in response to P-deprivation seems to be independent of auxin transport and BIG.

Another study on the role of auxin redistribution in responses to the root system architecture (RSA) during phosphate starvation was performed by Nacry *et al.* (2005). While the normal level of phosphate (P) in the root medium was 1 mM, the medium for P starvation contained 3 $\mu$M of P. These authors found that there is a great change in RSA during the 14 days after the planting of Arabidopsis seeds. During days 7 to 11, the RSA was changed by P starvation as reported previously: a shortening of the main primary root and an enhancement of lateral root formation. After day 11, the P starvation caused a general reduction not only of the growth of the main root but also of the lateral roots: growth and density were reduced. Low P had contrasting effects on lateral roots during the various stages of their development. Low P caused a strong inhibition of primordia initiation but a strong stimulation of the activation of the initiated primordia. The involvement of auxin signaling in morphological changes was investigated in wild-type seedlings of Arabidopsis treated with IAA or 2,3,5 triiodo-benzoic acid (TIBA) as well as using mutants (*axr4*, *aux1-7*, and *eir1-1*) instead of wild-type seedlings. *Aux1* and *eir1* as well as TIBA reduced IAA transport while *axr4* reduced IAA sensitivity. Most effects of low P on RSA were dramatically modified in the mutants or hormone treated with wild-type seedlings. This already showed that auxin (IAA) plays a major role in the P starvation-induced changes of root development.

The authors assumed that several aspects of RSA response to low P are triggered by local modifications of auxin concentration. They proposed a model that postulates that P starvation results in: (1) an over-accumulation of auxin in the apex of the primary root and young lateral roots; (2) an over-accumulation of auxin or a change in sensitivity to auxin in the lateral primordia; (3) a decrease in auxin concentration in the lateral primordia initiation zone of the primary root and in "old" lateral roots. The results of measurements of local changes in auxin concentrations induced by low P — either by direct quantification or by biosensor expression pattern (DR5:GUS reporter gene) — are in line with this postulated model.

Furthermore, the observation that low P availability mimicked the action of auxin in promoting lateral roots development in the *alf3* mutant confirmed that P starvation stimulates primordia emergence through increased accumulation of auxin or change in sensitivity to auxin in the primordia. Both the strong effect of TIBA and the phenotype of the auxin transport mutants (*aux1*, *eir1*) suggest that low P availability modifies local auxin concentration within the root system through changes in auxin *transport* rather than auxin *biosynthesis*. Clearly, the two studies by Lopez-Bucio *et al.* (2005) and Nacry *et al.*

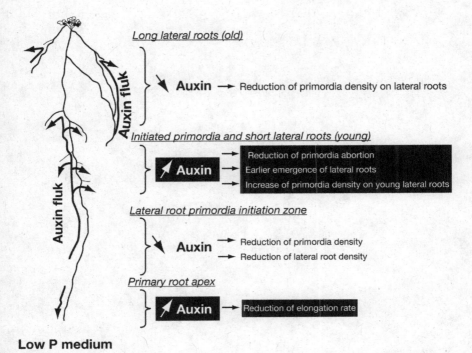

**Fig. 61.** Schematic model of auxin redistribution within the root system in response to P starvation. (From Nacry *et al.*, 2005.)

(2005) indicated that auxin is involved in the development of lateral roots. Nacry *et al.* (2005) proposed a scheme (Fig. 61) for their model of auxin redistribution within the root system, in response to P starvation. Are additional phytohormones also involved in root patterning? The answer may be provided in future studies.

# CHAPTER 12

# Patterning of the Shoot

The shoot-apex meristem (SAM) is the "fountain" of the above-soil organs and tissues of land angiosperms. These organs, with the exception of the shoot itself (e.g. flowers and carpels), are modifications of leaves. This assumption was already included in the doctoral thesis of Caspar Friedrich Wolff (1733–1794) that was submitted (in Latin) to the University of Halle in 1759 (the German translation was published a few years later). The botanical literature tended to ignore Wolff and usually mentioned Johann Wolfgang von Goethe (1749–1832) as the source of the idea of the metamorphosis of leaves, although Goethe proposed this idea many years after Wolff.

The patterns of SAMs differ among angiosperms, but within each species, this pattern changes only very little during ontogeny. The plants developed mechanisms that maintain the same basic SAM in their indeterminate main shoot. This chapter will contain a section that describes these mechanisms and discusses the genes and phytohormones that are involved in shaping the SAMs.

Plants developed some interesting means to avoid shade ("shade avoidance"). A special section in this chapter will be devoted to this subject.

Until recent years, the control of the formation of lateral branching from the main shoot was considered to be by "Apical Dominance"; meaning that the basipetal flow of auxin from the tip of the shoot prevents branching. In recent years, it was revealed that a new type of phytohormone participates in the control of branching: the strigolactones. Thus, another section of this chapter will be devoted to the initiation of laterals from the main shoot.

Leaves that are initiated from the SAM are arranged in a way that is typical for each species. This leaf arrangement is termed *phyllotaxis*. A considerable number of studies and mathematical models have been published to explain the phyllotaxis of angiosperms, and these will be discussed in one of the sections of this chapter.

Finally, shoots undergo a transition from vegetative growth (producing leaves in the axils of which shoots may develop) to reproductive growth: producing inflorescence and flowers. The last section of this chapter will handle this transition.

## Genes and Phytohormones Affecting the Patterning of SAM

The structure of the SAM already started to be investigated during the 18th and 19th centuries. However, this field of investigation progressed rapidly only in the first half of the 20th century. This progress was due to an improvement of the "tool kit" of methodologies, such as induction of mutation, induction of polyploidy, refined histological methods, and surgery. The early history of SAM research was summarized up to the end of the 1950s by several good reviews (Snow and Snow, 1931; Gifford, 1954; and Wardlaw, 1957). Later, important updates of SAM research included: Steeves and Sussex (1989), Sussex (1989), and Kerstetter and Hake (1997). In my book (Galun, 2007), the literature on SAM is updated till the end of 2006. Since the latter book contains a whole chapter on SAM, attention in the present book is focused on more recent studies and details the roles of phytohormones in the patterning of the SAM. Repetitions of information provided in the previous book will appear in the present book only to serve as background.

The role of the SAM is to equip cells with three main purposes: (1) to maintain a niche of dividing cells that may be regarded as "stem cells"; (2) to supply cells that will initiate lateral organs as new leaves with their axils from where auxillary shoots and/or flowers may emerge; and (3) to contribute cells that will differentiate into the different tissues of the mature shoot as epidermis cells and vascular tissue.

We will see that the SAM is regulated by interactions of gene products and phytohormones. Phytohormones were discussed in Part I of this book and a list of important genes (mainly of Arabidopsis) are listed in Table 4.

The publications of two very good reviews (Shani *et al.*, 2006; Tucker and Laux, 2007), as well as Chapter 8 of Galun (2007) update the knowledge on the roles of phytohormones and genes in controlling the pattern and functionality of the SAM. Briefly, the SAM can be divided into different histological zones, as shown for Arabidopsis in Fig. 62. Figure 63 provides additional information on shape, phytohormones, and genes that interact in the SAM. Lateral organs are initiated from the peripheral zone (PZ) which is located at the meristem flanks. The central zone (CZ), at the summit of the SAM, contains self-maintaining, slowly dividing cells, which provide initials for the PZ. The relative stage of the primordium development is described in terms of plastochrome (P), whereby the latest emerging primordium is termed $P_1$, the next primordium is termed $P_2$ and so on. The region within the meristem from which the next lateral organ primordium will be formed is termed $P_o$.

Proper SAM function requires the maintenance of a delicate balance between the production of lateral organs from the flanks and indeterminate growth at its center. The roles of phytohormones in controlling SAM function were revealed many years ago. Details of these roles were provided by several reviews, such as the roles of auxin by Benjamins and Scheres (2008) and the role of GAs by Schwechheimer and Williger (2009).

As for phytohormones, hormones such as auxin (AUX) and cytokinin (CK), appear to display a dynamic concentration gradient in the SAM. How can the plant sense a gradient and how are these gradients "translated" into the specification of unique tissues that are separated by clear boundaries, are major questions. Shani *et al.* (2006) summarized answers to the second question, but no satisfactory answers were given to the question of how plant cells sense gradients. Can they sense the different levels of auxin in two adjacent cells or in different intracellular locations?

**Table 4. Genes Involved in Meristem Establishment and Maintenance**

| Full name | Abbr. | Protein type | Role |
|---|---|---|---|
| Names of genes are from At* unless indicated to be from other species | | | |
| *AGAMOUS* (*PLENA* Am) | *AG* | MADS TF | AC function floral identify gene. In addition, active in floral meristems to repress WUS. When mutated, floral meristems are indeterminate and flowers are formed within flowers. |
| *ASYMMETRIC LEAVES1* (*PHANTASTICA*, Am *ROUGH SHEETH2*, Zm) *ASYMMETRIC LEAVES2* *CLAVATA* group | *AS1* | MYB TF | Restricts KNOX proteins such as BP and KNAT2 from leaves of many plants, where KNOX level is kept at bay. |
| | *AS2* | LOB TF | Forms a complex with AS1. |
| | *CLV* | | |
| *CLAVATA1* (*THICK TASSEL DWARF1*, Zm) *CLAVATA3* | *CLV1* | LRR receptor kinase | Part of a membrane receptor and signal transduction complex with CLV2. |
| | *CLV3* | CLE peptide | Small polypeptide ligand for the CLV1/CLV2 complex which restricts WUS activity. |
| *CUP-SHAPED COTYLEDON 1/2/3* (*NO APICAL MERISTEM* — Ph *CUPULIFORMIS*, Am) | *CUC1/2/3* | NAC TF | Specify organ boundaries and help to define STM embryonic expression. CUC1&2 are negatively regulated by miR164. |
| *HAIRY MERISTEM* Ph | *HAM* | GRAS TF | Expressed "outside" of the meristem proper and probably required for cellular responses to WUS and STM. |

*(Continued)*

**Table 4.** (*Continued*)

| Full name | Abbr. | Protein type | Role |
|---|---|---|---|
| KNOTTED-LIKE group SHOOT MERISTEMLESS (KNOTTED Zm) | KNOX STM KN | Homeodomain TF | Required for establishment and maintenance of the SAM. Promote cytokinin biosynthesis and negate GA. Expressed in the shoot meristem, but expression disappears when leaf primordial are specified. Inhibit leaf differentiation, and thus are not expressed in simple leaves but are expressed in compound leaves. Proteins are able to move through plasmodesmata. |
| BREVIPEDICELLUS/ KNAT1 KNAT2/6 | BP/KNAT1 KNAT2/6 | Homeodomain TF | Have redundant role with STM. BP regulates pedicel elongation and lignin synthesis. |
| LEAFY (FLORICULA, Am FALSIFLORA Sl, ABERRANT LEAF AND FLOWER Ph, UNIFOLIATA Ps) | LFY | LEY TF | Specify the fate of floral meristems as flowers and not as leaves. |
| PHABULOSA, PHAVOLUTA REVOLUTA, CORONA | PHB PHV REV, COR | ClassIII HD-ZIP TF | Maintain the SAM, specify the adaxial side of lateral organs and promote xylem formation in vascular strands. Negatively regulated by miR165/6. |
| MONOPTERUS | MP | ARF TF | An auxin response factor that facilitates root meristem pole and flower formation. |
| STIMPY/WOX9 | STIP | WOX TF | Promote cell division in the vegetative SAM but can be replaced by sucrose. |

(*Continued*)

**Table 4.** (*Continued*)

| Full name | Abbr. | Protein type | Role |
|---|---|---|---|
| WUSCHEL (ROSULATA Am) | WUS | WOX TF | Specify a sub-domain of the central zone, just below the CLV1 domain. Required for the SAM maintenance by a regulatory loop that involves the removal of CLV3 expression. |
| ZWILLE/PINHEAD ARGONAUTE1 | ZUPNA AGO1 | AGO member of the RISC | Function as a "slicer" in the RISC complex. Acts together with AGO1 to establish and maintain the apical meristem through proper regulation of miRNAs and their target genes. |

*At, *Arabidopsis thaliana*; Am, *Antirrhinum majus*; Ph, *Petunia hybrida*; Sl, *Solanum lycopersicum*; Ps, *Pisum sativum*; Zm, *Zea mays*; TF, transcription factor.

This table was compiled on the basis of numerous publications. The references to these publications were provided in Table 4 of Galun (2007) from where this table was copied. This table does not include genes involved in the biosynthesis of phytohormones; information on such biosyntheses is presented in Part I of this book. I am grateful to Professor Yuval Eshed, from the Department of Plant Science, Weizmann Institute of Science, Rehovot, Israel, for bringing this table to its final form.

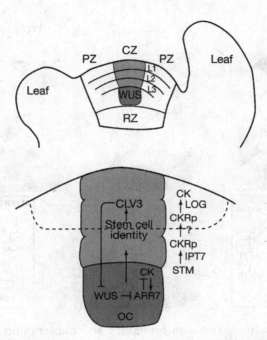

**Fig. 62.** Stem cell niches in the shoot apical meristem. The Arabidopsis shoot meristem is divided into three tissue layers (L1, L2, L3) and into various zones that separate slowly dividing cells in the central zone (CZ) from rapidly dividing cells in the peripheral zone (PZ). There is also an underlying rib zone (RZ), which contributes the inner tissues of the stem. The shoot meristem stem cells reside in all three layers of the CZ above the organizing center (OC) and are maintained as undifferentiated by positional information. The regulatory framework below shows interactions between meristem regulators, CK and its ribophosphate precursor (CKRp), and stem cell function. The dashed line indicates the approximate expression domain of the *LOG* gene, as inferred from studies in rice. Note that expression and functional details of the Arabidopsis *LOG* homolog are yet to be determined. *STM* is expressed in all cells of the shoot meristem (but not in incipient organ primordia). (From Tucker and Laux, 2007.)

Let us look at three types of gene-encoding transcription factors that seem to interact with plant hormones in the SAM (AUX, GAs, and CK). One group of these genes encodes KNOTTED-like homeobox (KNOXI) proteins that are located in specific sites of the SAM

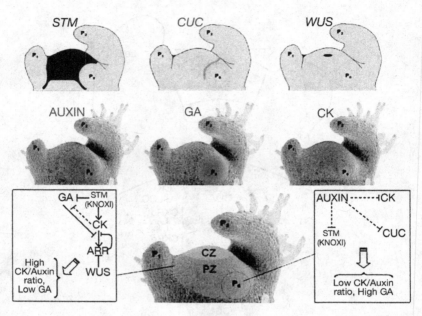

**Fig. 63.** Interactions between hormones and transcription factors in the shoot apical meristem (SAM). Top row: the expression patterns of the Arabidopsis genes *STM* (a *KNOXI* gene), *CUC*, and *WUS* are shown on a schematically drawn SAM. Middle row: predicted distribution of auxin, gibberellin (GA), and cytokinin (CK), shown over a scanning electron microscopy (SEM) image of a tomato SAM. The distributions are hypothesized on the basis of interpretations of the presented data. $P_1$ is the primordium of the youngest initiating leaf; $P_2$ is the next oldest leaf; and $P_0$ designates the region recruited to produce the next leaf primordium. Bottom row: a model of the generation of a distinct boundary between the SAM and the initiating primordia. In the STM expression domain, STM induces CK biosynthesis and represses GA biosynthesis and activity. WUS enhances CK activity locally by repressing ARR. The resulting high CK:auxin ratio and low GA concentration promotes indeterminate growth. In the $P_0$ region, high auxin concentrations restrict the expression of STM and CUC, and repress CK biosynthesis. High GA concentrations further decrease CK activity. The result is a low CK:auxin ratio, and a high level of GA, which together promote lateral organ initiation. The SEM image was provided by Arnon Brand. (From Shani *et al.*, 2006.)

(see Fig. 63). By using loss- and gain-of-function mutants of KNOXI have been implicated with the KNOXI proteins in the maintenance of the indeterminate nature of these meristems and in the formation of organ boundaries. Members of the NAC group of transcription factors, including genes that encode CUP-SHAPED COTYLEDON1 (CUC1), CUC2, and CUC3 (from Arabidopsis) are also essential for meristem establishment and for the formation of boundaries between organs and meristems, and between adjacent organs. In vegetative apices, members of the *NAC* gene family are expressed in strips in the SAM, which correspond to the future organ-organ and meristem-organ boundaries.

The WUSCHEL (WUS) encodes a homeodomain protein that is expressed in a defined group of cells below the CZ (see Fig. 62). WUS is involved in the maintenance of meristem indeterminacy ("stem cells," division without differentiation). WUS activity is non-cell autonomous, as the protein is expressed (i.e. transcribed) below the CZ but is required to maintain the CZ. On the other hand, CLAVATA3 (CLV3) is a ligand that acts non-cell autonomously through the CLV1-CLV2 receptor complex, and the combined CLV1-CLV2-CLV3 regulate WUS and thus restrict the size of the CZ. Details of the biology of the CLV1-CLV2-CLV3 were presented in Chapter 8, where polypeptide phytohormones were mentioned. The transcript of CLV3 is translated into a long protein that matures into a 96 amino acid protein and is then further processed to a peptide that contains a conserved ~14 amino acid box that is essential for the complexing with the CLV1-CLV2 receptor. As the processed CLV3 has to travel through several layers of cells, it deserves to be termed a "hormone."

In two publications that were published consecutively and in the same issue of *Current Biology* (Jasinski *et al.*, 2005; Yanai *et al.*, 2005), interactions between KNOXI protein and CK and between CK, GA and KNOXI were reported. Yanai *et al.* (2005) focused on the regulation of CK biosynthesis of *KNOXI* genes. They found that three different KNOXI proteins resulted in a rapid increase of the CK biosynthesis: gene *IPT7* and the activation of *ARR5* encoding a cytokinin response factor. They also found that a rapid and dramatic increase in CK levels followed the activation of the KNOXI protein SHOOT MERISTEMLESS (STM). The application of exogenous CK or the expression of a CK biosynthesis gene, through the STM promoter, partially rescued an *stm* mutant. The authors thus suggested that activation of CK biosynthesis mediates *KNOXI* function in meristem maintenance. *KNOXI* genes have been shown to interact with phytohormones, directly repressing the expression of the GA biosynthesis gene *GA20oxidase*. Hence, KNOXI factors negatively regulate GA biosynthesis; *KNOXI* gene probably also regulates CK biosynthesis. The authors thus suggested that KNOXI proteins are central regulators of hormone levels in plant meristems. The authors' model for the KNOXI/phytohormone interaction is shown in Fig. 64.

Jasinski *et al.* (2005), in their publication that was adjacent to the publication of Yanai *et al.* (2005), started with the interaction between KNOX transcription factors and the biosynthesis of GAs at the SAM. They noted that KNOXs promote meristem function partially through the repression of the synthesis of GAs, although the regulation could not fully account for KNOX action. Thus, the authors

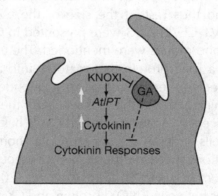

**Fig. 64.** A model of the interaction between *KNOXI* genes and the growth regulators cytokinin and GA in balancing meristem function. In the central zone of the SAM, KNOXI proteins induce the expression of *AtIPTs* among other targets, causing the accumulation of cytokinin. In parallel, KNOXI proteins repress GA biosynthesis. During leaf initiation, *KNOXI* genes are down-regulated, causing a reduction in cytokinin levels and elevations in GA levels. (From Yanai *et al.*, 2005.)

claimed (as was also claimed by Yanai *et al.*, 2005) that the function of KNOX is also mediated by the level of CK. The authors found that KNOX activity is sufficient to rapidly activate both CK biosynthesis gene expression and a SAM-localized CK response regulator. CK signaling was found to be necessary for SAM function in the presence of a weak hypomorphic allele of the *KNOX* gene *SHOOT MERISTEMLESS (STM)*. Additionally, the authors presented evidence that a combination of constitutive GA signaling and reduced CK levels is detrimental to normal SAM function. They then claimed that CK activity is both necessary and sufficient for stimulating GA catabolic gene expression, by that the latter genes reinforce low-GA regime establishment by KNOX protein in the SAM. The authors thus concluded that KNOX proteins may act as general orchestrators of growth regulator homeostasis at the SAM (of Arabidopsis) by simultaneously activating CK and depressing GA accumulation. The publications of Jasinski *et al.* (2005) and Yanai *et al.* (2005) support each other.

Several research reports (e.g. Tucker *et al.*, 2008; Braun *et al.*, 2008; Schuetz *et al.*, 2008; Gomez-Mena and Sablowski, 2008) and reviews (e.g. Tucker and Laux, 2007; Sablowski, 2007; Kepinski, 2006; Schwechheimer and Willige, 2009; Benjamins and Scheres, 2008) that were published in recent years further clarified the gene–phytohormone interactions that participate in the patterning of plant SAM (most of the studies were performed with Arabidopsis). The main findings and conclusions of these publications will be summarized below.

Tucker *et al.* (2008) analyzed the transition from the developing Arabidopsis embryo to the early functions of the Arabidopsis SAM. They focused on *ZWILLE (ZLL)* that is a member of the ARGONAUTE (AGO) family. The authors showed that the stem cell-specific expression of the signal peptide gene *CLV3* (see above for its description) is not maintained in the embryo despite increased levels of the homeodomain transcription factor WUS. WUS is expressed in the organizing center (OC) of the niche and normally promotes stem cell identity. The authors found that during embryonic patterning, *ZLL* acts from the emerging vascular primordium, located below the OC. The *ZLL* is specifically required for embryonic shoot meristem development. Whereas both ZLL and AGO1 appear to be required for normal expression of WUS, their effects on CLV3 expression appeared to be temporarily separated. One plausible model, suggested by Tucker *et al.* (2008), is that AGO1 and ZLL act sequentially in stem cell development with AGO1 being essential for initiation of the stem cell program, until the heart/torpedo stage, and ZLL being essential for its maintenance during embryo maturation. The authors also suggested that AGO1 has additional functions that cannot be rescued by late ZLL expression.

Kepinski (2006) updated the information on the interactions in the SAM of genes and phytohormones up to the end of 2005. This update was accompanied by a detailed scheme of these interactions, as shown in Fig. 65. The update already included summaries of the publications by Jasinski *et al.* (2005) and Yanai *et al.* (2005), who revealed the central roles of KNOX and its interactions with GA, IAA, and CK.

The review by Sablowski (2007) focused on the regulation of stem cell population in the SAM and discussed the interactions of WUS and the stem cell niche with several phytohormones. WUS activity strongly affects the increase of the stem cell niche while CLV3 suppressed the WUS (after complexing with CLV1/CLV2). Auxin suppresses the *STM* gene, while *STM* induces the biosynthesis of CK by up-regulation of *IPT7*. The increased level of CK promotes the cell division of stem cells. STM, in addition to promoting cell division, through IPT7, also inhibits cell differentiation. Also, there is a role of the *ARR7*: it antagonizes the CK-induced cell division. Sablowski also suggested that the retinoblastoma protein is a key regulator of cell differentiation, but further study of this issue should verify the retinoblastoma protein's role in the SAM because until now, the control of differentiation was reported only in the root apical meristem.

Another recent study, this time by Gomez-Mena and Sablowski (2008), focused on the development of the interface between the stem and lateral meristems. For understanding this interface differentiation, they looked at the *ARABIDOPSIS THALIANA HOMEOBOX*

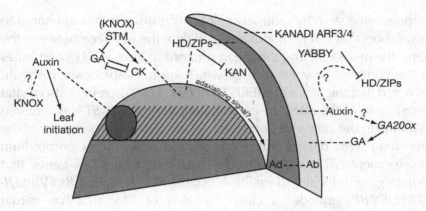

**Fig. 65.** Interactions between hormones and transcription factors that mediate SAM function and adaxial–abaxial patterning in the leaf. Arrows indicate positive and T-bars, negative regulation. In the Arabidopsis SAM, the KNOX protein STM suppresses GA biosynthesis and promotes CK biosynthesis. At the flanks of the meristem, in the PZ (hatched area), the accumulation of auxin prompts the initiation of a leaf primordium (circle), possibly by down-regulating *KNOX* expression. In the leaf (right-hand side), adaxial (Ad) and abaxial (Ab) identities are specified by the mutual antagonism between HD-ZIP and KANADI/YABBY transcription factors, apparently under the influence of an adaxializing signal that emanates from the meristem. The repression of HD-ZIP expression also involves ARFs and possibly asymmetric auxin accumulation. Auxin in the leaf might also promote GA levels by positively regulating the GA biosynthesis gene, *AtGA20ox1*. (From Kepinski, 2006.)

*GENA1* (*ATH1*) that is a BELL-type gene. They found that this is required for the proper development of the *boundary* between the stem, and both vegetative and reproductive organs. The role of *ATH1* partially overlaps with the role of *CUP-SHAPED COTYLEDON* (*CUC*) genes (see Fig. 63). During the vegetative phase of the shoot, *ATH1* also functions redundantly, with light-activated genes, to inhibit growth of the region below the shoot meristem. *ATH1* is down-regulated at the start of inflorescence development (see below in the section of the transition of the shoot from vegetative to reproductive growth). Ectopic *ATH1* expression prevents growth of the inflorescence stem by reducing cell proliferation. It thus appeared to the authors that *ATH1* modulates growth at the interface between the stem, the meristem, and the organ primordia: thus, *ATH1* contributes to the compressed vegetative growth habit of Arabidopsis. It should be noted that the SHOOT MERISTEMLESS (STM) protein is also a vital component of the SAM. In combination with *WUS*, *STM* is required to maintain the stem cell population in the CZ. In the periphery of the meristem, *STM* delays differentiation and antagonizes primordium development. *STM* also functions, together with CUC genes that repress growth locally to establish boundaries. The gene *BREVIPEDI-CELLUS* (*BP*) encodes a close homolog of *STM* that has partial redundant functions in meristem maintenance. In spite of the reports (mentioned above) by Jasinski *et al.* (2005) and Yanai *et al.* (2005), it is still not clear how KNOX proteins, such as STM and BP, control the behavior of meristem cells, although the latter proteins are now known to have a local control capability of some phyto-hormone levels. STM represses the biosynthesis of GAs and activates catabolism of GAs. Hence, STM is clearly involved in meristem functions.

The multiple roles of KNOX proteins in the meristem raise the question of whether these proteins function together with localized factors to control specific aspects of meristem function. Candidate cofactors are homeodomain proteins of the BELL family which form heterodimers with STM/BP and have been proposed to control distinct aspects of meristem function. The only example that has been functionally characterized is the BELL-type protein that is most often referred to as BELLRINGER (BLR), and is also termed PENNYWISE, REPLUMLESS, VAAMANA, LARSON, and BLH9. *BLR* is required for correct phyllotaxis and it interacts with BP to promote inflorescence stem development. The roles of other KNOX-interacting BELL proteins in meristem development remain unknown.

The *ATH1* gene encodes a BELL-type homeodomain protein that was initially described as a light-regulated transcription factor and more recently assumed to be an activator of the flowering suppressor *FLOWERIN LOCUS C* (*FLC*) (see below in the section on the transi-tion from vegetative to reproductive growth, as well as Chapter 10 of

Galun, 2007, Fig. 51). The ATH1 protein was found to form heterodimers with STM and BP, and to interact synergistically with ectopically expressed STM. This led to the assumption that ATH1 functions as a partner of STM and BP in the meristem. The authors further investigated the role of *ATH1* in the SAM and showed that *ATH1* is required for the development of the boundaries between shoot organs and the stem, and that this function partially overlaps with that of *CUC* genes. Also, ATH1 functions redundantly with light-activated genes to prevent stem growth during the vegetative phase. At the transition to reproductive development, when stem growth is activated (e.g. in Arabidopsis), ATH1 is down-regulated; conversely, constitutive ATH1 inhibits growth of the inflorescence stem by inhibiting cell proliferation. The strategy of plants to evolve two different means to control an important structural change is reminiscent to what was attributed to a Chinese emperor: he used yellow suspenders and a yellow belt to ensure that his trousers will not fall down ... in short, the authors suggested that ATH1 represses growth at the interface between the meristem, the stem, and the lateral organs (i.e. leaf initials) to establish the basal boundaries of shoot organs and the compressed rosette habit of Arabidopsis.

A recent update on the role of auxin in plant growth and development was written by Benjamins and Scheres (2008) from the Utrecht University in the Netherlands, who termed auxin as "the looping star of plant development." The components involved in the transcriptional loop of auxin are schematically (and clearly) presented in Fig. 66, but note that this scheme is not restricted to the SAM but rather also deals with the root apical meristem.

Benjamins and Scheres summarized their review on the roles of auxin in plant patterning in the following "points":

— Auxin is involved in many aspects of plant growth and development. Many molecular components have been identified and their biological function is at least partially understood.
— Members of the TIR1 family of proteins are able to bind auxin, and therefore represent auxin receptors for which a downstream signaling pathway is known. TIR1 proteins are involved in the

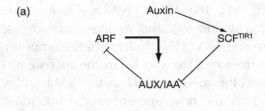

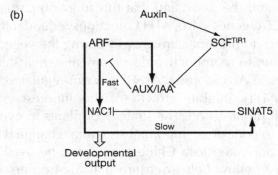

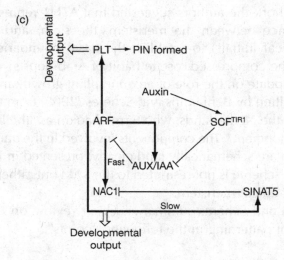

**Fig. 66.** Schematic representation of the components involved in the transcriptional loop. (a) The expression of *AUXIN/INDOLE-3-ACETIC ACID* (*Aux/IAA*) genes is induced by auxin, mediated by the SCF$^{TIR1}$ (AKP1, Cullin and F-box protein, in this case TRANSPORT INHIBITOR RESPONSE1) ubiquitin ligase complex and AUXIN RESPONSE FACTOR (ARF) proteins, which in turn feeds back on the sensitivity of cells toward auxin, in most cases

regulation of the stability of AUXIN/INDOLE-3-ACETIC ACID (AUX/IAA) proteins, which are transcriptional regulators, acting in concert with AUXIN RESPONSE FACTOR (ARF) proteins.

— The currently known regulatory mechanisms stress the importance of feedback loops in auxin action. At least two such regulatory loops exist: a transcriptional loop and a polarity loop.

— The transcriptional loop involves auxin regulation of the activity of SCF$^{TIR1}$ (SKP1, Cullin, and F-box protein, in this case TRANS-PORT INHIBITOR RESPONSE1; see Chapter 1 of this book for details), thereby regulating the transcriptional activity of ARF proteins via the proteolysis rate of AUX/IAA proteins. PLETHORA (PLT) proteins directly or indirectly regulate the transcription of *PIN FORMED* (*PIN*) genes that feedback on auxin by regulating the distribution of auxin, thereby completing the loop. The relative timing and balance between auxin-mediated expression of *AUX/IAA* genes and the auxin-dependent degradation of the

---

**Fig. 66. (*Continued*)**   making the cell less sensitive. ARF proteins interact with Aux/IAA proteins and together regulate the transcription of auxin-responsive genes, including *Aux/IAA* genes. (b) The stability of Aux/IAA proteins is dependent on the activity of the SCF$^{TIR1}$ complex in response to changing auxin levels. High auxin levels will lead to the degradation of Aux/IAA proteins, resulting in the derepression of ARF proteins, inducing the expression of auxin-responsive genes. In case of lateral root development, this regulatory loop is fine-tuned by the action of NAC1, a transcription activator of the NAM, ATAF, CUC family, depending on the activity of SINA of *Arabidopsis thaliana* 5 (SINAT5), a RING-finger protein with ubiquitin ligase activity, which acts downstream of TIR1. The induction of expression by auxin is faster in the case of NAC1 compared to that of SINAT5, allowing fine-tuning of the loop. (c) Transcription of *PLETHORA* (*PLT*) genes is dependent on ARF action and PLT proteins induce the expression of *PIN FORMED* (*PIN*) genes, which in turn affects cellular auxin concentrations. The NAC1 and PLT proteins mediate the developmental outputs in response to auxin. Black arrows indicate regulation at the transcriptional level. Thin lines indicate positive (arrows) or negative (bars) regulation at the protein level. Multiple arrows indicate regulation at the transport level. (From Benjamins and Scheres, 2008).

AUX/IAA proteins balance the auxin response and auxin transport.
— The polarity loop involves the regulation of the subcellular local-ization of PIN efflux facilitators by affecting the phosphorylation status through the activity of PINOID (PIN) (or PID-like) kinases and PROTEIN PHOSPHATASE 2A (PP2A) phosphatases. It is likely that GTPase also perform a regulatory function in polarity determination.

Studies on the auxin–gene interactions (e.g. as shown in Fig. 65) suggest a situation in the SAM in which at a certain site of the PZ the auxin level is increased. At this site, a lateral organ (a leaf in the vegetative shoot or a flower in the reproductive shoot) is initiated. Once initiated, there is a flow of auxin to this site, depleting the neighbor cells from auxin. This depletion of IAA in the neighborhood requires auxin transport. The depletion is weakened in cells that are far away from the site of lateral initiation, and at another site, far removed from the previous site, there is sufficient IAA to initiate another (younger) lateral organ. The *MONOPTEROS* (*MP*) gene that encodes an auxin response factor is a positive regulator of *PIN* (efflux) genes and auxin transport. Schuetz *et al.* (2008) focused on the role of initiating new lateral organs. They formed a double mutant: *mp, pin1*. This mutant had a leaf-less dome at the SAM because it was double defective in IAA signaling and IAA transport. Notably, *mp* mutants could not be restored to normal phenotypes by the addition of auxin. Previous studies already suggested that *mp* mutants have a strongly reduced IAA sensitivity. The authors found that *mp,pin1* double mutants, as well as *mp* mutants in which auxin efflux was inhibited, displayed strong synergistic abnormalities: they failed to develop any lateral organs, and their SAM developed into a leafless dome. The authors concluded that the role of MP in shoot meristems is not limited to the regulation of auxin transport: the novel pheno-type suggested that auxin *transport* and *signaling* are essential for the regulation of the meristem architecture. The experimental results of Schuetz *et al.* (2008) suggested a "novel" role for MP in the SAM: the induction of actual outgrowth of leaves and flowers from the PZs. In a further section, phyllotaxis will be further looked at and there will also be an examination of the model suggested by Reinhardt *et al.* (2003) that was formulated in association with C. Kuhlenmeier and others. According to this model, leaf primordia are formed at sites of elevated epidermal auxin concentration. Pre-existing primordia are thought to influence the position of the new primordia by depleting the vicinity of the existing primordia through auxin transport. Thus, new primordia will only form at sites that are far enough from the existing primordia to allow new auxin maxima to form. This "distance" limitation will lead to the proper phyllotaxis pattern. As auxin transport in the *mp* mutants is strongly reduced, the *mp* mutants are not only defective in new lateral formations, but also have a vastly expanded SAM.

Intensive studies during recent years increased the understanding of how auxin is perceived by plant cells. In particular, these studies led to the identification of the TRANSPORT INHIBITOR RESPONSE PROTEIN (TIR1) F-box factor as the receptor of auxin whose binding with auxin leads to the repression (by ubiquitination) of the AUX/IAA transcriptional repressor. However, there were indications that TIR1 is not the only receptor of auxin (IAA). One such potential receptor is AUXIN BINDING PROTEIN1 (ABP1). The ABP1 was found to be active in several cell responses, but the study in whole plants was hampered by the lack of appropriate *ABP1* mutants. This was because the lack of ABP1 in the embryos is lethal: they are aborted at the transition from the globular to the heart-shaped stage. Recently, Braun *et al.* (2008) were able to overcome this difficulty by finding a way to suppress ABP at a given phase of the development of the pant. They termed their method "conditional repression of ABP1" and used this method to investigate the function of ABP1 during the vegetative shoot development. The essence of this conditional repression is the use of either a cellular immunization approach or an inducible antisense. The activities of the immunization or the RNA antisense to *ABP1* were achieved by constructing plasmids that contained promoters that either activated the formation of the appropriate antibody against ABP1 or the appropriate antisense RNA against the *ABP1* mRNA. The promoters could be activated at will by a simple signal, such as alcohol vapor. Arabidopsis was genetically transformed by this construct. In a previous study, these authors used this approach successfully with cultured tobacco cells (BY2), so they used it again with Arabidopsis seedlings. They found that ABP1 is required for the maintenance of shoot and leaf growth. They also found that ABP1 is required for the coordination of cell division and cell expansion. Especially during very early stages of leaf formation, even a transient loss of ABP1-mediated coordination of cell plate formation and cell expansion had drastic consequences for the whole plant's growth. This showed that ABP1 is required during postembryonic development of Arabidopsis. The authors thus concluded that the cells at the site of presumptive leaf initiation are more sensitive to ABP1 repression than other regions of the meristem and suggested a model in which ABP1 acts as a coordinator of cell division and expansion with local auxin levels influencing ABP1 effectiveness. This model is shown in Fig. 67.

Gibberellic acid (GA) promotes a range of developmental and growth processes in plants, such as germination, elongation, and flowering time. The involvement of GA in the development of SAMs

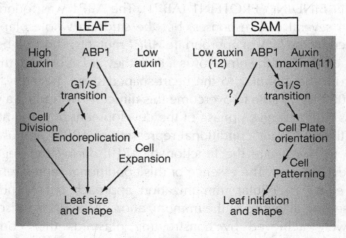

**Fig. 67.** Model for a context-dependent role of ABP1 in the shoot. Depending on the local auxin concentration (gray shading) and developmental context of the tissue, ABP1 may mediate different cellular outputs of division frequency, orientation, and expansion. (From Braun *et al.*, 2008.)

is mentioned above. DELLA repressors constitute key players in the impact of GA on growth and differentiation. The term DELLA is derived from a conserved motif in DELLA proteins that consist of a sequence of five amino acids (Asp-Glu-Leu-Leu-Ala). The presence of the DELLA motif is essential for DELLA protein interaction with GIBBERRELIC ACID INSENSITIVE DWARF1 (GID) receptors. The DELLA repressors were found to function as repressors of the PHY-TOCHROME INTERACTING FACTOR3 (PIF3) and PIF4 transcriptional activators in light-regulated seedling development. Schwechheimer and Williger (2009) used the interaction of PIFs with the active (Pfr) form of phytochrome B that leads to PIF degradation. These authors studied the evolution of GA signaling from a moss (*Physcomitrella*) through a lycophyte (*Selaginella*) to angiosperms. The authors found that while the moss did not have the GA signaling that is found in angiosperms, the lycophyte did contain this signaling. The absence of GA binding or of GA-dependent interactions between the GID1 and DELLA protein candidates from *Physcomitrella* however, does not necessarily allow to conclude that there are proteins unrelated to the GA signaling pathway: they may only have some protein functions in common with the established pathway. GID1–DELLA repressor inter-actions were therefore also tested between species, and two studies showed that there is no interaction between an angiosperm GA signaling component and the presumed interacting protein of the moss. The authors noted that it was previously found that there is a strong but GA-independent interaction between the moss GID1 candidate protein and a lycophyte DELLA protein. The evolutional history of the GA signaling is still far from clear but molecular data indicate that lycophytes are GA-responsive, but the physiological function of GA signaling in lycophytes is still elusive.

## Shade Avoidance

It is obvious to farmers and botanists that plants planted in high density (a great number of plants per unit area) tend to have long shoots with only few side branches. This elongation of shoots has an advantage to the respective plant because it will be able to compete successfully with a neighboring plant in getting more sunlight for its optimal development. This long and slender shoot characteristic is actually controlled by the quality of light. Light in which the ratio between red light and far-red light is low will lead to this phenomenon. There is a change going on in this ration, in densely grown plants: red light is absorbed by the canopy leaves of the neighbor plants and there is a reflection of far-red light from the plants' leaves (see Roig-Villanova *et al.*, 2007). Due to the fact that this plant behavior is leading to seeking direct sun illumination and avoiding shade, the phenomenon was termed "shade avoidance."

Two consecutive papers in the journal *Cell*, by Tao *et al.* (2008) of the laboratory of Joanne Chory and of other laboratories (a total of 17 investigators), and by Stepanova *et al.* (2008) of the laboratories of Gerd Jurgens and Jose Alonso (a total of 9 investigators) found that an increase in auxin biosynthesis was involved in shade avoidance. Friml and Sauer (2008) were very impressed with the findings of Tao *et al.* (2008) and Stepanova *et al.* (2008), and a month after the publication of the two papers, Friml and Sauer published a review on them.

The increase of auxin was attributed to the promotion of a "short-cut" biosynthesis of auxin, from tryptophan. It was previously assumed that the bulk of auxin synthesis from tryptophan goes through trp-decarboxylase to tryptamine and from there several additional steps, one of them, processed by YUCCA protein (e.g. CYP79B2, CYP79B3) leads to indole-3-acetaldehyde (IAAld). The "short-cut" pathway is from tryptophan to indole-3-pyruvic acid (IPA) and from there directly to IAAld (see Chapter 1 and Fig. 3 for background of pathways of auxin biosynthesis). The gene *TAA1*, that was detected by Tao *et al.* (2008) and by Stepanova *et al.* (2008), encodes the TAA1 enzyme that converts tryptophan to indole-3-pyruvic acid. The "shade avoidance" syndrome includes not only stem elongation but also fewer branches and early flowering. Friml and Sauer (2008) claimed that it is a change in the auxin gradient that induces the "shade avoidance." The "gradient" concept is problematic because in order to sense a gradient, the cell has to be able to compare the auxin levels at the cell sites of auxin influx and auxin efflux. The other possibility is that the cell can just sense the level of auxin in it. In the latter case, cells with different levels of auxin trigger different developmental programs. The level of auxin can be changed by different means. The auxin source may be greater by different transport effectivity, and by changes in the auxin influx and/or efflux.

Tao *et al.* (2008) found the gene that affects "shade avoidance" by screening for mutants that are not able to perform "shade avoidance." They identified the *TAA1* gene, the mutant of which (*taa1*) had no shoot elongation caused by shade. The mutant also had lower auxin levels than w.t., and the auxin content of *taa1* stem was not elevated by shade. However, furnishing external auxin restored the "shade avoidance" of *taa1* plants. Further experiments assured that indeed, *TAA1* encodes the enzyme of the short-cut that converts tryptophan to pyruvic-3-acid. They also found that TAA1 is mainly confined to leaf margins and is increased in shade.

Stepanova *et al.* (2008) had basically the same experimental results as Tao *et al.* (2008). The former investigated the developmental relevance of the enzyme that is encoded by *TAA1*. They also found two related genes that partially compensate for the loss of *TAA1*, but when all three genes are inactive, the consequences are disastrous: the triple mutants are already defective during embryo development and have no roots. Stepanova *et al.* (2008) also found that the *taa1* mutants were not sensitive to ethylene, but external supply of auxin restored ethylene sensitivity, confirming the cross-talk between auxin and ethylene.

The "shade avoidance" syndrome may not only concern the elongation of shoots in a field of densely planted plants. Every plant has a typical ramification of its branches so that a plant can be recognized from quite a distance (for example, whether a tree at a distance of a few hundred meters is a cypress or an oak). The distance between laterals on a main stem and the relative elongation of the laterals may be regulated similarly as the "shade avoidance." Hence, similar developmental controls may occur inside one plant as they occur between plants.

# Phyllotaxis

When phyllotaxis is mentioned, it means the process and mechanisms that regulate the arrangement of new lateral organs (mainly the development of new leaves) around the shoot apex of a plant. Phyllotaxis has interested scholars since ancient times. Some botanists attributed the study of phyllotaxis to Leonardo da Vinci who claimed, about 500 years ago, that the proper arrangement of plant organs is crucial for plant growth and survival. Some patterns of phyllotaxis are shown in Fig. 68. Much of the research on the control of phyllotaxis was performed with shoots that have a "spiral" phyllotaxis (such as Arabidopsis). Interestingly, until 1994, phyllotaxis was investigated without any reference to phytohormones (e.g. Callos and Medford, 1994). Much effort was devoted recently to the study of phyllotaxis by Cris Kuhlemeier (from the University of Berne, Switzerland) (Reinhardt, 2005; Reinhardt *et al.*, 2000, 2003, 2005;

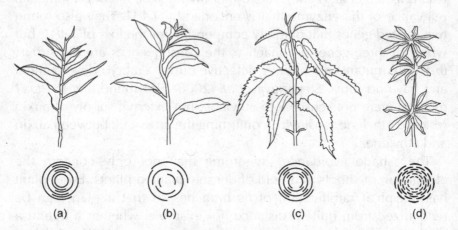

(a)        (b)        (c)        (d)

**Fig. 68.** Phyllotactic patterns. (a) Distichous or alternate: one leaf per node with 180° divergence angles (*Trisetum distichophyllum*). (b) Spiral: angle is ~137.5° (*Lysimachia dethroides*). (c) Decussate: two opposite leaves per node at 90° angles between the pairs (*Urtica dioica*). (d) Whorled: multiple leaves per node (*Galium odoratum*). Drawings by Peter Leuthold. (From Kuhlemeier, 2007.)

Smith *et al.*, 2006). These investigations — as well as investigations of other laboratories — on phyllotaxis were reviewed by Kuhlemeier (2007), and Benjamins and Scheres (2008). The subject was also reviewed (until the end of 2006) by Galun (2007). The number of types of phyllotaxis is great but only four of them are shown in Fig. 68. Of these, the *distichous* (or alternate) type is typical to maize although there are maize mutants that have a different type of phyllotaxis (see below). The *spiral* type is common in dicots. Some dicots, such as coleus, have *decussate* phyllotaxis. The spiral phyllotaxis may have different divergence angles. These angles are defined as shown in Scheme 1 in which the leaves of the shoot are schematically seen from above. In the scheme, the divergence angle is ~137 which is a rather common angle among plants having a spiral phyllotaxis.

The currently accepted model of new leaf initiation at the SAM of dicots has its root in a book by W. Hofmeister (1868) that claimed that each new leave at the SAM arises in the largest gap between the edges of the previous leaf primordium. Only in recent years was the auxin issue added to the model. In the current model, the initiated leaf primordium serves as a sink of auxin: the auxin in its vicinity is pooled into this primordium and by that, causes a depletion of auxin in the surrounding epidermal cells. Then, beyond the depleted zone, a new leaf primordium can be initiated because such an initiation requires a high level of auxin. How is the auxin transported from cell

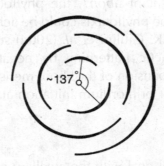

**Scheme 1.** Phyllotaxis with divergence of ~137°.

to cell? Benjamins and Scheres (2008) wrote a detailed review on this subject. One important issue is the influx and efflux of IAA. It should be noted that IAA is a weak acid, thus IAA can also enter cells without a strong transporter because the inside of the plant cell has a relatively high pH. However, once IAA is inside the cells, it will not exit without a transport protein. Thus, an active efflux is required to "push" IAA out of the cell. This latter refluxing is mediated by PIN-FORMED (PIN) proteins. There are several genes (in Arabidopsis) that encode PIN proteins. These proteins are predicted to contain 6 to 10 transmembrane domains (similar to bacterial transporters), and they are located on the cell membrane (inside of the cell wall). As noted above, IAA can enter the cytoplasm of the cell by diffusion, but an active influx of IAA is also possible. The protein AUXIN RESISTANT1 (AUX1) is such an influx carrier. The involvement of auxin flow in phyllotaxis became obvious when *pin* mutants were used: initiation of new leaves was then suppressed. It was suggested that normally, there is a co-expression of *PIN* genes and *AUX* genes that together facilitate the polar transport of auxin.

The abovementioned model of phyllotaxis has been tested in computer simulations, and generally these simulations generated realistic phyllotactic patterns (e.g. Jönsson *et al.*, 2006; Smith *et al.*, 2006).

Besides the main impact of auxin on the regulation of phyllotaxis, it was reported by Giulini *et al.* (2004) that CK can induce a change in phyllotaxy. As mentioned above, w.t. maize has a distichous phyllotaxis, but in the mutant *abph1*, the phyllotaxis is decussate! A change to the wild-type phyllotaxis could be achieved by treating the *abph1* mutant with CK. Giulini *et al.* (2004) suggested that *ABPH1* controls the phyllotactic patterning by negatively regulating the cytokinin-induced expansion of the shoot meristem, thereby limiting the space available for primordium initiation at the apex.

## Lateral Branching

In this section, we will deal with the auxillary outgrowth from angiosperm shoots. This outgrowth becomes evident when we describe the shoot as a chain of phytomers. Each phytomer consists of a node from which a leaf develops, followed by an internode. The axil of the leaf may be devoid of an auxillary bud (AB) outgrowth or an AB does develop. The history of revealing the mechanisms that lead to AB development or AB inhibition was detailed in Chapter 9 of this book, where the relatively new phytohormones, the strigolactones, were discussed at length. Napoli *et al.* (1999) reviewed the subject, and it appears that up to about a dozen years ago, the model of AB outgrowth was simple and coined as "Apical Dominance." According to this model, auxin that flows basipetally from the shoot tip inhibits the AB outgrowth and CK flowing in the stem acropetally promotes the AB outgrowth. The apical dominance model emerged from numerous studies in which the removal of the tip of the shoot caused ample outgrowth of AB. This also fitted the experience of horticulturalists over the past thousands of years (e.g. pruning of trees). However, there were doubts ... there were several indications that the apical dominance model is far from being able to explain the control of AB outgrowth (see Beveridge, 2000). One obvious finding was that auxin does not flow into the axil in which the AB was inhibited. As written above, the full "history" of the present model of AB control is provided in Chapter 9. Here, only a summary of its history is repeated. Beveridge (2000) suggested a model that she described schematically (see Fig. 48). Her grafting experiments suggested roles for two graft-transmissible signals while they gave only little direct evidence that auxin and CK are the main regulators of branching in peas. Using mutants, the Beveridge model included four *Ramosus* genes (*Rms1*, *Rms2*, *Rms3*, and *Rms4*). Of these, the product of *Rms1* has an inhibitory effect on the outgrowth of the AB. In further studies, Beveridge and her associates (Foo *et al.*, 2005; Morris *et al.*, 2005) revealed further details of the model. The *Rms1* seemed to encode a carotenoid cleavage enzyme that acts with *Rms5* to control the level of an unidentified mobile branching inhibitor. Van Norman *et al.* (2004) used Arabidopsis in their study and found that the gene *BYPASS1* (*BPS1*) encodes a protein that is required to prevent the function of a root-derived, graft-transmissible signal. These investigators suggested that this signal is likely to be a "novel" carotenoid-derived

molecule. The latter molecule (or its precursor) is apparently transported acropetally from the root.

As carotenoid cleavage takes place in the plastid, the investigators
looked for genes that encode intra-plastid enzymes. Hence, it
became evident that the elusive apocarotenoid is generated, at least
in part, by CCD7/MAX3 and/or CCD8/MAX4. Similar findings were
reported by studies with petunia (Snowden *et al.*, 2005), where
*Dad1/PhCCD8* were revealed. The petunia mutants *dad1*, *dad2*, and
*dad3* had increased branching. Simons *et al.* (2007) continued the
studies with petunia and found that the phenotype of increased
branching in *dad1* and *dad3* mutants could be changed to the w.t.
phenotype, when grafted on root stocks that contained DAD1 and
DAD3. It should be noted that the double mutant *dad1,dad3*
had additional phenotypic changes: decreased shoot length, delayed
flowering, and reduced germination. Hence, the respective genes
*DAD1* and *DAD3* probably have other "tasks," in addition to the
control of branching.

The interest in branching went to China... where Zou *et al.* (2006)
studied tillering. Tillering is an important issue in rice cultivation. The
correct amount of tillers provides the highest yield of grain. Rice
contains a gene termed *HIGH-TILLERING DWARF1* (*HTD1*) that
encodes a protein that is orthologous to MAX3 of Arabidopsis. HTD1
is probably a carotenoid cleavage dioxygenase. Auxin induces the
expression of HTD1 suggesting that in rice, the tillering is also controlled via auxin. Excessive tillering reduces the height of rice and
removal of tillers restores normal height.

After it was established that three phytohormones are involved in
the control of shoot branching (auxin, CK, and the recently discovered strigolactones), Ongaro and Leyser (2008) reviewed in some
detail the hormonal control of shoot branching. These authors
claimed that the auxillary meristems have the same developmental
potential as the primary shoot apical meristem. Therefore, each of the
former can form an entire secondary shoot. However, the auxillary
meristems frequently form only a few leaves before arresting to form
a dormant auxillary bud. The buds can subsequently be reactivated

and produce a branch. This flexibility makes it possible to change the architecture of the shoot system in response to environmental changes, such as the light conditions and nutrition. These environmental conditions are probably relayed through phytohormones. The roles of auxin and inhibitors (though not "direct inhibition") was well documented, and also the direct promotion of branchings by CK was verified. The authors thus focused on the "novel" phytohormones: the strigolactones. The mutants *max1*, *max2*, *max3*, and *max4* of Arabidopsis, and *rms1*, *rms2*, *rms3*, *rms4*, and *rms5* of pea, as well as *dad1*, *dad2*, and *dad3* in petunia served in genetics and graft studies to implicate the strigolactones in the repression of branching, as noted above. Of these, the Arabidopsis gene *MAX3* clearly encodes a carotenoid cleavage dioxygenase (CCD) that is essential in the production of the final hormone. Also, the two other Arabidopsis genes, *MAX1* and *MAX4*, encode enzymes that are required to produce the final hormone; while *MAX2* encodes an F-box protein LRR family and thus it probably involves a specific ubiquitination and protein degradation. It should be noted that the numbering of the genes in different plants might be confusing: so that *MAX2* is orthologous to *RMS4* and *MAX4* is orthologous to *RMS1* and *DUD1*.

As for interactions between phytohormones, there is good evidence for the down-regulation of CK by auxin. The latter probably reduces the CK biosynthesis enzyme adenosine-phosphate-isopentenyl transferase (IPT). The interaction between auxin, CK, and strigolactones is described in Fig. 69.

As indicated in Chapter 9 earlier, the chemical structures of active strigolactones were revealed only recently in two sequentially printed papers in *Nature* by Gomez-Roldan *et al.* (2008) and Umehara *et al.* (2008). The structures are shown in Fig. 50.

It was mentioned in Chapter 9 and reported by Besserer (2006) that strigolactones released from plant roots serve as attractants to arbuscular symbiotic fungi and as facilitators of the germination of parasitic plants, such as *Striga* and *Orobanche*. These subjects are beyond the scope of this book and will not be discussed further.

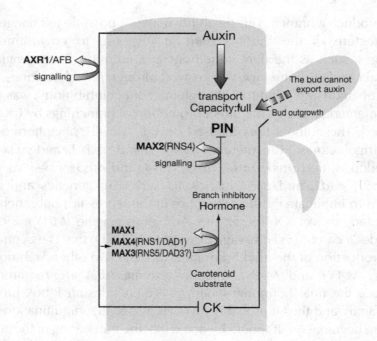

**Fig. 69.** Model for the hormonal control of shoot branching in Arabidopsis. Auxin is transported down the stem in the polar transport stream. The capacity of the stream is regulated by the MAX pathway, in which MAX3, MAX4, and MAX1 act in the synthesis of novel upwardly mobile hormones that regulate PIN1 levels through MAX2. For bud activation, buds must be able to export auxin and thus MAX-limited auxin transport capacity in the stem prevents bud outgrowth. In addition, the concentration of auxin in the polar transport stream is monitored by the AXR1/AFB pathway to regulate cytokinin synthesis. High auxin down-regulates cytokinin synthesis, inhibiting bud activation. The model predicts, rather paradoxically, that because of the increased capacity for auxin transport in the stem of *max* mutants, there will be high auxin and thus low cytokinin in the stem, but nonetheless, increased branching because of the high transport, capacity. Orthologs of the *MAX* genes in pea (*RMS*) and petunia (*DAD*) are written in brackets. (From Ongaro and Leyser, 2008.)

## Transition from Vegetative Growth to Flowering

About 45 years ago, just before Arabidopsis became a model dicot plant outside of Germany and Hungary, my colleagues and I published a paper (Galun *et al.*, 1964) on the induction of flowering in very young seedlings of *Pharbitis nil*. Our results indicated that the transition to flowering — that is induced by the appropriate photoperiod — requires mRNA synthesis that takes place in the leaves (cotyledons). The methodologies for the identification of specific mRNAs were, at that time, not available, and we stopped this research. Since then and up to the end of 2006, a great number of studies on the genetics and molecular biology of the transition from vegetative growth of the shoot to flowering were conducted. The results of these studies were discussed in Chapter 10 of Galun (2007). In the following, I will focus on the recent publications in this field. Figure 70 provides a scheme of the flowering induction as reviewed by He and Amasino (2005). A more updated scheme was provided by Jan A.D. Zeevaart (2008) and shown in Fig. 71. The latter scheme shows the transmission of the flowering signal from the leaf of a photo-induced Arabidopsis up to the differentiation of the SAM. The signal coming from the leaf is the FT. According to this model, the CONSTANT (CO) protein is produced and stabilized on the photo-induced leaf and then induces the expression of *FT* in phloem companion cells. The FT protein is transported in the sieve tubes, to the SAM. After arriving at the tip of the shoot, the FT protein is complexed with the transcription factor FLOWERING LOCUS D (FD) — a heterodimer (FD/FT) promotes the flowering by activating *SUPPRESSION OF OVEREXPRESSION OF CO1* (*SOC1*) and the floral meristem gene *APETALA1* (*AP1*). Hence, the site of the production FT mRNA is in the leaves and the site of FT action is in the shoot apex. Therefore, FT (the mRNA or the protein) or its product is a good candidate for the *florigen* signal. While this proposed sequence of events was mainly based on studies with Arabidopsis, there is now evidence for the role of FT from several other species, including monocots. As can be seen in Fig. 70, there are three pathways for floral induction: (1) induction by photo induction; (2) induction by vernalization; and (3) induction by an autonomous pathway. We will deal with these pathways sequentially.

Jiao *et al.* (2007) reviewed the light-regulated networks in angiosperms. The light-regulated plant development includes several issues, such as germination of seeds, photomorphogenesis (e.g. elongation of the hypocotyl), shade avoidance, phototropism, and the

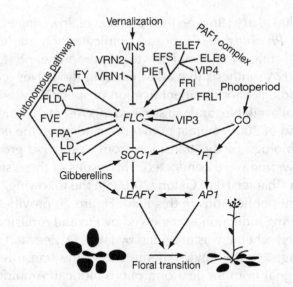

**Fig. 70.** A scheme of the induction to flowering. The autonomous pathway regulators and vernalization repress *FLC* expression. FRI and RFLI up-regulate *FLC* expression, and the H3-K4 trimethylation mediated by the PAF1 complex (ELF7, ELF8, and VIP4) activates *FLC* expression. *FLC* represses expression of the "flowering-time integrators" *SOC1* and *FT*, whereas the photoperiod pathway promotes expression of these integrators. *SOC1* and *FT* expression leads to the induction of floral-meristem-identity genes such as *LEAFY* and *AP1*, and thus of flowering. Lines with arrows indicate up-regulation (activation) of gene expression and lines with bars for gene repression. *AP1, APETALA1; CO, CONSTANS; EFS, EARLY FLOWERING IN SHORT DAYS; ELF7, EARLY FLOWERING 7; ELF8, EARLY FLOWERING 8; FLC, FLOWERING LOCUS C; FLD, FLOWERING LOCUS D; FLK, FLOWERING LOCUS K; LD, LUMINIDE-PENDENS; PIE1, PHOTOPERIOD INDEPENDENT EARLY FLOWERING 1; SOC1, SUPPRESSOR OF OVEREXPRESSION OF CONSTANS 1; VIN3, VER-NALIZATION INSENSITIVE 3; VIP3, VERNALIZATION INDEPENDENCE 3; VIP4, VERNALIZATION INDEPENDENCE 4; VRN1, VERNALIZATION 1; VRN2, VERNALIZATION 2.* (From He and Amasino, 2005.)

induction to flowering by the appropriate photoperiods. Each of these involves specific photo receptors, such as phytochromes and cryptochromes. Moreover, a huge number of transcription factors is required for each of these issues, and in total there are dozens of genes involved in the respective regulations. The interaction with diurnal changes and the light regime (light/dark periods) are especially complicated issues that are outside the scope of this book. Principally, the transition to flowering may be imposed by diurnal cycles that have dark periods that are shorter than a given number of hours (e.g. a "night" shorter than 11 hours), while other plants require short days (long nights) for the induction of flowering. The plants are thus regarded as Short Day (SD) plants or Long Day (LD) plants. There are also plants that have the autonomous pathway and do not require a specific photoperiod for the induction of flowering. Some of the latter do require a cold treatment: vernalization. As for vernalization (yarowization), this term was probably first used by Trofim Lysenko, the Soviet experimentalist who was active in putting the great Russian botanist, Nicolei Vavilov to jail, where he died of hunger and mistreatment. In the very same species, there may be variants that require LD for flowering while others require SD or are independent of light regimes. The same is true for vernalization in the very same species (e.g. wheat, *Triticum aestivum*). Some varieties require vernalization of their seeds or seedlings while others do not require cold treatment for induction of flowering.

As for the photoperiod pathway, between 2006 and 2009, there were numerous publications concerning various plants that support the model of floral transition, as schematically shown in Fig. 71.

The concept that a flowering signal that originates in the leaves is transferred in a long range and then has its impact on the SAM is rather old. This transfer was revealed by Knott (1934) who used the LD spinach. He found that exposure of a leaf to LDs caused flowering while exposure of the shoot tip to LDs did not cause flowering. That means there should be a "flowering hormone." Chailakhyan (1936) also claimed that such a hormone must exist and termed it *Florigen*. Anton Lang spent some time in the laboratory of Chailakhyan and found support to the Florigen system of floral induction (see Lang, 1965). The abovementioned work of Galun *et al.* (1964) also found in seedlings of the SD plant *Pharbitis nil* that cotyledons perceived the photo-induction and send a long-range signal to the shoot tip, causing floral bud initiation. However, the

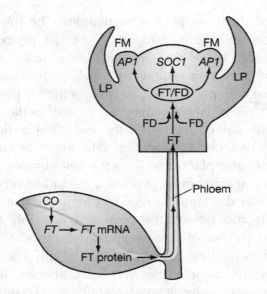

**Fig. 71.** FT protein as a transmissible signal for flowering. In long days, CO protein accumulates in the leaves and induces expression of *FT* in the phloem companion cells. FT protein is transported in the sieve tubes to the shoot apex, where it forms a heterodimer with FD. The FD/FT complex activates expression of *SOC1* and *AP1*, which leads to floral initiation. FM, floral meristem; LP, leaf primordium. (From Zeevaart, 2008.)

chemical structure of Florigen remained elusive for many years. Although several genes were found in the pathway from light perception to flowering, such as *CONSTANT (CO)*, *FT*, and *AP1*, the function of these genes, the location of their expression and especially whether the transported signal is the mRNA of *FT* or this encoded protein, were not clear until a few years ago. Below, important publications that led to the present understanding of photo-induced flowering are mentioned.

E. Lifschitz, J.P. Alvarez, Y. Eshed, and their associates (Lifschitz *et al.*, 2006a,b) worked with tomato, Arabidopsis, and tobacco. They reported that, contrary to previous reports, the mRNA *SFT* (*SFT* is the tomato ortholog of the *FT* of Arabidopsis) could not be found in the shoot tips of tomato. Moreover, the mRNA of *SFT* is not graft-transmissible. The graft-transmissible *SFT* complemented all developmental defects of the tomato mutant *sft* and it substituted for LD in Arabidopsis as well as for SD of "Maryland Mammoth" of tobacco (a SD cultivar). Furthermore, the SFT protein facilitated flowering and could move systematically in the plant.

Jaeger *et al.* (2006), and Jaeger and Wigge (2007), working with Arabidopsis found that the photoperiod pathway acts predominantly through the gene *CO* in the leaves, to activate the *FT* gene and cause the transcription of this gene, and this causes the production of the (systemic) FT protein that arrives at the shoot apex, where it interacts with the transcription factor protein FLOWERING LOCUS D (FD) that is a bZIP that is continuously maintained in the apical meristem.

Support for the claim that the FT protein acts as the long distance flowering signal (i.e. Florigen) came from a study by Lin *et al.* (2007) who studied cucurbits (*Cucurbita moschata* and *C. maxima*). *C. moschata* is a SD species. When *C. moschata* was grown in LD, but the expression of FT was induced in these plants (mediated with the zucchini yellow mosaic virus, ZYMV), it was found that the expression of *FT* under LD did cause flowering, although the location of this expression was far removed from the shoot apex. The authors concluded that the cause of floral induction was the FT protein, rather than the mRNA of *FT*. Also, grafting flowering *C. maxima* stocks with *C. moschata* caused flowering in LD maintained *C. moschata* scions. The phloem saps of *C. maxima* did not contain FT mRNA. The signal that crossed the graft connection was found to be the FT protein and not the *FT* mRNA.

Support for the scheme reported by Zeevaart (2008) also came from an independent study by Mathieu *et al.* (2007). The results of these investigators' study provided evidence for the following pathway. In the phloem, companion cells interact after photo-induction with CO and produce FT mRNA in the leaves. This mRNA produces the FT protein that is a globular protein of about 20 kDa. Such small proteins can cross plasmodesmata. After arriving at the SAM, the FT interacts with FD. Further, it was found with both Arabidopsis and rice, that the moving hormone is FT. This flow of FT was needed and sufficient to induce flowering.

Further support for the scheme in Fig. 71 came from a recent study with *Pharbitis* (now termed *Ipomoea nil*) by Hayama *et al.* (2007), published 43 years after the publication by myself and associates on the transfer of a photo-induced signal from *Pharbitis* cotyledons to the shoot tip. The former investigators found that orthologs of FT (PnFT1, PnFT2) promoted flowering. The endogenous levels of the FT orthologs were raised only when a continuous darkness of at least 11 hours was maintained. The authors suggested that the SD response in *Pharbitis* is controlled by a dedicated light-sensitive clock, set by dusk that activates Pn transcription in darkness, in a different mechanism for measuring day length than described for Arabidopsis and rice.

Evidence for the pathway in which FT moves a long way up to the SAM and there interacts with FD to produce AP1, causing transition to flowering in Arabidopsis, came from another study (Notaguchi *et al.*, 2008). Yang *et al.* (2007) reviewed the "drama" of first claiming that the FT mRNA was the florigen and then finding that the florigen is the FT protein.

A recent review on FT and the induction of flowering was written by George Coupland and associates from the Max Plank Institute of Cologne, Germany (Turck *et al.*, 2008). In this review, the main claims of the role of the FT protein were discussed. Also, Turck *et al.* (2008) integrated in their review the circadian expression of the key components in the photoperiod pathway, such as CO mRNA, CO protein, and FT mRNA. They also discussed the roles of specific photoreceptors, such as PhyA, PhyB, and CRY in the regulation of genes involved in the photo-induction of flowering. It should be noted that about 20 years ago, C. Somerville and associates found that GA is required for flowering in Arabidopsis plants that are maintained in SD (remember that Arabidopsis is considered to be an LD plant although different Arabidopsis lines react differently to day lengths).

A review of vernalization in wheat and barley, and its comparison with photoperiod-induced transition of flowering in Arabidopsis, was written by an Australian team of authors (Trevaskis *et al.*, 2007). Vernalization of these two cereals is regulated by three genes: *VRN1*, *VRN2*, and *VRN3*. Of these, *VRN1* is induced by vernalization and it accelerates the transition to reproductive development at the shoot apex. *VRN3* is an ortholog of *FT* and is induced by long days and its encoded protein, the Florigen, further accelerates reproductive development in the apex.

The *VRN2* is a floral repressor that integrates vernalization and day length responses by repressing FT (VRN3) until the plant is vernalized. The authors claimed that the vernalization response is controlled by different MADS-boxes (see Chapter 12 of Galun, 2007, a discussion on MADS-box genes), but the integration of vernalization and LD response occurs through similar mechanisms. The schematic description of the influence of seasonal cues on SAMs of cereals and the basis of floral induction in cereals and in Arabidopsis is provided in Fig. 72 (from the publication of Trevaskis *et al.*, 2007).

## The Architecture of Inflorescences

In recent years, numerous research teams published studies that were devoted to provide answers to the question of how the various forms of inflorescences are patterned. Among these is the team of R. Koes of Amsterdam (e.g. Angenent *et al.*, 2005; Koes, 2008; Souer *et al.*, 2008; Rebocho *et al.*, 2008), the teams of P. McSteen and S. Hake of California (e.g. McSteen *et al.*, 2007; Hake, 2008; Barazesh and McSteen, 2008), the team of E. Coen of the John Innes Centre in Norwich, UK (Prusinkiewicz *et al.*, 2007), and the teams of Y. Eshed and D. Zamir of Rehovot and E. Lifschitz of Haifa (e.g. Lippman *et al.*, 2008; Shalit *et al.*, 2009).

The investigators focused commonly on specific model plants, such as maize and rice of the Gramineae, petunia, tomato, and pepper of the Solanaceae, snapdragon of the Scrophulariaceae, and Arabidopsis of the Brassicaceae. It was claimed (e.g. Angenent *et al.*, 2005) that distinct plant (and inflorescence) architectures arose from each other or from a common ancestor, by evolution, therefore it is likely that many of the genes that dictate this architecture are conserved and that diversification resulted from alterations in a few of these genes. Recent studies on inflorescence were mainly conducted to reveal and characterize these genes and to investigate their origin.

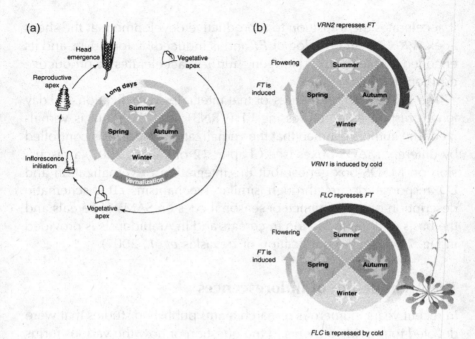

**Fig. 72.** (a) The influence of seasonal cues on shoot apex development in the temperate cereals. Varieties that require vernalization are sown in late summer or autumn. The shoot apex develops vegetatively until winter, when vernalization occurs. This promotes inflorescence initiation, which occurs as temperatures increase in spring. Long days in spring promote subsequent stages of reproductive apex development; head emergence occurs in late spring or early summer. (b) Molecular basis of vernalization-induced flowering in cereals vs Arabidopsis. In the temperate cereals (top), *VRN2* represses *FT* and blocks long-day promotion of flowering before winter. *VRN2* is not expressed in the short days of winter, when *VRN1* is induced by prolonged exposure to cold. After winter, *VRN1* expression remains high. This promotes inflorescence initiation and represses *VRN2*, to allow long-day induction of *FT* to accelerate reproductive development. When flowering occurs, *VRN1* expression is reset to establish the vernalization requirements in the next generation. In Arabidopsis (bottom), *FLC* is expressed before winter and represses *FT*. Vernalization represses *FLC*, and this allows long-day induction of *FT* (and *SOC1*) to promote flowering in spring. *FLC* expression is reset during meiosis to establish the vernalization requirement in the next generation. (From Trevaskis *et al.*, 2007.)

Angenent *et al.* (2005) reviewed the studies on meristems of shoots (including patterning of inflorescences) and advocated the choice of petunia in these studies because its body architecture is different from the architecture in other model plants such as maize, snapdragon, and Arabidopsis. Also, petunia lends itself well to molecular genetic studies. This review updated the processes in the inflorescence meristem (IM) of petunia that lead to its inflorescence in wild-type and mutants, up to the end of 2004. The authors listed several useful definitions, as you can see below.

*Cymose* is an inflorescence structure in which the IM is transformed into a floral meristem after which a secondary IM is formed on the flank of the apical dome, resulting in a zigzag structure. *Determinated inflorescence* is an inflorescence that is terminated by a flower; the IM is transformed into a flower and meristematic activity is lost. *Indeterminated inflorescence* is an inflorescence that is not terminated by a flower and the inflorescence meristem maintains its meristematic activity. *Raceme* is an inflorescence structure with an IM that remains indeterminate and produces floral meristems on its flanks.

Angenent *et al.* (2005) listed 11 genes in petunia that are involved in meristem function. Most of these have orthologs in Arabidopsis but only six of them have orthologs in snapdragon as listed below. This requires an explanation, because phylogenetically petunia is closer to snapdragon than to Arabidopsis.

The development of distinct inflorescence architecture, according to Angenent *et al.* (2005), as provided in Fig. 73, is that it schematically describes the development of an indeterminate raceme (in Arabidopsis), the single apical flower (in tulip), and a cymose inflorescence (in petunia). The apical meristem in this scheme may go from a vegetative meristem (VM) to an inflorescence meristem (IM), and from the latter, a part develops in the bract or leaf axils into a flower meristem (FM). This will result in a raceme as in Arabidopsis. In some plants (such as tulip), the VM is converted into an IM and then to a FM, resulting to a single apical flower. A third sequence can start with VM that is converted into an IM and from this meristem, a FM is formed that still retained a second IM at its flank.

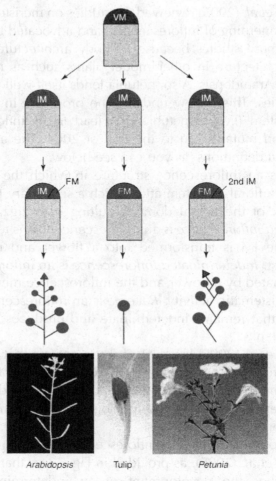

**Fig. 73.** Development of distinct inflorescence architecture. After germination, the shoot apical meristem (SAM) initially has a vegetative nature (VM; green; indicates the vegetative meristem) and, upon the switch to flowering, transforms into an inflorescence meristem (IM; blue). In racemes (left), this IM maintains its indeterminate character indefinitely and forms new floral meristems at the periphery, resulting in the formation of a main axis that is topped with the IM and flowers placed on the side, as exemplified by the Arabidopsis inflorescence (bottom left). In other species, the apical IM is determinate and undergoes a transition into a flower (middle). Since no meristematic cells remain, only a single flower is formed, as exemplified

**List of petunia genes involved in meristem function and their putative orthologs from Arabidopsis and snapdragon**

| Petunia gene name | Proposed function | Arabidopsis orthologs | Snapdragon orthologs |
|---|---|---|---|
| *TFR* | Stem cell maintenance | *WUS* | ? |
| *PhSTM* | Preventing meristem differ | *STM* | ? |
| *HAM* | Meristem maintenance | ? | ? |
| *NAM* | Initiating (auxillary) meristems | *CUC1,CUC2, CUC3* | ? |
| *DAD1* | Controls auxillary branching | *MAX4* | ? |
| *ALF* | Meristem identity | *LFY* | FLO |
| *PIE7* | Meristem identity | *AP1* | SQUA |
| *PMADS3* | Specification stamen, carpel | *AG* | PLE |
| *FBP11* | Specification ovule identity | *STK* | DEFH9 |
| *FBP2* | Specification floral organ identity | *SEP* | DEFH 72,84,200 |

From Angenent *et al.* (2005), where references were provided.

While there are numerous types of inflorescences, Koes (2008) described three of them: *Raceme, Panicle,* and *Cyme* (Fig. 74). In a more recent publication, the Koes team (Rebocho *et al.*, 2008) focused on the role of one gene, *EVERGREEN* (*EVG*), in patterning

**Fig. 73. (*Continued*)** by the tulip inflorescence (bottom center). Cymose inflorescences develop as solitary flowers, except that they can form a new (secondary) sympodial meristem that will again terminate with the formation of a flower (right). The reiteration of this program results in a zigzag structure carrying multiple terminal flowers, as exemplified by the *Petunia* inflorescence (bottom right). Notice that the more vigorous growth of the sympodial shoot tends to push the flower to a more lateral position, weakening the zigzag shape of the inflorescence. (From Angenent *et al.*, 2005).

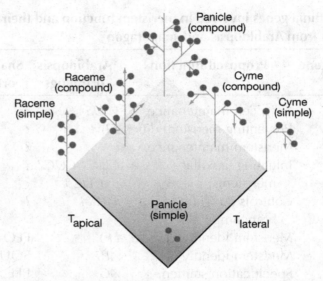

**Fig. 74.** Structure of inflorescences and position in morphospace. Flowers are indicated by red circles and meristems by green arrows. The inflorescence types are positioned in a 2D morphospace defined by the time required for apical ($T_{apical}$) and lateral ($T_{lateral}$) meristems to acquire floral fate. (From Koes, 2008.)

the cymose inflorescence of Petunia. Rebocho *et al.* (2008) noted several genes in Petunia that affect inflorescence architecture. There is the *ABERRANT LEAF AND FLOWER* (*ALF*) that is the homolog of *LEAFY* (*LFY*) of Arabidopsis and the *DOUBLE TOP* (*DOT*) that is the homolog of *UNUSUAL FLORAL ORGANS* (*UFO*). *DOT* encodes an F-box protein that interacts with ALF and is required for transcription activation of all known target genes of *ALF*. The time and place where flowers are formed is largely determined by the transcriptional activation of *DOT*. The proteins ALF and FLY, as well as DOT and UFO, are functionally similar proteins: they acquired widely divergent expression patterns that are probably a key factor in the evolution of inflorescence architectures. An important feature of the cymose inflorescence (e.g. of Petunia) concerns the initiation and development of the IM by which growth of the inflorescence continues after the apex

terminates by forming a flower. In Petunia, the gene *EVG* was identified and was found to be required for the development of the lateral
inflorescence shoot, the activation of *DOT*, and the specification of
floral identity in the apical FM. The authors therefore regarded *EVG*
as a key factor in the evolution of inflorescences. A model of explaining the phenotype of *evg* and various inflorescences is provided in
Fig. 75. We should recall that according to the suggestions of Souer
*et al.* (2008), the FM identity is regulated in Arabidopsis via the
transcription of *LEAFY* (*LFY*) that encodes a transcription factor that
promotes FM1. In Petunia the regulation is via the transcription of a
distinct gene, *DOUBLE TOP* (*DOT*), that is a homolog to *UNUSUAL
FLORAL ORGANS* (*UFO*) from Arabidopsis. Mutation of DOT or its
tomato homolog *ANANTHA* abolishes FM1. Ubiquitous expression
of *DOT* or *UFO* in Petunia causes very early flowering and transforms
the inflorescence into a solitary flower and leaves into petals. Ectopic
expression of *DOT* or *UFO*, together with FLY or its homolog *ABBER-
ANT LEAF AND FLOWER* (*ALF*) in petunia seedlings, activates genes
required for identity or outgrowth of organ primordia. DOT seems to
interact physically with ALF. Thus, according to Souer *et al.* (2008),
DOT and UFO have a wider role in the patterning of flowers than
that attributed to them previously. The authors claimed that the
different roles of LFY and UFO homologs in the spatio-temporal control
of flower identity in distinct plant species results from their divergent
expression patterns.

The team of Enrico Coen (e.g. Prusinkiewicz *et al.*, 2007) has
many years of experience with floral patterning of snapdragon
(*Antirrhinum majus*) that has a typical racemous inflorescence. They
wished to know the constraints on the biological diversity of inflorescences. They suggested that a single developmental model accounts
for the restricted range of inflorescence types that exist in nature. The
model predicts association between inflorescence architecture, climate,
and life history. The Coen team claimed that paths, or evolutionary
"wormholes," (an astrophysical term for a hypothetical space-time
tunnel connecting a "black hole" with another universe) link different
architectures in a multi-dimensional fitness space, but the rate of evolution along these paths is constrained by genetic and environmental

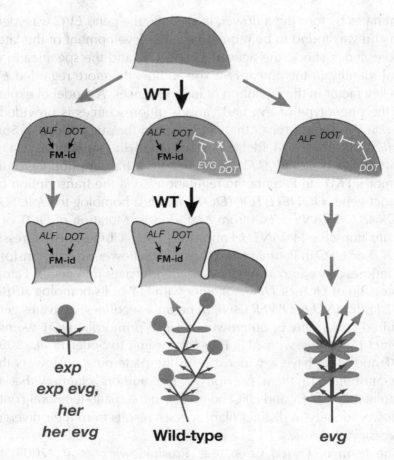

**Fig. 75.** Model explaining the phenotype of *evg* in distinct backgrounds. During development, FM identity is specified by *ALF* and *DOT*. Initiation and identity of the lateral IM is controlled by *EXP* and *HER*. The model assumes that an unknown mobile factor, designated "X," synthesized in the lateral IM inhibits *DOT* expression in the lateral IM and in the neighboring FM and thereby specification of floral identity. *EVG* disrupts the inhibitory effect of X on *DOT* expression in the FM, possibly indirectly by promoting proliferation of the lateral IM and separation from the apical FM. In *exp* and *her* mutants, development of the lateral IM is compromised and X is not made; hence, EVG is no longer required for *DOT* expression. (From Rebocho et al., 2008.)

factors, which explains why some evolutionary transitions are rare between closely related plant taxa.

A team composed of scientists from the Weizmann Institute of Science (Y. Eshed, J. Pekker, and J.A. Alvarez), the Faculty of Agriculture of the Hebrew University in Rehovot (D. Zamir, Z.B. Lippman), and the ARO of the Volcani Center in Bet Dagan (I. Paran, O. Chen, M. Abu-Abied) studied the formation of inflorescence in tomato and related members of the nightshade family (Lippman *et al.*, 2008). Tomato (*Solanum lycopersicum*) and other related species of the Solanaceae have various branchings in their inflorescence that determine flower number and distributions. Generally, there is a "sympodial" program in the "wild-type" tomato cultivars, in which the few-flowered inflorescence is formed: a branch terminates with a flower and a side meristem takes over to produce a short side shoot, which again produces a few phytomers that terminate with a flower. This process continues to produce a compound cyme (see Fig. 74). There are two mutants that change the regular inflorescence pattern. One is compound inflorescence (*s*) and the other is anantha (*an*). These have a highly branched phenotype. *S* and *AN* encode a homeobox transcription factor and an F-box protein, respectively, controlling inflorescence architecture by promoting successive stages in the progression of an IM to floral specification. In wild-type tomato, *S* and *AN* are sequentially expressed during this gradual phase transition, and the loss of either of these genes delays flower formation but resulting in additional branching. The inflorescences of such mutants can reach a phenotype with a cauliflower-like inflorescence. To describe the phase changes that lead to the regular cyme of tomato, let us start with a shoot apical meristem. The tip of this apex is converted into a FM but on its side, another meristem is formed (with S expression). The tip meristem loses the S expression and is converted into a flower bud. This process is repeated so that a few vegetative phytomers develop between two flowers. The reproductive phytomers may have more than one flower. In "wild-type" tomato, the expression of *S* and *AN* transforms the tomato "vine" into a highly branched inflorescence with hundreds of flowers. The *S* encodes an

F-box protein ortholog to a protein encoded by *UFO*. The *S* encodes a transcription factor that is related to a gene called WUSCHEL HOMEOBOX 9 (WOX9). The team found that *S* is a major determinant of inflorescence architecture in tomato. Interestingly, *S* and *AN* have little, or no effect, on plants without a racemose (indeterminate) inflorescence (such as Arabidopsis). However, the transient and sequential expression of *S* followed by *AN* promotes branch termination and flower formation in plants with a determinate shoot. In pepper, there is a single auxillary flower below the newly developing shoot apex. This architecture can be converted to a compound inflorescence (compound cyme) upon mutating its *AN* ortholog. Thus, the authors suggested a developmental mechanism whereby inflorescence elaboration can be controlled through temporal regulation of floral fate.

In a recent publication, Shalit *et al.* (2009) studied the *SINGLE FLOWER TRUSS* (*SFT*) that encodes the tomato precursor of *florigen* and *SELF-PRUNING* (*SP*) that encodes a potent SFT-dependent SFT inhibitor that were found to be prime targets of mobile florigen. The impact of florigen on organ-specific traits is imported by SAMs. Analyzing the graft-transmissible impacts of florigen in perennial tomato, the authors found that in addition to it being imported by SAMs, the florigen is also imported by organs in which SET is already expressed. By modulating local SFT/SP balances, in tomato, florigen confers differential flowering responses of primary and secondary apical meristems, regulates the reiterative growth and termination cycles typical of perennial plants, accelerates leaf maturation, and influences the complexity of compound leaves, the growth of stems, and the formation of abscission zones. The florigen is thus established as a plant protein that functions as a general growth hormone. Developmental interactions and a phylogenetic analysis suggest that the SFT/SP regulatory hierarchy is a recent-evolutionary innovation that is unique to angiosperms. Accepting the roles attributed by the authors to florigen as multi-purpose phytohormones, we should recall that these multiple effects of florigen are not unique to florigen. Most phytohormones have several impacts on patterning in angiosperms, such as the several roles of GA (in promotion of flowering and modifying growth) and auxin (in controlling numerous changes in growth and differentiation).

While many studies were performed in order to better understand the shaping of inflorescence of dicots, fewer investigators focused their attention on the shaping of inflorescence of monocots. However, two researchers, S. Hake of the USDA/ARS in Albany, CA, and her associates, as well as P. McSteen of Penn State University and her associates did devote research efforts to the shaping of inflorescences of grasses (Gramineae), especially of the inflorescence of maize and rice (e.g. McSteen and Hake, 2001; McSteen *et al.*, 2007; Barazesh and McSteen, 2008). It should be noted that already more than 50 years ago, J. Heslop-Harrison performed experiments on the impact of phytohormones on the sexuality of maize, and I, together with my associates, investigated, at about the same time, the impact of phytohormones on the sexuality of cucumber (see Frankel and Galun, 1977, for old references). However, these "old" studies were performed much before molecular methodologies became available. It is also noteworthy that McSteen, Hake and associates were dealing with the hormonal control of shaping grass inflorescences while there is only little information from recent studies on the impact of phytohormones on inflorescences of dicots, not taking into account the recently discovered florigen as a wide-spread phytohormone (see Shalit *et al.*, 2009). Schemes of the rice inflorescence, the male (tassel) inflorescence of maize and the inflorescence of Arabidopsis are shown in Fig. 76.

McSteen's and Hake's teams (McSteen *et al.*, 2007) cloned *bif2*, a mutant in maize, that is a co-ortholog of *PINOID* (*PID*), in Arabidopsis. *PID* regulates auxin transport in the latter plant. Further phenotypic analyses of *bif2* showed that it is expressed in all auxillary meristems and lateral primordia during inflorescence and vegetative development in maize and rice. The analyses of maize *bif2* mutants illustrated that this gene is involved in numerous impacts on vegetative development. The authors suggested that *bif2/PID* sequences and expressions are conserved between grasses and Arabidopsis, indicating the importance of *bif2/PID* sequences in the differentiation of angiosperms. The authors claimed that their findings provide support to the idea that

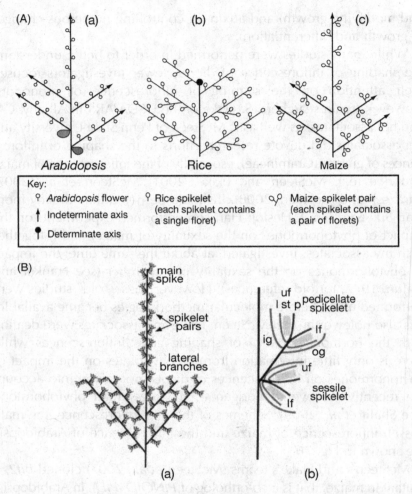

**Fig. 76.** (A) The structures of the Arabidopsis, rice, and maize inflorescences. (From Barazesh and McSteen, 2008). (B) Diagram of a normal tassel and spikelet pair. (a) Diagram of a normal tassel (male inflorescence). The tassel consists of a central main spike with long lateral branches at the base. Short branches called spikelet pairs cover the main spike and the lateral branches. (b) Diagram of a spikelet pair from a normal tassel. The pedicellate spikelet is borne on a pedicel while the sessile spikelet is attached at the base. Each spikelet contains two florets, the upper floret (uf) and the lower floret (lf) enclosed by two glumes, the inner glume (ig) and the outer glume (og). Each floret consists of lemma (l), palea (p), two lodicules (not shown), and three stamens (st). (From McSteen and Hake, 2001.)

*bif2* and *PID* are required for the initiation of both auxillary meristems and lateral primordia. Clearly, because *PID* regulates auxin transport, *bif2* may act through this transport to regulate maize inflorescences.

Barazesh and McSteen (2008) recently reviewed the recent progress in understanding the hormonal control of grasses (mainly in maize and rice). Much of this progress was due to the completion of the sequence of the rice genome and a draft sequence of maize. The uniqueness in the studies of McSteen and Hake was that these investigators not only dealt with the genetic regulation of maize and rice, but also investigated the impact of phytohormones on inflorescence development. Barazesh and McSteen claimed that while monocot and dicot inflorescences appear to have vastly different morphology, this divergence in architecture could be attributed to simple differences in the activity of auxillary meristems. The basic unit of grass inflorescence architecture is the spikelet that is a short branch bearing leaf-like structures, termed glumes, that enclose one or more florets (see Fig. 76). The highly branched inflorescence of maize (the male inflorescence) and the florescence of rice are produced from multiple types of auxillary meristems. The auxillary meristems produced from the apical inflorescence meristem form in the axils of bract leaf primordia. The latter are suppressed from further growth. Rice inflorescences have a central inflorescence stem that aborts after the production of several branches. Both primary and secondary branches bear spikelets singly and each spikelet contains a single floret (Fig. 76B). Maize (male) inflorescence (tassel) is more highly branched than the rice inflorescence. The female inflorescence of maize (the "ear") is unbranched and consists of an inflorescence stem bearing pairs of spikelets that produce pistillate florets. Interestingly, the early development of the ear follows a similar pathway as the tassel except that spikelet meristems (SMs) are not produced. Hence, selective organ abortions give rise to pistillate florets in the ear and staminate florets in the tassel. Decades of maize genetics established a huge number of genes, and recently available DNA sequences in maize have also shown that phytohormones are important regulators of inflorescence development. However, even before this accumulated information, it was known that auxins and CKs control the development of maize sexual expression (e.g. Heslop-Harrison, 1975, 1961). The auxin effects were related to auxin transport and the latter issue was studied in detail, in Arabidopsis, but the monocots have homologous transport systems as the *PIN* family of auxin efflux carriers and the serine/threonine protein kinase *PINOID* (*PID*) genes. The mechanism of inflorescence development in maize is more complicated than in Arabidopsis; some controlling elements of this development seem to be conserved and an *OsPID* was revealed in rice.

CKs are also involved in the structuring of rice. The gene *LONELY GUY* (*LOG*) affects an enzyme in the pathway of CK and *log* mutants fail to maintain meristematic cells in the SAM. LOG maintains a large apical meristem on the SAM as well as in the auxillary meristem of the inflorescence.

Clearly, GA is also involved in inflorescence architecture. Recently, three genes in the *ramosa* pathway were identified in maize. The *ramosa* pathway promotes the determining of the SPM (Spikelet Pair Meristem). *Ra1* encodes the "zinc finger" putative transcription factor and is expressed at the boundary of the SPM and the main axis, implying that it interacts with a mobile signal to impose determinacy on the SPM. It seems possible that this "mobile signal" is actually GA, because exogenous application of GA has been reported to suppress the *ra1* phenotype, but direct evidence is lacking. Further understanding of the molecular interaction of the ramosa genes with phytohormones will probably provide important progress on understanding the impact of hormones on the shaping of inflorescences. It is worthwhile to mention some old results of D. Atsmon and myself: when GA is applied to female cucumber plants, male (staminate) flowers develop on the treated female plants; this change in sex expression is not because the pistillate floral bud changes its sex but rather because GA suppressed the development of the pistillate, young floral buds. This will replace the female (pistillate) flowers with staminate flowers.

# CHAPTER 13
# The Leaf

Angiosperms, with the exception of a very few plants, such as the angiosperm parasites dodder (*Cuscuta*), witchweeds (*Striga*), and broomrapes (*Orobanche*), produce leaves that contain a similar basic structure. The leaves contain a peduncle to which a lamina is connected. The laminae are flattened and are organs with very efficient photosynthesis. Most of the photosynthesis of terrestrial angiosperms happens in the laminae. The laminae can attain very different shapes. The spectrum of laminar shapes of present-day angiosperms is relatively "new." In the late-Silurian era, the ancestors of angiosperms, termed "tracheophytes," of about 420 million years ago (MYA), had neither lateral branches nor leaves. These plants consisted of naked dichotonized branching stem systems. However, after about 40 MYA there was a drastic change so that from the close of the Devonian era, leaf laminae started to evolve. Presently, the laminae produce annually about $5 \times 10^{16}$ g of fixed carbon (C) from $CO_2$, water, and light energy. It should be noted that during the early evolution of leaves the levels of atmospheric $CO_2$ was rather high. Zimmermann (1952) proposed the "Telome Theory." He envisioned the transformation of a stem ("telome") into a leaf by modifications of existing organs rather than through a major change in the body plan. The rather old assumption stated in the PhD Thesis of Caspar Friedrich Wolff (1759), and a few years later phrased by J.W. von Goethe, says that all the above-soil plant organs consist of stems and leaves that have undergone metamorphogenesis. If we accept the Zimmermann hypothesis that basically suggests that leaves are derived from stems, then the above-soil organs of plants consist only of stems that have undergone metamorphosis. The principle of Zimmermann's Telome Theory was elaborated by several subsequent studies and was recently reviewed by Beerling and Fleming (2007). A summary of the latter review is presented below.

## The Telome Theory

This section describes the origin of the present-day megaphylls by critically evaluating Zimmermann's Telome Theory, which is mainly based on the review by Beerling and Fleming (2007). This review starts at the late-Silurian era when there existed the earliest tracheophytes that lacked side branches and leaves but consisted of dichotomized stems. During this era, the atmosphere of our globe was rich in $CO_2$. The weathering of Ca-Mg silicate rocks (between the early and the late-Devonian era) that stored $CO_2$ caused further increase of $CO_2$ to the atmosphere as well as causing a high temperature. The release of $CO_2$ to the atmosphere probably led to a rather high $CO_2$ level in the air. After this late-Silurian era and around a further 40 MYA later, there was probably a vast diversification of land plants and there was apparently an evolutionary selection of plants that withstood the changing (harsh) environmental conditions (Hao *et al.*, 2003). Larger megaphylls evolved from late-Devonian to early-Carboniferous eras, when atmosphere $CO_2$ dropped massively. In his review, Zimmermann (1952) formulated the Telome Theory, suggesting a three-step process for the conversion of a stem into a megaphyll leaf: *overtopping, planation,* and *webbing.*

*Overtopping*: this step was a change from a dichotomous pair of branches into one "dominant branch" and another "side branch." This overtopping occurred between 410 MYA and 390 MYA. The dominant stem (or telome) was indeterminate while the side branch was determinate. This overtopping is a recognized innovation within the early-Devonian era. It should be noted that branching in angiosperms requires lateral secondary meristems, coined "auxillary meristems." Hence the overtopped branch in angiosperm ancestors may have evolved similarly to present auxillary meristems of angiosperms. Once an auxillary meristem has been initiated, it may either grow out or remain dormant. The dormant situation was explained in the past as "Apical Dominance," meaning that the out-growth is suppressed by auxin that flows from the tip of the stem and CK flowing acropetally. As discussed in a previous chapter of this book, now the consensus process includes a role for a novel carotenoid-associated signaling system that has a role in inhibiting auxillary branch outgrowth. Nevertheless, the role of auxin flow is not eliminated in spite of the evidence that auxin does not enter the auxillary initial. The involvement of auxin flow in non-angiosperm species, with respect to branching has not yet been investigated. Note that there is a difference between the angiosperm leaf that is determinate and the angiosperm SAM that is indeterminate (at least during differentiation of some nodes). This difference in determi-nancy between the leaf and the SAM in angiosperms is controlled by genes encoded by some transcription factors (e.g. *KNOX* and *ARPs*). In most angiosperms, the *KNOX* genes are expressed in the SAM but the SAM lacks the expression of *ARPs*. On the other hand, leaves are characterized by the expression of *ARPs* and consequently do not express *KNOX*. However, the situation is more complex because the timing of *KNOX*-off and *KNOX*-on is crucial for the patterning of the angiosperm leaf, as will be mentioned below in this chapter. Here, I will mention the role of the KNOX/ARP module in the closely related plants of Arabidopsis and *Cardamine hirsuta* (Hay and Tsiantis, 2006; Barkoulas *et al.*, 2007). Arabidopsis has simple leaves while *Cardamine* has dissected leaves with individual leaflets. When KNOX-off is delayed this probably leads to the compound leaves of *Cardamine*. The latter situation of change in *KNOX* also affects auxin flow (by changing the auxin efflux transporter, PIN1, as revealed by Barkoulas *et al.*, 2007).

*Planation*: this term was used to describe the flatting of the three-dimensional terate stem segment into a single-plane organ. There is no experimental evidence that favors the idea that the conversion of the three-dimensional laminal leaf precursor into a two-dimensional structure (e.g. the process of planation) had selective advantages, but computer models suggested such advantages in spore dispersal and light interception. Unfortunately, there are no present examples that represent planated branching stems. Auxin was found to play a central role in the control of leaf initiation and its patterning. The auxin flux through plant tissue is directed by the pattern of expression of a series of auxin efflux (*PIN*) and influx (*AUX*) carriers. Local accumulation of auxin at discrete sites within the meristem dictates the position of leaf formation. This is probably also true for the initiation of leaflets during the formation of a compound leaf.

*Webbing*: the webbing leads to flat-photosynthesizing laminae. Such broad laminae were formed in some plants of the mid-Devonian and more frequently in the late-Devonian forest trees (progymnosperms). The flat-formed and large laminae were established worldwide in floras during the late Carboniferous. The webbing of telomes to produce a laminate leaf blade first requires the production of lateral outgrowth; secondly, either congenial or post-genial fusion of adjacent branches. The lateral outgrowth of angiosperm leaves, to form a lamina, is under the control of a complex transcriptional and signaling network that requires the juxtaposition of fields of cells that express either adaxial or abaxial identity genes. The interaction of these two domains leads to a gradual restriction of growth to a plane that is centered around the primary vascular element. It should be noted that the adaxial leaf domain in angiosperms is characterized by the expression of *ARP* genes, suggesting a close mechanistic interplay between determinate lateral organ formation, proximal-distal growth, and lamina formation.

In modern angiosperms, fusion events between adjacent primordia can be observed in certain genetic contexts. For example, the *CUP-SHAPED COTYLEDON* (*CUC*) genes of Arabidopsis, that encode transcription factors, prevent the fusion of cotyledons and mutations of the *CUPULIFORMIS* gene in *Antirrhinum* and lead to the dramatic formation of fused organs throughout the plant. These genes must be expressed during the formation of primordia to prevent fusion, and fusion occurs in their absence. In addition to these transcription factors, auxin has also been implicated in the control of lateral organ fusion. Hence the disruption of auxin flux often leads to fused organs, and it is possible to induce collar-type leaves by manipulating auxin patterns around the SAM. With respect to postgenital fusion, the cuticle composition of the cell wall is a potential target. Thus, manipulations that lead to the loss or disruption of cuticle lead to post-genial fusion events, as was the case with the angiosperm ancestors' webbing.

Palaeobotanical studies (see Beerling and Fleming, 2007) lead to estimates of the time scale during which these three major leaf transformations took place:

**Overtopping     Planation     Webbing**

410 MYA $\rightarrow$ 390 MYA $\rightarrow$ 380 MYA $\rightarrow$ 370 MYA

These estimates provide a time length of about 40 MYA from the dichotomeous branching of plants to the early formation of flat leaf laminae.

## Some General Remarks on Leaf Patterning

The patterning of leaves — as affected by various genes and phytohormones — was studied by several research teams. A considerable part of these studies was conducted by the Israeli teams of Y. Eshed, E. Lifschits, N. Ori, and D. Zamir (Hareven *et al.*, 1996; Ori *et al.*, 2000; Emery *et al.*, 2003; Eshed *et al.*, 2004; Yanai *et al.*, 2005; Pekker *et al.*, 2005; Lifschits *et al.*, 2006a,b; Ori *et al.*, 2007; Efroni *et al.*, 2008; Berger *et al.*, 2009). Teams of investigators from other countries, such as those headed by M. Tsiantis, S. Hao, A. Hay, A.J. Fleming, S. Jasinski, C. Ferrandiz, M.C.P. Timmermans, M. Barkoula, M.E. Byrne, N. Sinha, and R.A. Kerstetter also wrote very significant publications on leaf patterning as affected by genes and phytohormones (Tsiantis *et al.*, 1999; Hao *et al.*, 2003; Tsiantis and Hay, 2003; Piazza *et al.*, 2005; Fleming, 2005; Jasinski *et al.*, 2005; Hay and Tsiantis, 2006; Ballanza *et al.*, 2006; Chitwood *et al.*, 2007; Kinder *et al.*, 2007; Barkoulas *et al.*, 2007, 2008; Pinon *et al.*, 2008; Kimura *et al.*, 2008; Jasinski *et al.*, 2008; Wu *et al.*, 2008; Blein *et al.*, 2008). I shall briefly review some of these publications. Readers who are interested in additional details may receive them from the listed publications.

The team of E. Lifschits from the Technion in Haifa, in collaboration with Y. Eshed (Hareven *et al.*, 1996) who have been studying the differentiation of tomato leaves for many years, used the *KNOTTED-1* (*Kn1*) gene (of maize) and its orthologs (from other species) to study the differentiation of simple and compound laminae. They found that misexpression of *Kn1* conferred different phenotypes on simple and compound leaves. When tomato plants were manipulated to express *Kn1*, up to 2000 leaflets were formed on these tomato plants. However, the *Kn1* did not elicit the leaf ramification in plants with inherent simple leaves. The tomato *Kn1* ortholog, unlike the *Kn1* of Arabidopsis, is not expressed in the leaf primordia. The authors suggested that simple and compound leaves are the result of different patterns of meristematic activities. They also proposed that simple leaves are morphogenetically rigid, while compound leaves (such as those of W.T., tomato) are developmentally more flexible. By the use of mutations it was possible to obtain a wide range of leaf-types in tomato: from simple Lanceolate leaves to super-compound leaves with a vast number of primary and secondary leaflets. There seem to be inherent fundamental differences in the development of simple and compound leaves. Simple and compound leaves are probably determined by two different developmental programs, and the gene systems that condition them are conserved among species with simple and compound leaves, respectively.

Tsiantis *et al.* (1999) from the University of Oxford (UK) looked at the disruption of auxin transport and its association with aberrant leaf development in maize. The authors based their study on previous findings, that the maize mutant *rough sheat2* (*rs2*) has ectopic expression of *knox* genes, causing a range of phenotypes as aberrant vascular development, ligular displacement, and dwarfism. These investigators showed that *rs2* mutants display decreased polar auxin transport in the shoot and that germination of wild-type maize seedlings on agents known to inhibit polar auxin transport mimic aspects of the *rs2* mutant phenotype, but the phenotype which elaborated inhibitor-treated plants is not correlated with ectopic KNOX protein accumulation.

Naomi Ori (now in Rehovot, Israel) collaborated with the investigators Y. Eshed, G. Chuck, J.L. Bowman, and S. Hake (Ori *et al.*, 2000) in a research on the mechanisms that control *knox* gene expression in the Arabidopsis shoot. Misexpression of the *knox* genes *KNAT1* and *KNAT2* in Arabidopsis produces a variety of phenotypes, including lobed leaves and ectopic stipules and meristems in the sinus (the region between the lobes). These investigators wished to determine the mechanisms that control *knox* gene expression in the shoot by examining recessive mutants that share phenotypic characteristics with transgenic *35S::KNAT1* plants. It was found that double mutants of *serrate* (*se*) with either *asymmetric1* (*as1*) or *as2* showed lobed leaves, ectopic stipules in the sinuses, and defects in the timely elongation of sepals, petals, and stamens, similar to *35S::KNAT1* plants. Ectopic stipules, and in rare cases ectopic meristems, were detected in the sinuses on plants that were mutants for *pickle* and either *as1* or *as2*. *KNAT1* and *KNAT2* were misexpressed in the leaves and flowers of single *as1* and *as2* mutants and in the sinuses of the different double mutants, but not in the *se* or *pickle* single mutants. The authors claimed that these results suggest that AS1 and AS2 promote leaf differentiation through the repression of *knox* expression in leaves, and that SE and PKL globally restrict the competence to respond to genes that promote morphogenesis.

The Oxford team of Tsiantis and Hay (2003) reviewed the studies of comparative plant development that concern leaf differentiation, until the end of 2002. They summarized that a key problem in the developmental biology is understanding the origin of morphological innovations, and stated that comparative studies in plants that have different leaf morphologies indicate that the developmental pathway defined by KNOTTED1-type homeodomain proteins could be involved in generating different leaf forms. The differential expression of regulatory proteins emerged as an important factor in deriving morphological innovations in the plant kingdom — an idea that is supported by quantitative trait locus analyses.

A team of investigators (Emery *et al.*, 2003) — who at the time worked at J.L. Bowman's laboratory in the University of California, Davis — and its members who are now spread in three continents

(America, Asia, and Australia), focused on the "*Class III HD-ZIP*" genes: *PHAVOLUTA* (*PHV*), *PHABULOSA* (*PHB*), and *REVOLUTA* (*REV*), as well as on *KANADI*. They found that gain-of-function in *REV* mutants caused adaxialization of lateral organs (i.e. leaves). Such an effect was based only on sequence change in the mRNA of *rev-σmi*: encoding a CCA triplet (that codes for proline) rather than the wild-type triplet CCG (that also codes for proline). This "mute" mutation, that does not change the translated peptide, does change, apparently, the binding of a MIR 165/166 to the conserved binding sequence of *PHV*. When investigators find a case in which a mutant is producing the same protein as in the wild-type but has a "mute" exchange of one nucleotide in their respective mRNA that encodes a polypeptide — they look for a MIR that binds differentially to the mutated mRNA vs the wild-type mRNA. It should be noted that among the *Class III HD-ZIP* genes, *PHB* and *PHV* are most important for patterning in lateral organs, whereas *REV* is more important for vascular patterning (the xylem of the stem is adaxial while the phloem is abaxial). It was previously found that in Arabidopsis gain-of-function alleles *PHB* and *PHV* result in adaxialization of lateral organs, while loss-of-function alleles of *KANADI* also cause adaxialization. There are parallel effects of gain-of-function of *REV* alleles and loss-of-function of *KANADI* alleles, with respect to vascular patterning. Simultaneous loss-of-function of *PHB*, *PHV*, and *REV* has severe effects: it causes abaxialization of cotyledons, abolishes the formation of the primary apical meristem, and in severe cases eliminates bilateral symmetry. The investigators proposed that a common genetic program dependent upon miRNAs governs the adaxial–abaxial patterning of leaves (and of radial patterning of stems) in the angiosperm shoot. They also suggested that a common mechanism of patterning is shared between apical and vascular meristems. Y. Eshed, J.L. Bowman and associates (Eshed *et al.*, 2004) further studied the asymmetric blade expansion in Arabidopsis. As noted above, asymmetric development of lateral plant organs (e.g. leaves) is initiated by a partitioning of organ primordia into distinct domains along their adaxial/abaxial axis. The two main determinants of abaxial cell fate are members of the *KANADI* and *YABBY* gene families. Progressive loss of KANADI activity in loss-of-function mutants results in progressive transformation of abaxial cell types into adaxial ones, as well as to a correlated loss of laminae formation. The authors noted that already previous members of the *YABBY* gene family have been proposed to promote abaxial cell fates so that the activity of genes, such as *FIL*, *YAB3*, and *CRC*, caused the formation of abaxial tissue differentiation in adaxial positions of cotyledons and floral members. Conversely, loss-of-function alleles of *CRC*, when in combination with *kan1* mutations, result in adaxial tissues developing in abaxial positions in the gynoecium. During leaf development *FIL* and *YAB3* are expressed in abaxial regions of Arabidopsis leaves and their expression patterns parallel to that of the progress of leaf differentiation. The authors presented lines of evidence that *YABBY* gene activity is associated with laminae expansion and proposed that boundaries of *YABBY* gene expression marking the abaxial/adaxial boundary are intimately linked to the proposed communication between the abaxial and adaxial domains during leaf development. The authors also proposed that the initial asymmetric leaf development is regulated primarily by mutual antagonism between *KANADI* and *PHB-like* genes that is "translated" into polar *YABBY* expression. Thereafter, polar *YABBY* expression contributes both to abaxial cell fate and to abaxial/adaxial juxtaposition-mediated lamina expansion.

In the previous chapter (on the patterning of the shoot), the two sequentially published papers of Jasinski *et al.* (2005) from Oxford (UK) and of Yanai *et al.* (2005) from Rehovot were discussed at some length. Both papers demonstrated that the SAM proteins promote the synthesis of CK. CK facilitates cell division but CKs were also activating the enzyme GA2ox that degrades GA. These interactions may confine active GA to the abaxial region of the leaf. Both the abovementioned papers agree that KNOX1 proteins are central regulators of hormones (CK and GA) in the meristem and in the emerging leaves.

In an additional publication, from Rehovot and Davis, Pekker *et al.* (2005) analyzed the effects of *Auxin Response Factors* (*ARFs*) on leaf symmetry in Arabidopsis. As indicated above, the *KANADI* gene family in Arabidopsis regulates abaxial identity. The authors isolated *ARF3* and *ARF4* and found that in the double mutants, the ectopic *KANADI* is enhanced throughout the plant. The abaxial tissues were transformed into adaxial ones in all aerial parts, resembling mutations in *KANADI*. The experimental results led to a model for the role of auxin in mediating abaxial/adaxial partitioning of organ primordia. The model suggests that initiating organ primordia cells co-express abaxial and adaxial factors. The partitioning of lateral organs into abaxial, *KAN*-expressing domain, and the adaxial *PHB*-like-expressing domain, is gradual and evolves by mutual antagonism between the two types of factors and external morphogenic input. With the rapid expansion of the growing primordium, auxin concentrations form a slight gradient via asymmetric auxin influx carrier distribution, and due to the conversion from being a sink to a new source of auxin synthesis, this gradient of auxin is translated into differential action of specific subsets of ARFs, enabling KAN to override PHB-like activities at the abaxial domain. Subsequently, gradients of these ARFs help differentially translate auxin presence to maintain abaxial fate, leading to the abaxial/adaxial partitioning. We should recall that this partitioning of adjacent domains is required to pattern the flat leaf lamina.

The Oxford team (Piazza *et al.*, 2005) wrote an extensive review (a *Tansley Review*) on the evolution of leaf development mechanisms in which the relevant literature was considered (up to the end of 2004). The review traced these mechanisms from multicellular algae (*Chara*) up to angiosperms, with Arabidopsis and *Antirrhinum* as dicot model plants and maize as a monocot model plant. An overview of the involvements of AUX, CK, and GA, as well as numerous genes, in shaping the asymmetric leaf, was presented in Fig. 77. The tomato and the pea served as model plants for the formation of compound leaves. In pea there is a *UNI* gene that encodes a transcription factor (that is an ortholog of the *LFY* of Arabidopsis, and *FLO* of *Antirrhinum*). The *UNI* may be required to maintain the leaves' transient phase of indeterminacy (ability to generate novel structures, hence to lead to leaflet initiation). *UNI* and *STP* of pea act synergistically to promote leaflet formation. The interactions of *FLY* and *UFO* of Arabidopsis and

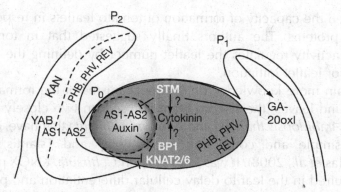

**Fig. 77.** Cartoons depicting some of the factors controlling SAM and leaf development in Arabidopsis. (From Piazza *et al.*, 2005.)

between *FLO* (of pea) and *FIM* (of *Antirrhinum*) also promote leaflet formation. Piazza *et al.* (2005) supplied a lot of genes that are involved in the formation of dissected (compound) leaves. An apparently important issue in this process is the longer maintenance of a transient indeterminacy in the developing leaf, but the overall mechanisms by which the leaf is dissected in an acropetal manner are still not clear. However, it is clear that *KNOX* genes play an important role in this dissection process. As Hareven *et al.* (1996) reported many years ago, the overexpression of KNOX in tomato leads to super-dissected leaves. On the other hand, GA appears to reduce the dissection process in tomato leaves. In pea there seems to be a *KNOX*-independent mechanism of leaf dissection. In a recent publication the Oxford team, Jasinksi *et al.* (2008), returned to the involvement of GA in the shaping of the dissected leaves of tomato. They noted the well-documented observation that KNOX proteins are positive regulators of leaflet formation, while GA can be a negative regulator of leaflet formation. The authors showed that the tomato mutant *pro* mimics the situation of GA application. This suggested that *PRO* causes the formation of leaflets in tomato. Support for this suggestion was provided by the finding that *PRO* encodes a DELLA-type growth repressor. *PRO* was found to be required for the early stages of growth of leaf primordia, and reduced GA biosynthesis increased the capacity of formation of tomato leaflets in response to KNOX proteins. The authors finally suggested that in tomatoes, DELLA activity regulates the leaflet number by defining the correct timing for leaflet initiation.

To gain more knowledge on the mechanisms of the formation of compound leaves, the Oxford team focused on two closely related plants: *Arabidopsis thaliana* and *Cardamine hirsuta* that have, respectively, simple and compound leaves (Hay and Tsiantis, 2006; Barkoulas *et al.*, 2008). It was found that in *C. hirsuta*, KNOX proteins are required in the leaf to delay cellular differentiation and produce the dissected (compound) form. In contrast, in *A. thaliana* the exclusion of KNOX from leaves resulted in a simple leaf form. The differences in KNOX expression arised through changes in the activity of upstream gene-regulatory sequences. The function of ARP proteins repressed *KNOX* expression conserved between the two species, but in *C. hirsuta* the ARP-KNOX regulatory module controlled new developmental processes in the leaf. Thus, according to Hay and Tsiantis (2006), evolutionary tinkering with KNOX regulation constrained by ARP function, may have produced diverse leaf forms by modulating growth and differentiation patterns in developing leaf primordia. Barkoulas *et al.* (2008) reported that lateral leaflet formation in *C. hirsuta* requires the establishment of growth foci that are formed after leaf initiation. These growth foci are recruited at the leaf margin in response to active maxima of auxin. As noted above, KNOX proteins promote leaflet initiation in *C. hirsuta*. The authors found that this action of KNOX proteins is contingent on the ability to organize auxin maxima via the PIN1 auxin efflux transporter.

Another approach to understand the mechanism that leads to compound leaf formation in tomato was taken by the Rehovot team (Ori *et al.*, 2007) of 14 investigators that included Y. Eshed, J.P. Alvarez, N. Ori, and D. Zamir. These authors focused on *LANCE-OLATE* (*LA*) and *miR319*. In the partially dominant mutation of *LA*, the large compound leaf of a tomato is converted into a small simple leaf. The investigators found that in the W.T. of tomato (*LA*) there is a binding sequence to *miR319*. Thus, the expression of this gene is reduced and W.T. compound leaves are formed. However, in mutants (such as *La*, *La2-4*, and *La5*) there are single changes in the nucleotide sequence probably preventing the binding of mRNA to the *miR319*. The *La* isolates thus have point mutations that reduce the sensitivity of the respective mutations to *miR319*. The LA is elevated in the very young leaf primordia causing the precocious differentiation of leaf margins. Using ectopic expression of *miR319* resulted in larger leaflets and continuous growth of leaf margins. The results implied that the regulation of LA amounts by *miR319* defines a flexible window of morphogenetic competence along the developing leaf margin that is required for leaf pattern elaborations. A Rehovot team (Efroni *et al.*, 2008) came up with a unique idea regarding the shaping of plant organs that they termed "digital differentiation index" (DDI). DDI was based on a set of selected markers with informative expression during leaf ontogeny. The leaf-based DDI reliably predicted the developmental state of leaf samples from diverse sources. The details of this interesting approach are beyond the scope of this book.

Another contribution of the Rehovot team to understand the shape of compound tomato leaves was provided by N. Ori, Y. Eshed, J.P. Alvarez and associates (Berger *et al.*, 2009). In tomato, the prolonged leaf patterning enables the elaboration of compound leaves by reiterative initiation of leaflet with lobed margin. In the *globlet* (*gob*), loss-of-function mutants primary leaflets are often fused, secondary leaflets and marginal serrations are absent, and SAMs often terminate precociously. The investigators found that *GOB* encodes a NAC-domain transcription factor expressed in narrow stripes at the leaf margins flanking the distal side of the future leaflet primordia and at the boundaries between the SAM and leaf primordia. Leaf-specific over-expression of *miR164* (a negative regulator of *GOB*-like genes) also leads to loss of secondary leaflet initiation and to smooth leaflet margins. Plants that have a dominant *gob* allele with an intact ORF but disrupted miR164 binding site produce more cotyledons and floral organs, have split SAMs and surprisingly simpler leaves. The overexpression of a form of *GOB* with an altered *miR164* binding site in leaf primordia leads to delayed leaflet maturation, frequent improperly timed and spaced initiation events, and a simple mature leaflet form owing to secondary leaflet fusion. The *miR164* also affects leaflet separation in *C. hirsuta*, a species that has complex (compound) leaves but is a close relative of the simple-leaved Arabidopsis. The investigators suggested that GOB marks leaflet boundaries and that its accurate spatial temporal and quantitative activity affects leaf elaboration. The results of Berger *et al.* (2009) hinted that the boundary specification of GOB interacts with additional factors in the context of compound leaf patterning. Future studies may identify additional factors.

The Oxford team, in collaboration with investigators from Versailles, continued the study on the shaping of compound leaves (Blein *et al.*, 2008). They noted that the different leaf shapes are the result of alterations of the leaf margins. For example, deep dissection leads to leaflet formation and less pronounced incision results in the serration of lobes. They selected five species that have compound leaves: *Aquilegia caerulea, Solanum lycopersicum, Solanum toberosum, Cardamine hirsuta,* and *Pisum sativum.* They found that there are two pathways that promote the formation of compound leaves in seed plants. One of these pathways involves expression in the primordia of compound leaves of class homeodomain KNOXI transcription factors that were initially identified for their role in maintenance of meristem activity. This pathway is active in a wide range of angiosperms that include *S. lycopersicum* and *C. hirsuta.* A second pathway involving the *UNI* gene was revealed in *P. sativum*, which does not express *KNOXI* genes in the leaf primordium. *UNI* encodes a member of LFY family transcription factors. The latter were initially identified for their role in floral meristem identity. The authors also noted that the generation of activity maxima of auxin is a mechanism that facilitates initiation and separation of both leaves at the SAM and leaflets from the rachis. Other key regulators of organ initiation and delimitation are *NAM/CUC3* which are members of a large evolutionary conserved family of plant transcription factors that are subdivided into *NAM* (*NO APICAL MERISTEM*) and *CUC3* (*CUP-SHAPED COTYLEDON3*) clades. They are expressed in the boundary of organ primordia, where they repress growth to allow organ separation. In addition, they are involved in meristem establishment via their activation of KNOX1 expression. The authors isolated 11 *NAM/CUC3* genes from five different plants (mentioned above) and searched the sites of their expression. These genes were grouped into either *NAM* or *CUC3* clades. The *NAM/CUC3* genes had a typical expression in the boundary domain at the base of organ primordia. The investigators examined the expression of these genes during leaf development. The expression was in a narrow strip of cells at the distal boundary of leaflet primordia, while no expression was observed in the proximal region. The conserved pattern of expression of *NAM/CUC3* suggested to the authors that this expression reflects a fundamental mechanism of leaflet formation.

A quite different approach to study leaf patterning was taken by investigators from the John Innes Centre in Norwich (UK) and the Cold Spring Harbor Laboratory in New York (Pinon *et al.*, 2008). These investigators used mutants of ribosomal proteins, termed PIGGYBACK1 (PGY1), PGY2, and PGY3 as well as mutants in the MYB domain transcription factor *ASYMMETRIC LEAVES1* (*AS1*), and *HD-ZIPIII* genes (*PHB, PHV, REV*) in various combinations. They revealed interesting leaf phenotypes. The *pgy1*, *pgy2*, and *pgy3* mutants encoded ribosomal proteins that were altered in the large subunits of the cytoplasmic ribosomes L10a, L9, and L5, respectively. The authors proposed several models to explain the causal correlation between the genotypes of the plants they constructed and the leaf phenotypes. However, the exact role of the altered ribosomal proteins and the specific leaf phenotypes awaits further clarification; the investigators claimed that it is timely to address the role of ribosomes as regulators of development.

A relatively recent publication (Wu *et al.*, 2008) handled the interaction between the *ASYMMETRIC LEAVES2* (*AS2*) gene, the *KAN1* gene, and the adaxial/abaxial polarity. The investigators of this publication showed that the adaxial regulator *AS2* is a direct target of the abaxial regulator *KAN1*, and that KAN1 represses the transcription of *AS2* in abaxial cells. Mutation of a single nucleotide in a KAN1 binding site of the *AS2* promoter causes *AS2* to be ectopically expressed in abaxial cells resulting in a dominant adaxialized phenotype. Also, the abaxial expression of KAN1 was found to be mediated (directly or indirectly) by *AS2*. The authors concluded that KAN1 acts as a transcriptional repressor and mutual repressive interactions between *KAN1* and *AS2* of Arabidopsis contribute to the establishment of adaxial/abaxial polarity of plant laminae.

## The Formation of Trichomes in Leaves

Trichomes are single-celled epidermal hairs that are formed from epidermal cells in various plant organs, such as leaves and stems, but are commonly absent in roots, hypocotyls, cotyledons, petals, stamens, and carpels of the model plant Arabidopsis. Because leaf trichomes of Arabidopsis have several sharp-pointed lobes, it is believed that trichomes constitute a defense organ against insect herbivores. Indeed, many plant species respond to insect damage by increasing the number of trichomes. Trichomes constitute favorable organs in genetic studies because they are not essential for reproduction, and mutants that were defective in their trichomes can easily grow to sexual maturity.

A comprehensive discussion on the morphology and the genetics of trichome development was provided in Chapter 12 of Galun (2007), but the role of phytohormones in trichome formation was not reported in the latter chapter. Below, I shall report on two studies that handled the roles of phytohormones in the formation of trichomes.

Traw and Bergelson (2003) were probably the first to report in detail on the roles of phytohormones (e.g. JA, salicylic acid, and GA) on the formation of leaf trichomes in Arabidopsis. One previous publication though, did report that GA promotes the *GLABROUS1* gene of Arabidopsis. The investigators Traw and Bergelson (2003) found that artificial physical damage and JA (jasmonic acid) caused significant increase in trichome production in leaves. Also, the *jar1-1* mutant exhibited normal trichome induction following treatment with JA, suggesting that adenylation of JA is not necessary. Salicylic acid had a negative effect on trichome production and consistently reduced the effect of JA, suggesting negative cross-talk between the JA and salicylate-dependent defense pathways. Interestingly, neither JA nor salicylic acid affected the number of epidermal cells in Arabidopsis leaves. The interaction between GA and salicylic acid was interesting. In the absence of applied salicylic acid, the application of GA caused a 72% increase of trichomes, but when salicylic acid was applied, then GA caused only about 29% increase of trichomes. The change in the trichome number was scored in the newly formed leaves. Could it be that JA up-regulates trichome formation, as hinted by some observations in the corollas of *Petunia*? In any case, the study by Traw and Bergelson (2003) suggested an important link between the GA and JA pathways in Arabidopsis.

There are DELLA repressors of growth: *GA-INSENSITIVE1* (*GA1*) and *REPRESSOR OF ga1-3* (*RGA*). Gan *et al.* (2007) have probed the relative roles played by RGA, GA1, and two homologs — *RGA-LIKE1* (*RGA1*) and *RGA2* — in the GA-induced phase changes. These investigators found that the DELLA acted collectively to regulate trichome initiation in all aerial organs, and that the onset of their activity is accompanied by the repression of most genes known to regulate trichome production. These effects are consistent with the results of genetic analysis, which conclusively place these genes downstream of the DELLA. The authors found that repression of trichome regulatory genes is rapid, but involves an indirect, rather than a direct, molecular mechanism, which requires *de novo* protein synthesis. DELLA activity also influenced post-initiation events, and the authors claimed that GA1 is a major repressor of trichome branching, a role in which it is antagonized by RGL1 and RGL2. The investigators also suggested that, in contrast to most other effects, the repression by GA application of flower trichome initiation is not dependent on RGA, GA1, RGA1, or RGL2. In summary, Gan *et al.* (2007) believe that DELLA proteins are central to trichome development, as schematized in Fig. 78, and that their effect can be largely explained by their transcription influence on trichome initiation activators.

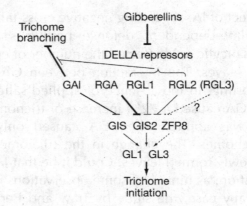

**Fig. 78.** Model of DELLA-mediated control of trichome initiation and branching. Dotted arrows indicate relationships that have not been fully characterized. Thicker lines indicate a more predominant effect in the pathway. GIS, GLABROUS INFLORESCENCE STEM; GL, GLABROUS; ZFP, ZINK FINGER PROTEIN. (From Gan *et al.*, 2007.)

## Tendrils

Several climbing plants use tendrils to climb on their neighbors. Tendrils have already been examined by Charles Darwin (1809–1882) in his book (Darwin, 1875) devoted to climbing plants, as well as by Jaffe and Galston (1968) in a review devoted to tendrils. Some of the tendrils are searching for support by circular movements, and when the support is sensed, the tendril will grasp the support and entwine it. Plant tendrils may be derived from a variety of plant organs, such as leaf components and stems. In grapevine, the tendrils are considered to derive from inhibited inflorescences. In cucumber, the tendril is an auxillary entity. The latter tendrils are strongly elongated by GA treatment, and I used cucumber tendrils, many years ago, to develop a bioassay for GA levels (Galun, 1959d). In a recent publication, Hofer *et al.* (2009) studied the genetics of tendril formation in peas. Peas are legumes. The latter are commonly divided into three subfamilies: Caesalpinioideae, Mimosoideae, and Papilionoideae. These three subfamilies together comprise over 19,000 species; hence they are one of the largest flowering plant families. Interestingly, it appears that tendrils evolved several times in legume species. In Caesalpinioideae (e.g. in the genus *Bauhinia*), tendrils arise from the base of the leaf, while in Mimosoideae and Papilionoideae, the tendrils are derived from the distal region of the leaves. The term "pea leaves" used by Hofer *et al.* (2009) means plants of two different kinds of "pea": (1) garden pea (*Pisum sativum*) and (2) sweet pea (*Lathyrus odorantus*). In botanical terms *Lathyrus odorantus* is neither a pea nor is it sweet, although the flowers of *L. odorantus* do have a "sweet" scent. The garden pea (W.T.) has a compound leaf with (commonly) two pairs of leaflets that evolve acropetally. Distal to the leaflets the tendrils are differentiated. A key regulator of the compound leaves of legumes is the meristem identity gene *Unifoliata* (*Unl*), the ortholog of *LEAFY* (*LFY*) of Arabidopsis. Mutants of garden pea and sweet pea that lacked tendrils were termed *tl* (*tendril less*). These mutants are semidominant. The *tl* (homozygous) converts tendrils to leaflets. Hofer *et al.* (2009) used a systematic marker screen of fast neutron-generated *tl* deletion mutants to identify the *Tl* as a class homeodomain leucine zipper (HD ZIP) transcription factor. Hence, the authors confirmed the tendril-less phenotype as a loss-of-function by targeting induced local lesions in genomes of garden pea and by analyzing the tendril less phenotype of the *t* mutant in sweet pea. The conversion of tendrils into leaflets in both mutants (*tl* and *t*) demonstrates that the pea tendril is a modified leaflet, inhibited from completing laminar development by Tl. According to the authors, there is evidence that shows that laminae inhibition requires *Unifoliata/LEAFY*-mediated *Tl* expression in organs emerging in the distal region of the leaf primordium. Phylogenetic analyses showed that Tl is an unusual class I HDZIP protein and that tendrils evolved either once or twice in Papilionoid legumes. The authors also suggested that tendrils arose in the Fabeae clade of Papilionoid legumes through acquisition of the *Tl* gene.

## Venation and Polar Auxin Transport

Venation in angiosperm leaves leads to a transport system composed of vascular bundles made of phloem and xylem tissues, which are the channels for the translocation of photoassimilates and minerals (and water), respectively. Studies since many years ago (see Esau, 1977; Foster and Gifford, 1959) showed that all vascular cells differentiate from a vascular meristematic tissue, the procambrium. The procambrial leaf cells become apparent as narrow, cytoplasmic dense cells emerging from the subepidermal tissue of the leaf primordium, meaning that the initial procambrial cells are formed very early in the development of the young leaf, and they continue to be formed with the maturation of the leaf. There is a considerable variability in the pattern of venation (and a great difference between the venation pattern in dicots and monocots). However, the functionality of the venation pattern in all angiosperms is very similar: each leaf cell is only a few cells away from a vascular bundle.

In Chapter 11 of Galun (2007), I provided a detailed discussion of vein differentiation in leaves. This discussion included information on the role of auxin in patterning venation. Not much relevant information was added since this Chapter 11 was phrased (end of 2006), but the detailed analysis by Scheres and Xu (2006) of the publication of Scarpella *et al.* (2006) is an important contribution for understanding the role of auxin in vein patterning in leaves. The article ("Perspective") by Scheres and Xu will therefore be narrated in the following text. The authors first stated that for a considerable time biologists and mathematicians were trying to find regularities in the various patterns of leaf venations. The vein organization is actually fulfilling — in all leaves — the same principal task of coming close to all leaf cells in order to supply the transport into-and-out of these cells, the products that are required to render the leaf cells efficient photosynthetic organs. However, there could be different means to achieve the optimal venation. Genetic input into patterning systems can be responsible for predictable gross changes in venation networks, although it appears that considerable variation in the venation pattern exists even among leaves of the very same plant. The latter fact suggests that there must be stochastic elements in the details of the venation patterning. The publication of Scarpella *et al.* (2006) provided insights into how genetic and stochastic vein patterning processes might be related. The latter publication also connects the initiation of the venation patterns to models of other auxin-dependent patterning processes in other plant organs, such as roots and shoots.

Observations at the cellular level indicated that vein progenitor cells, termed "pre-procambrial" cells, become incorporated into veins by selection from equivalent subepidermal leaf cells. During this process, pre-precambrial cells divide and elongate along a common axis which is essential for the formation of a continuous vein network that can carry out its transport role. Therefore, any model to explain venation network formation will first have to explain how narrow rows of precisely connected cells can be specified from a uniform field of cells. Then it will also have to explain how different regularities can be programmed to give rise to the characteristic venation classes. In addition, it will have to deal with observed variability in the final patterns.

In the past, two major models have been proposed in order to start explaining vein patterning; both models incorporate self-organizing properties that can give rise to the variable features of venation networks. Model 1 elaborates on Turing's reaction-diffusion principle (Turing, 1952, in his publication on differentiation and morphogenesis Hydra). Turing headed the successful Enigma Project in England during World War II to decipher German coded military messages. The basic idea in Turing's reaction diffusion principle was that auto-catalytic production of an activator triggers a faster diffusing inhibitor that keeps a new activator at a distance. Depending on parameters, this system can form connected or disconnected net-like prepatterns, as well as freely ending networks, if the activator in addition removes its own substrate.

Model 2, the "canalization hypothesis," specifically proposes the phytohormone auxin as a patterning agent. This second model is rooted in the observations on the vein-inducing capacity of auxin, and proposes that the positive feedback between auxin flow through a cell, the capacity of this cell to transport auxin leads to preferred conduits that will differentiate into vascular tissue. This model is traced back to the many studies of the late T. Sachs from the Hebrew University in Jerusalem (Sachs, 1981, 1991).

In recent years, the debate on the correct model for leaf venation turned to the model plant Arabidopsis, in which it was relatively easy to isolate and characterize venation mutants. The question then was whether or not such mutations could help to decide which one of the two venation models is the correct one. The critical distinction of the two models is that discontinuous ("patchy") initial specification of venation networks is allowed by the reaction-diffusion model but not by the canalization model. Therefore, observations of discontinuous and unconnected regions of vein specification, the so-called "vascular islands," were thus proposed to refute the canalization model in favor of the reaction-diffusion mechanism, for vein precursor specification. However, discontinuities in the *initiation* of vein pattern can only be defined by the earliest markers available. Scarpella *et al.* (2006) supplied such early markers. These investigators used a functional fusion of GFP with one of Arabidopsis pin-formed (PIN) transmembrane polar auxin transport proteins, PIN1. The PIN1 protein marks incipient vein cells earlier than any other pre-procambrial marker described as well as marked polarity. Thus PIN1:GFP monitoring can provide a direct observation of the basic mechanism proposed to act in the canalization hypothesis. The tracing by PIN1:GFP provided a strong case for canalization mechanisms operating during vein development. The observations based on the fusion protein led to the following claims:

— The first expression of PIN1:GFP (PIN1) is *epidermal* with a convergence point containing cells with opposite PIN1 polarity at the leaf primordium tip. This position foreshadows the formation of the mid vein.
— The formation of branches of the mid vein is foreshadowed by new epidermal convergence points at the leaf margin of the growing leaf primordium, again defined by cells with opposing PIN1 localization.
— New vein extensions start out as free ends (toward the existing veins) and connect to previously formed veins with PIN1 polarity directed toward them.

— PIN1 localization identified closed veins as a composite expression domain with opposite polarity, connected by a single internal bipolar cell. Such expression domains start out as free-ending veins.
— PIN1 expression domains start out as wide domains that subsequently narrow down.

These observations suggested a "leaf margin-guided" venation model composed of two separate processes: (1) specification of a PIN1 convergence domain and auxin accumulation at the convergence point lead to elevation of PIN1 expression and polarization correlated with the gradual selection of a narrow strand of pre-procambrial cells; (2) pre-existing veins polarize surrounding cells, that lead to "backward" recruitment of pre-procambrial cells that meet and connect to form a closed network. The second principle is used over and over again to specify the many higher-order veins in the network, and bipolar cells are the likely meeting points of these growing vein cells. The last added veins can remain "open" because the process is halted at the time when mesophyll cells, the alternative choice for internal leaf precursor cells, start to differentiate. There were also earlier observations that implicated auxin accumulation and the regulation of this process by polar auxin transport in vein formation (Mattsson *et al.*, 1999, 2003). However, the study of Scarpella *et al.* (2006) had the advantage of tracing the flow of auxin by PIN1 that enabled the tracing to a very early stage of vein differentiation. Also, auxin application, in lanolin paste, by the latter authors, led to more epidermal PIN1 convergence points, while it did not interfere with self-restriction of the size of internally projected procambrial PIN1 expression domains. The polar auxin transport inhibitor NPA, on the other hand, provoked more convergence points but also delayed self-restriction.

The strongest argument against the canalization hypothesis of vein formation was, until the Scarpella *et al.* (2006) publication, that there are "vascular island" mutants in which the veins are patchy. The latter authors thus focused on an extreme mutant, *van3*, where the venation is very patchy. The authors went on and studied pre-procambrial cell selection as monitored by PIN1 accumulation and polarity in *van3* plants. In support of the canalization-related model early leaf vascular development, as marked by PIN1, was found to be initially continuous in the *van3* background. This continuity "breaks up" when later markers are examined. "Islands" therefore appear to result from inappropriate stabilization of cell fate that isolates previously connected pre-procambrial cells. It seems plausible that vascular island formation results from inappropriate regulation of auxin sensing or polar auxin transport. Taken together, the data provided by Scarpella *et al.* (2006) are consistent with the canalization hypothesis. The data also add an extra factor: initiation of branching can be controlled at the level of the specification of PIN1 convergence points. While the canalization concept is reinforced as a stochastic mechanism that specifies vein tracks in a field of growing cells, the question of the different classes of venation patterns in nature can now be related to genetic (unknown) factors that contribute to the initiation of PIN1 convergence points and the associated auxin response maxima that suggest auxin accumulation at these points. In other words, there is progress in understanding how venation is induced in plant leaves, but we are still far away from understanding the genetics of specific venation patterns (e.g. what are the genes that cause different vein patterns in oak leaves vs the ones in maple leaves?). However, we can agree that variation in venation patterns, in different species, can now be viewed as a different balance between self-organizing and genetically determined positioning of epidermal PIN1 convergence points. Stochasm is involved in leaf venation but it is limited by genetic factors.

A relatively recent publication (Fukuda *et al.*, 2007) reports on a study of vascular development. This development was found to be involved in a plethora of phytohormones. It was not performed with leaves but rather with isolated *Zinnia elegans* cells, cultured *in vitro*, that have a xylogenesis capability. The many phytohormones involved in this xylogenesis justify the presentation of a summary of this research. Recent previous studies, some of them reported earlier in this book, showed that *CLAVATA3/ENDOSPERM SURROUNDING*

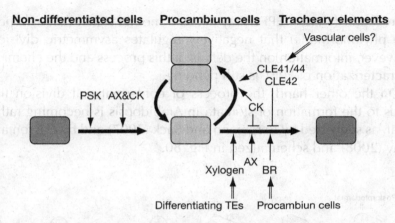

**Fig. 79.** An illustration showing cell-cell communication through plant hormones and peptides during xylem development. AX, auxin; BR, brassinosteroids; CK, cytokinin; PSK, phytosulfokine. (From Fukuda *et al.*, 2007.)

*REGION* (*CLE*) genes that encode peptide ligands, function in various developmental processes (as keeping in bay the amount of stem cells in the SAM). The authors of the aforementioned publication characterized TDIF, a dodeca-CLE peptide that is suppressing tracheary element differentiation. This indicated that there is a regulation of vascular organization by cell-cell communication through a CLE peptide. It was also found that extra-cellular peptides, such as phytosulfokine (PSK) and other phytohormones (such as AUX, CK, and BR), participate in the formation of vascular tissue. This suggestion is illustrated in Fig. 79.

## The Formation of Stomata

During the past few years from the end of 2006, only little additional information was accumulated on stomata formation that was not yet covered in Chapter 11 of Galun (2007). One recent review (Nadeau, 2009) mentioned an intercellular-moving peptide that may participate in the spacing between stomata: *EPIDERMAL PATTERNING FACTOR1* (*EPF1*). The encoded peptide is a candidate for a ligand for receptor kinases. EPF1 may act as a mobile signal (i.e. as a peptide phytohormone) that negatively regulates asymmetric division. However, information on the details of this process and the chemical characterization of EPF1 is still pending.

On the other hand, the process of epidermal cell division that leads to the formation of stomata in Arabidopsis is becoming rather clear, as suggested by Bergmann and Sack (2007) and by Casson and Gray (2008) and schematized in Fig. 80.

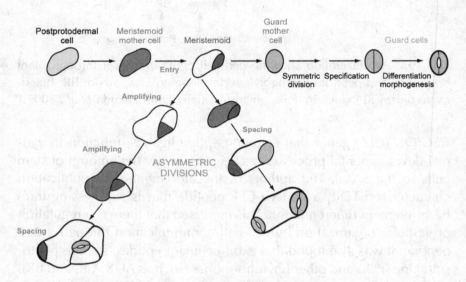

**Fig. 80.** Diagram of key stages and divisions in *Arabidopsis thaliana* stomatal development. Protodermal cells in the epidermis are converted into meristemoid mother cells (MMCs) through an unknown process. MMCs undergo an asymmetric entry division to create a meristemoid. Meristemoids may undergo additional asymmetric amplifying divisions, or convert into a guard mother cell (GMC). The GMC will divide a single time, symmetrically, to form the two guard cells. Later, morphogenesis and pore formation create the mature stoma. The division process is reiterative. Cells next to meristemoids, GMCs, and guard cells can become MMCs and undergo spacing divisions to create new meristemoids. The plan of this division is oriented so that the new meristemoid is placed away from the pre-existing stoma or precursor cell. (From Bergmann and Sack, 2007.)

Frederick Meins Jr's team and his associates at the Friedrich Miescher Institute in Basel, Switzerland, studied the involvement of a recently isolated microRNA, miR824, on the distribution of stomata in Arabidopsis leaves. The target of miR824 is a member of the MADS box protein family. MiR824 binds to a specific region of *AGAMOUS-like16* (*AGL16*) mRNA. Hence *AGL16* mRNA is targeted for sequence-specific degradation by miR824. A single exchange of the binding site of *AGL16*, even when such a change does not alter the amino acid sequence of the encoded protein, will render the *AGL16* resistant to miR824 (Kutter *et al.*, 2007). The latter authors noted that primary stomatal complexes could give rise to higher-order complexes derived from satellite meristemoids. The expression of a miR824-resistant *AGL16* mRNA (by the change of a single nucleotide, as noted above) in transgenic plants increased the incidence of stomata in higher-order complexes. Reduced expression of *AGL16* mRNA in the *agl16-1* deficiency mutant and in transgenic lines overexpressing miR824, decreased the incidence of stomata in higher-order complexes. The authors therefore concluded that the *miR824/AGL16* pathway functions in the satellite meristemoid lineage of stomatal development.

There are recent important reports on stomatal development, published after Galun (2007), such as Bergmann and Sack (2007), Nadeau (2009), and Serna (2009). These reports deal with the genetic-developmental aspects of stomatal patterning but not with the possible involvement of defined phytohormones on the differentiation of stomata. Bergmann and Sack (2007) stated specifically: "The roles of plant growth regulators and environmental signals in regulating stomatal number and development are largely unexplored." Hence, these reports will not be discussed in this book.

We will finish this section of stomatal patterning in leaves with two remarks. One is that, while the role of phytohormones in patterning of stomata in leaves was not explored, there is information on the impacts of ET, ABA, GA, and CK on stomatal development in hypocotyls. The other remark is that the impact of genes, such as *TOO MANY MOUTHS* (*TMMs*), on stomatal development is different in leaf stomata than in stomatal development of other organs, such as stems and siliques.

Studies on the differentiation of stomata in leaves have several advantages over other differentiation mechanisms. This is because the patterns of stomata can be easily evaluated by microscopical analyses, and because most phenotypes resulting from mutations are not lethal.

# CHAPTER 14

# Patterning of Flowers:
# Genes and Phytohormones

Since the mid 1980s, various schemes have been put forward to explain flower patterning on the basis of homeotic genes. These schemes were based on the principle of metamorphosis of leaves, which was suggested independently by Caspar Friedrich Wolff (in his PhD thesis to the University of Halle in 1759), and a few years later, by J.W. von Goethe. Both these scholars claimed that all above-soil plant organs (except for the shoot) are leaves that underwent metamorphosis. Hence, Haughn and Somerville (1988), working on Arabidopsis, presented a scheme for the homeotic transformation of flower organs that is caused by specific mutations. Leaves, sepals, petals, stamens, and carpels were included in this scheme. A couple of years later, the laboratory of Saedler (in Cologne, Germany), working on *Antirrhinum* (Schwarz-Sommer *et al.*, 1990), published a scheme for three types of morphogenetic genes that control floral organ identity in *Antirrhinum* (their own studies) and in Arabidopsis (based on information received from the Laboratory of Elliot M. Meyerowitz at Cal Tech). Only one year later, E. Coen (from the John Innes Centre, UK) and Meyerowitz wrote in *Nature* an article on "The War of the Whorls: Genetic Interactions Controlling Flower Development" (Coen and Meyerowitz, 1991). The scheme proposed by the latter investigators became known as the ABC model of floral patterning. The ABC was then modified and adopted to flowers of many genera. I have already devoted a detailed discussion on these models (Galun, 2007). The models did not elaborate on the impact of phytohormones on the development of floral members and the discussion will not be repeated in the present book.

I am starting this chapter with old studies on the sex expression of plants, meaning on the differentiation of stamens and pistils. My associates and I performed some of the early research on this subject. I was hired by Dr Oved Shifriss who headed the Plant Genetics Laboratory at the Weizmann Institute of Science in Rehovot, Israel. My task was to study the physiology and genetics of sex expression and to utilize these studies for the breeding of cucumber and melon. We found very early on in our screening of cucumber cultivars that there was a great variability between cucumber cultivars with respect to sex expression. Some cultivars produced (in LDs) almost only sta-minate flowers (most cucumber cultivars are monoecious, producing staminate and pistillate flowers), while the ratio of pistillate flowers to staminate flowers was much higher in other cultivars. Moreover, under conditions of short days (SDs) and lower temperatures, the same cultivars had very different ratios between pistillate to staminate flowers (Fig. 81). Furthermore, among the over 100 cultivars of cucumber, some cultivars (Shogoin-type, that we obtained from Japan, but were likely originally from Korea and/or China) had only pistillate flowers under SDs and relatively low temperature. This led to the idea to use the different types of sex expression for the mass production of hybrid-seed cucumber cultivars. Such a process was not known in cucumber at that time, i.e. the end of 1957. I thought that the basis for this process could be a pistillate line of cucumber that would be pollinated by a regular monoecious cucumber line. Seeds collected from the pistillate line should all be hybrid seeds. However, there was a problem: how to propagate the pistillate line. Could treatment with a phytohormone help? There were previous reports on the change of sex expression in cucumber. Already in 1938, E.G. Minina then working in Kiev (in 1948 she was sent by T. Lysenko to Siberia to study the sex expression of forest trees; she was to join her husband who openly criticized the doctrine of Lysenko), found that some gases (probably containing ethylene) were causing the femalization of cucumber plants growing in closed green-houses (Minina, 1938). A few years later, Laibach and Kribben (1949)

APRIL SOWING                                    JULY SOWING

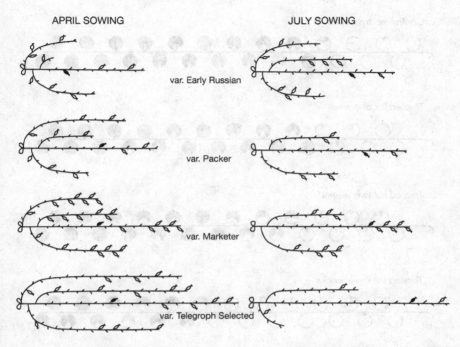

var. Early Russian

var. Packer

var. Marketer

var. Telegroph Selected

**Fig. 81.** Developmental patterns of pistillate flower distribution in four varieties of cucumber grown in two seasons in 1955, in Rehovot, Israel. The first pistillate flower on the main axis is shown in black. The nodes which are not marked by the presence of flowers are actually strictly staminate, with the exception of terminal nodes, on which information is lacking. (From Shifriss and Galun, 1956.)

reported that auxins would enhance femalization in cucumber. However, I needed to achieve the opposite: formation of staminate flowers. These were the years during which gibberellins (GAs) became popular because of their various effects on several plants. I therefore decided to make a "shot-in-the-dark" and sprayed pistillate (female) cucumber lines with GA. The effect was clear: spraying the plants with gibberellic acid caused the formation of staminate flowers. Also, spraying monoecious cucumber plants with gibberellic acid increased the ratio of staminate to pistillate flowers (Fig. 82). In future experiments I used different GAs and found that the $GA_{4+7}$

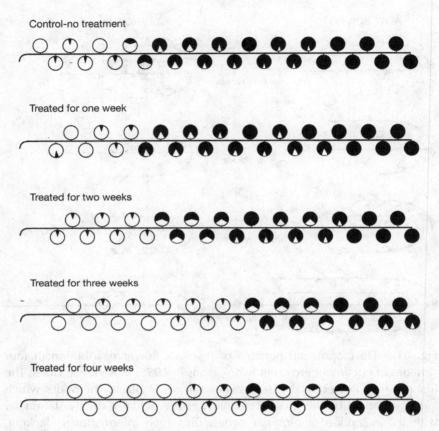

Control-no treatment

Treated for one week

Treated for two weeks

Treated for three weeks

Treated for four weeks

**Fig. 82.**    Change in sex expression in heterozygous-female ($st^{st}/st$) cucumber plants treated with gibberellic acid. Circles: mean sex at specific nodes; clear areas: proportion of nodes bearing staminate flowers; shaded areas: nodes bearing female flowers. (From Galun, 1959b.)

mixture is much more effective than $GA_3$. In practice, growing a field (in isolation) of pistillate cucumber and spraying every third row with GA solved the problem of maintaining the pistillate seed-parent. For a hybrid production field: one row of the monoecious plants was planted between several rows of (pistillate) seed parents and the hybrid seeds were collected from the pistillate rows. The first commercial hybrid cucumber seeds were thus available in 1960. The procedure was improved during the years (i.e. the pollen parent was converted into a staminate line that was maintained by spraying part of the field with Ethephon). Also, spraying the propagation field of the pistillate plants with 2-chloroethyl phosphonic acid ("Ethephon") that is converted by the plants to ethylene, will eliminate sporadic staminate flowers from the pistillate rows in the hybrid seed production field. A rather detailed description of hybrid seed production for cucumber was provided in Frankel and Galun (1977). The basic genetic system that controls sex expression was already formulated in Galun (1961a) and independently by Shifriss (1961); (Shifriss left the Weizmann Institute of Science in 1956). These two authors used different designations for the genes that affect sex in cucumber. Below, the designations of E. Galun will be used. In the following, three main genes will be described.

— $M$ is a gene that will cause the "trigger mechanism," meaning that a floral meristem will develop a unisexual flower: leading to either a staminate flower or a pistillate flower, but not a hermaphrodite flower. Hermaphrodite flowers will develop only in plants that are homozygous $mm$.

— $A$ is a gene that allows the development of pistillate flowers; hence $aa$ plants will produce only staminate flowers. $A$ is dominant over $a$, so that $Aa$ will allow the formation of pistillate flowers.

— $St^+$ pushes the floral buds toward the production of stamens, and $st$ pushes the floral buds to pistillate flowers. $St^+$ is partially dominant over $st$. Hence the genotype $stst$ will produce only pistillate flowers, while the $st^+$ plants can produce both pistillate and staminate flowers. The genotype $st^+ st^+$ will have mostly staminate flowers.

Based on these effects of the genes that control sex expression in cucumber, I rationalized that it should be possible to construct a genetic combination that would lead to a sex expression that does not exist in the species of cucumber (*Cucumis sativus*): hermaphrodite plants that bear only bisexual flowers. This should evolve from the combination of $mm$ and $stst$. Indeed, by crossing pistillate plants

(*MM stst*) with andromonoecious plants (*mm st⁺ st*), such hermaphrodite plants could be obtained in the F2 generation. It should be noted that in addition to the three major genes affecting sex expression, there are also modifying genes in cucumber cultivars. Therefore there is a variability in the monoecious cucumber cultivars as is shown in Fig. 81. The variability in sex expression also affected andromonoecious cultivars, and different types of hermaphrodite plants could be derived from the abovementioned crossing. Some plants were strictly hermaphrodite, while others also had staminate flowers (being andromonoecious), and the relative size of the pistils was variable: in some flowers the pistils were much smaller than the stamens. However, strictly hermaphrodite types could be maintained and even used in breeding programs. I sent seeds of the hermaphrodites to B. Kubicki in Poland and he found that these hermaphrodite lines were partially parthenocarpic and very useful in the production of hybrid seeds in cucumber (Kubicki, 1970).

Keeping in mind that there are modifying genes that affect sex expression, we may write down the assumed genetic composition of the main sex types of cucumber, as follows.

— Male (androecious):       *M/−, st⁺/−, a/a*
— Monoecious:               *M/−, st⁺/−, A/−*
— Female (gynoecious):      *M/−, st/st, A/−*
— Andromonoecious:          *m/m, st⁺/−, A/−*
— Hermaphrodite:            *m/m, st/st, A/−*

These genotypes of the *M/m, M/M, st⁺/st, st⁺/st⁺, st/st, A/A, A/a*, and *a/a*, combined with effects of day-length and two phytohormones (GA and ET as well as the anti-ET compound aminoethoxyvinyl glycine (AVG)) lead to different sex types of cucumber (Fig. 83). As mentioned above, most cucumber cultivars are monoecious, meaning they produce pistillate and staminate flowers. However, there are a few andromonoecious cultivars that produce staminate and hermaphrodite flowers. The latter hermaphrodite flowers have a round (apple-like) ovary in contrast to pistillate flowers that have an elongated ovary. In this respect, cucumbers differ from melons

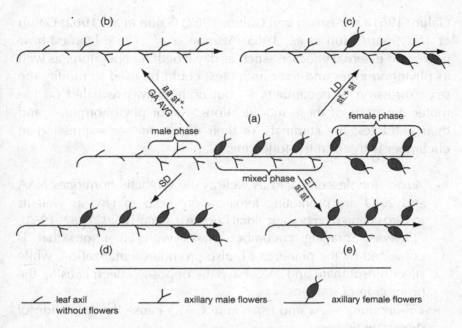

**Fig. 83.** Different sex expressions in cucumber genotypes derived from a hypothetical main shoot of cucumber. The [a] shoot can be converted to [b] by the *aa* genotype and by AVG (aminoethoxyvinyl glycine) or to [c] by LDs and by SDs to d. a is converted to e by ET.

(*Cucumis melo*), in which the ovary of hermaphrodite flowers can be either round or elongated and the ovaries of pistillate flowers can also be either round or elongated. Unlike most cucumber cultivars, most melon cultivars are andromonoecious (with round ovaries in the hermaphrodite flowers, but there is a group of melon cultivars that is used as (not sweet) vegetables (common in Egypt), termed "fakuss." Hence I shall note — as will be detailed below — that while Linnaeus grouped cucumber and melon in the same genus (*Cucumis*), these two species are actually not closely related phylogenetically.

At the end of about nine years of studying the sex expression of cucumber, my associates and I (Shifriss and Galun, 1956; Galun 1959a,b,c,d; Atsmon and Galun, 1960; Galun and Atsmon, 1960,

Galun, 1961a,b; Atsmon and Galun, 1962; Galun *et al.*, 1962; Galun *et al.*, 1963; Galun *et al.*, 1965; Atsmon *et al.*, 1965) learned how genes and external effectors (such as day length, temperature, as well as phytohormones and their antidotes) could be used to modify the sex expression of cucumbers — but nothing was revealed on the molecular basis of these modifications. As for phytohormones and their antidotes, the summary of their effects on sex expression in cucumber is listed in the following:

— *Auxin.* Indoleacetic acid as well as the synthetic hormones NAA and 2.4D are promoting femalization, both *in vivo* as well as *in vitro* cultured embryonic floral buds (e.g. Galun *et al.*, 1962, 1963).
— *Ethylene.* Spraying cucumber plants with Ethephon, that is converted by the plants to ET, also promotes femalization, while silverthiosulphate and AVG have the opposite effect: causing the promotion of stamens.
— *Gibberellins.* GAs and especially $GA_{4+7}$ cause the formation of staminate flowers.

As noted above, there is a difference in the mode of promoting maleness by the application of GA or by the application of the antidote to ET (by spraying AVG). The application of GA causes the suppresion of the pistillate flower; but acropetally of the first pistillate floral bud in the axil of the shoot, there is a dormant axially inflorescence with potential staminate floral buds. As long as the basal pistillate flower is allowed to differentiate, it will suppress the (potentially) staminate floral buds; so that only a pistillate flower will develop in this leaf axil. On the other hand, if the development of the basal (pistillate) floral bud is suppressed by GA, then the embryonal floral buds — which are acropetally of the suppressed pistillate bud — can develop into staminate flowers. This is schematically shown in Fig. 84. The application of AVG seems to act in a different way. The AVG changes the pistil/stamen ratio in the very same flower bud and suppresses the development of the pistil, allowing the stamens to reach maturity. Hence, gynoecious cucumber plants sprayed with AVG will frequently produce hermaphroditic flowers, while no such

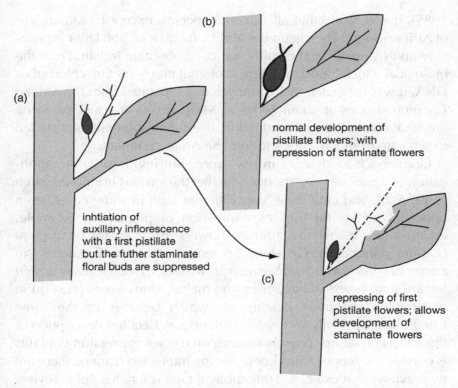

(a)

inhtiation of
auxillary inflorescence
with a first pistillate
but the futher staminate
floral buds are suppressed

(b)

normal development of
pistillate flowers; with
repression of staminate flowers

(c)

repressing of first
pistilate flowers; allows
development  of
staminate  flowers

**Fig. 84.** Change in sex expression of a pistillate flower. Without application of GA (or an anti-ethylene compound, AVG) the pistillate bud develops
into a pistillate flower, suppressing further (staminate) floral buds (b); when
the first (pistillate) floral bud is suppressed (by GA), the more distal staminate buds can develop into staminate flowers, several staminate flowers will
then evolve (c).

flowers will develop on gynoecious cucumber plants that are
sprayed with GA.

While treating cucumber plants with GA causes a significant
change in sex expression (enhancing the development of staminate
flowers), GAs have no effect on the sex expression of melon plants
although they were both included (by C. Linnaeus) in the same
*Cucumis* genus. In fact, from the studies of R. Perl-Treves and others
(in my laboratory, e.g. Perl-Treves and Galun, 1985; Perl-Treves *et al.*,

1985), it was found that all *Cucumis* species, except *C. sativus*, are of African origin. The cucumber and *C. hardwickii* (the latter is probably an escapee from the cultivated cucumber) are Indian. From the biological clock-based frequency of mutations in the chloroplast DNA, it was estimated that *C. sativus* was separated from the African *Cucumis* species at about 70 to 90 MYA. This is at about the same time that due to the continental drift, the Indian sub-continent started to separate from East Africa to join the Asian continent.

Going back to changes in sex expression imposed by the application of hormones, we should note that the various hormones, such as ET, AUX, and GAs, have very different (and in some cases, even opposite) effects on the sex expression of plants. For example, auxins cause a shift to staminate flowers in the castor bean plant (*Ricinus communis*). GA shifts sex expression toward maleness in cucumber but toward femaleness in maize. ET that causes a shift toward femaleness in cucumber and melon, shifts sex expression in watermelon (*Citrullus lanatus*) — which belongs to the same Cucurbitaceae family — toward maleness. A detailed description of the effects of various phytohormones on the sex expression of plants is outside the scope of this book. For the interested readers, there are two reviews that examined this subject. One is a rather "old" review but detailed and worth reading, by Heslop-Harrison (1957); the other is a more recent review by Perl-Treves and Rajagopalan (2006), which focused on floral members patterning of various monoecious species.

## Development of Floral Members in Monoecious Plants

The review on sex expression in plants by Heslop-Harrison (1957) detailed experimental work that showed by which means the sex expression of plants could be changed. This review handled external effectors such as phytohormones, day length, and temperature. Thereafter appeared a plethora of publications on the modification of sex expression in angiosperms. The book by Frankel and Galun (1977) updated these findings up to the end of 1976 on the sex expression in various monoecious, andromonoecious, and dioecious crops, such as asparagus (*Asparagus officinalis*), castor (*Ricinus communis*), cucumber (*Cucumis sativus*), hemp (*Cannabis sativa*), hop (*Humulus lupulus*), maize (*Zea mays*), melon (*Cucumis melo*), papaya (*Carica papaya*), and spinach (*Spinacia oleracea*). The choice of crop plants came about as a result of the book being intended for students of plant breeding. Additional information on sex expression and phytohormones was supplied by Zeevaart (1976), who reviewed the literature on the physiology of flower formation.

In this, Section I will mainly handle the sex expression of maize and of the two Cucurbitacean species, cucumber and melon. A further section on the roles of genes and phytohormones on the floret organs of grasses will be provided in a section at the end of this chapter.

Let us first look closely at the flowers of maize. Normally, the maize bears flowers in two parts of the mature plants. There are staminate (male) flowers on the indeterminate inflorescence (raceme) at the upper end of the plant. This is the tassel. Pistillate (female) flowers are produced in auxillary inflorescences, a few nodes below the tassel. The pistillate flowers have very long pistils termed "silk." The fundamental unit of maize flowers is the *spikelet* (see details in Irish, 1999). The spikelet has two subtending *glumes* that enclose two flowers, termed *florets* in the Gramineae. The florets lack the typical sepals and petals (that are the regular outer two whorls) as in the flowers of dicots, such as Arabidopsis. Instead, the maize florets develop a single *lemna*, a *palea*, three *lodicles*, three *stamens*, and a pistil that bears a single ovule. First, bisexual embryonal florets are developed, thereafter either the stamens or the pistil develop to maturity and, respectively, either the pistil or the stamens stop growing and degenerate. The mature floret is therefore normally unisexual. In the tassel, the pistil primordia of the two florets stop growing and eventually disintegrate, while the stamens of the two florets reach maturity. The abortion of the gynoecium was observed to go through a programmed cell death (PCD) process. This is manifested by vacuolization and loss of ribosomes and organelles in a group of cells. Also, localized DNA degradation and nuclear loss in the abortive tissue was observed while the full details of this abortion process are still pending (the details of PCD processes in maize florets, such as DNA fragmentation, expression of specific proteases, and classical apoptotic cellular morphology were not yet detailed at the molecular level).

In the auxillary, female inflorescence (the "ear"), rows of spikelet pairs are formed. In each spikelet, there are two florets. The second floret, that is the lower one, lags behind the first developing floret and is eventually aborted so that only one floret per spikelet matures. The upper floret first reaches the bisexual stage but thereafter its stamens' primordia are eliminated by an abortive process. The gynoecium of this single floret then develops a single ovule and a very much elongated pistil, the "silk."

There are maize mutants in which the stamens of the ear florets do not degenerate and bisexual florets reach maturity. The tassels of these mutants are normal (staminate). These mutants include recessive ones, such as *dwarf-1*, *dwarf-2*, *dwarf-3*, and *dwarf-5*, as well as dominant mutants, such ad *Dwarf-8* and *Dwarf-9*. It was found that the recessive mutants are deficient in GA synthesis and GA application will cause normal florets with no stamen in the ear. This strongly suggested that GA is inhibiting stamen development in the ear. Normally, the GA content in the ears is about 100-fold higher than in the tassel. Also, by transposon-tagging, another allele was found and the cloned respective gene encoded ent-kaurene-synthase A, an enzyme that catalyzes the first committed step in gibberellin synthesis, and it was also found that the maize *dwarf-3* gene encodes a cytochrome P450-type enzyme that is involved in a GA synthesis pathway (see Fig. 19 in Chapter 3). In the tassel, exogenous GA caused repression of stamens and promotion of pistil development, but in the ear the pistils were not affected; this probably means that in the ear GA is not required for normal pistil development. However, too little is known on the content of phytohormones in the individual florets and even less on these contents in the various organs of florets (stamens and pistil), which is insufficient to draw firm conclusions on the roles of GA in shaping the sex expression of the florets in maize. Moreover, it may be that changes in the sensitivities of the floret members to GA have a main role in the sex expression of maize.

Already in the early years of maize breeding, during the first half of the 20th century, it was realized that hybrid maize cultivars have "hybrid vigor." Investigators therefore looked for efficient means to produce such hybrids. The standard way was to grow several rows of regular monoecious maize plants, of a given cultivar, in between two rows of another cultivar that served as "pollen parent." The tassels of the rows between the pollen parents served as "seed parents." The tassels of the seed parents were removed before anthesis so that no self-pollination of the seed parents could take place. Another way to produce hybrid maize was to use cytoplasmic male-sterile (CMS) lines as seed parents. The maintenance of the CMS lines was performed by crossing them with maintainer lines that caused the restoration of male fertility. Numerous studies were then performed which aimed to isolate maize mutants that had various phenotypes of changed sex expression.

A group of these mutations was termed *tasselseed* (*ts*). In these mutations, there is no abortion of the carpel in the florets of the tassel so that seeds are formed on the tassel as well as in the ear. For this to happen, the plant should be homozygous for the recessive alleles (e.g. *ts1 ts1* or *ts2 ts2*). In such homozygous plants, the florets on the tassel commonly do not produce stamens: the tassel florets are pistillate. The *ts1 ts1* and *ts2 ts2* plants also lack the abortion of the secondary floret in the ear spikelets so that each spikelet produces two grains.

By studying a transposon-tagged *ts2* mutant, the *TS2* gene could be cloned. It was found that the gene encoded a factor that causes the abortion of the gynoecium of the florets in the wild-type tassel, as well as the pistil in the secondary ear floret. It was also found that in the plants with the transposon-tagged *ts2*, there could occur an excision of the transposon so that the section of the tassel where such an excision happened, the florets reverted to wild-type (with aborted gynoeciums). The expression of the *TS2* takes place when the spikelet is already quite advanced. The section with a reverted *ts2* to *TS2* could be very small. It was found that *TS2* encodes a protein that is homologous to short-chain alcohol dehydrogenases. The substrate of this enzyme is not yet assured. It could be GA or a BR. The *ts1* mutation has a similar phenotype as the *ts2* mutant. The *TS2* expression is abolished in the *ts1* mutants, indicating that *TS1* regulates *TS2* expression.

Another gene that has been known for many years to affect the sex expression of maize, is *silkless* (*sk*). This gene affects the ear so that in the genotype *sk sk* no ears are developing (such mutants can obviously be maintained through the heterozygotes *Sk sk* that produce normal ears). Thus it could be possible (theoretically) to convert maize into a dioecious species. This could be done by pollinating a silkless plant that also expresses tasselseed (*sk sk, ts2 ts2*) with pollen from a silkless plant (*sk sk, Ts2 ts2*); such a cross should result in two kinds of plants: (1) male plants that are *sk sk, Ts2 ts2* and (2) female plants that are *sk sk, ts2 ts2*. However, in practice, the *Ts2/–* plants are usually not "pure," as they do have a few florets that produce stamens in the tassel. Could these few stamens be eliminated by breeding? Possibly, because there seem to be several modifying genes that regulate the differentiation of maize florets. However, I do not know any commercial maize hybrid cultivars that are based on the *sk, st* genes.

The involvement of another phytohormone in the regulation of sex in maize was explored by Young *et al.* (2004). These investigators engineered maize with a gene (IPC) that encodes a cytokinin (CK)-producing enzyme. They put the coding region behind an Arabidopsis senescence-inducible promoter. The promoter was then activated and the gene was expressed. This expression rescued the second ear floret from abortion. The result was that the spikelets (each) produced two developing florets in the ear spikelet with two fertile ovules that led to fused seeds. This suggested that CK could prevent the entry of the second ear floret into the PCD pathway. On the other hand, from entering into PCD the high CK did not protect the gynoecium of male flowers, or the stamens of the female flowers. The reason for this change in florets in the ear was not fully explained. A converse experiment was also performed (Huang *et al.*, 2003) in which plants were engineered to overexpress cytokinin oxidase (a cytokinin degrading enzyme) in their anthers. This resulted in male sterility, suggesting that CK is required for the normal differentiation of anthers.

# Using Homologs of Maize Genes to Affect other Plant Species

Tassel seed homologs were cloned from the Arabidopsis and *Silene latifolia* genomes, by H. Saedler's laboratory (Lebel-Hardenack *et al.*, 1997); they had about 50% encoded amino acid identity with the original amino acid sequence encoded by *TS2*. The investigators asked if the *TS2* homolog is involved in pistil inhibition in male flowers of *Silene*. This was not the case because the homologous genes in both *Silene* and Arabidopsis are expressed only in the tapetal tissue of mature anthers. Thus the effect of the TS2 homologs in the two plants is not clarified. On the other hand, analyzing the effects of maize genes in other monoecious taxa of the Gramineae, strongly suggested that there is a common mechanism of floret differentiation in all these taxa. In the case of the wild genus *Tripsacum* (a relative of maize), maize *ts2* mutants could be crossed with *Tripsacum*. The results showed that the maize *ts2* and the respective *Tripsacum* homolog are "allelic."

# A New Wave of Studies of *Cucumis* Sex Expression

After the 1970s, a new wave of studies on sex expression in *Cucumis* emerged. These studies utilized improved methodologies and brought about a better understanding of the biosynthesis of phytohormones. An example of such studies is the work of Trebitsh *et al.* (1987). These investigators studied the production of ethylene by evaluating the level of crucial stage in its biosynthesis: 1-aminocyclopropane-1-carboxylic acid (ACC) in the apices of semi-isogenic cucumber cultivars (*Alma*, a gynoecious line and *Elem*, a monoecious line). The level of ACC was higher in the apices of the *Alma* plants than in the *Elem* plants. The application of IAA enhanced ethylene and ACC production in both the *Alma* and *Elem*, and the effect of IAA was stronger in *Alma* than in *Elem*. The induction of the higher levels of ethylene and ACC by IAA was blocked by the treatment with the anti-ethylene compound aminoethoxyvinylglycine (AVG), but the endogenous IAA content in the apices was lower in *Alma* than in *Elem*. The investigators concluded that although exogenous IAA enhances ethylene production, endogenous IAA might not have a major role in the control of sex expression in cucumber. It should be noted that when this study was performed, the methods for analyzing IAA contents were rather coarse, and shoot tip extract may not truly represent the IAA level in the developing floral bud. A similar test for IAA content was earlier performed by Galun *et al.* (1965). This work indicated that the IAA content of hermaphrodite cucumber plants (*mm, stst*) was higher than in andromonoecious plants (*mm, st⁺st⁺*).

In more recent studies, the involvement of ET in the sex determination was explored. The more advanced studies benefitted from the information on the pathway of ET biosynthesis in plants (see Chapter 2).

Yamasaki *et al.* (2003) further studied the role of ethylene on the sex expression of cucumber flowers. They recorded the arrest of either the stamens or the pistils in the early stages of floral bud development. To look for the effect of ET they treated the plants with either the ethylene-releasing ethephon or the ethylene antagonist AVG. Two types of cucumber plants were used. One type was monoecious plants (that produce mature pistillate and staminate flowers), while the other type was gynoecious plants that produce (without treatment) only pistillate flowers. They looked at the mRNA accumulation of the ACC synthase gene (*CS-ACS2*) and the ethylene receptor-like genes (*CS-ETR1, CS-ETR2,* and *CS-ERS*) by the respective *in situ* hybridization analyses. The *CS-ACS2* mRNA was located in the pistils' primordia of gynoecious cucumbers; but in flowers of the monoecious plants (that would probably develop into staminate flowers) this mRNA was detected *below* the pistil primordia and at the adaxial side of the petals. In flower buds of andromonoecious cucumber, only the *CS-ETR1* mRNA was located and expressed in the pistil primordia. There were some differences in the locations of the receptor's mRNA in flower buds of monoecious vs gynoecious plants but the meaning of these differences was not completely explained. The authors assumed that the tissues for both ethylene production and signal perception are the same and that ET induces pistil development. They also suggested that the receptor-related genes of cucumber floral bud accumulated in the stamens. Therefore, if *CS-ACS2* produces enough ET to inhibit stamen development, ET produced by *CS-ACS2* in the pistil primordia may diffuse to the primordia of stamens and inhibit stamen development in gynoecious cucumber plants. It is notable that Arabidopsis ACC synthase is encoded by a multigene family, of which 12 such genes were characterized. In cucumber four of the ACC synthase genes have been identified. Accumulation of *CS-AC2* mRNA is correlated with femaleness in cucumber, but further studies are required to clarify the positional relationship between cells producing and receiving ET in determining the sexuality of cucumber flowers.

Yamasaki *et al.* also addressed the question of why short days cause femaleness in most monoecious cucumber cultivars. They found that in the shoot tips of monoecious and andromonoecious cucumber plants there is a diurnal rhythm of the expression of *CS-ACS2*, *CS-ACS4* and *SC-ERS* genes with a peak in the middle of an 8 hour or a 16 hour light period, and that the expression of *CS-ACS2* was higher under short-day conditions than under long-day conditions. The authors suggested a feedback regulation of ET evolution in plants that have the *M* (rather than the *mm*) genetic constitution (i.e. monoecious plants).

Tova Trebitsh, now at the Department of Biology of the Ben Gurion University of the Negev in Beer-Sheva, is a veteran of sex expression studies in the Cucurbitaceae. Her work on sex expression in cucumber and its involvement with ET and auxin was mentioned above (Trebitsh *et al.*, 1987). In more recent studies, Trebitsh returned to this field of endeavor (Trebitsh *et al.*, 1997; Rimon *et al.*, 2006; Barak and Trebitsh, 2007); Salman-Minkov *et al.*, 2008).

In the first of the latter publications (Trebitsh *et al.*, 1997), the investigators focused on a gene that encodes ACC synthesis and isolated *CS-ACS1*. The expression of the latter gene was induced by auxin but not by ACC. The investigators then used two near-izogenic lines: *Alma*, a gynoecious line (*MM FF* by the Shifriss designation and *MM stst* by my designation) and *Elem*, a monoecious cucumber line (*MMFF*), and found an additional gene that encodes ACC (*CS-ACS1G*). Analyzing crosses, they found that in the $F_2$ population, *CS-ACS1G*, enhanced femaleness was 100% correlated with the *F* allele. There was no recombination between *CS-ACS1G* and *F*; both were located on the B linkage group. The authors proposed that *CS-ACS1G* plays a pivotal role in the sex determination in cucumber flowers, and that while *CS-ACS1* is present in a single copy in monoecious plants (*MM ff*), the gynoecious plants (*MM FF*) contain an additional copy: *Cs-ACS1G* of ACC synthase. When analyzing *CS-ACS1* and *CS-ACS1G* genes, Rimon Knopf and Trebitsh (2006) found that *CS-ACS1G* is the result of a recent duplication and recombination between *CS-ACS1* and a branched chain amino acid transaminase (*BCAT*) gene. It was therefore suggested that this duplication event gave rise to the *F* locus of the gynoecious cucumber plants. Computer analysis of the 1 kb region upstream of the transcription initiation site revealed several putative *cis*-acting regulatory elements, that can potentially confer the responsiveness of *CS-ACS1G* to developmental and hormonal factors, and thereby control female sex determination in cucumber. In an additional study, Barak and Trebitsh (2007) looked at a gene that encodes a putative GTP binding tyrosine, phosphorylated protein A (*CsTypA1*). The investigators found that the expression of *CsTypA1* is developmentally regulated. It is expressed in stamen primordia and its transcript is more abundant in the apices of monoecious plants, implying a role for *CsTypA1* in the early stages of male reproductive organ development. At later stages of flower development, a higher transcript level was observed in female flowers, in stigmatic papilla, nectary and in particular in the ovule/ovary tissue. The authors suggested that the differential expression of *CsTypA1* during male and female flower development indicates a role for *CsTypA1* in female flower development. However, what role exactly? This was not very clear. It was also suggested that *CsTypA1* may have a "dual" role: one in the early stages of flower development and the other, later, after sex determination during the development of the ovary/ovule.

The application of an ET releasing compound (ethephon) to cucumber and melon enhanced the femaleness and application of an anti-ET compound (AVG) or silverthiosulfate (that counteract ET) enhanced maleness. The reaction of watermelon (*Citrullus lanatus*) to

ET is very different. This was already reported in a review by J. Rudich and Zamski (1985) and by Sugiyama *et al.* (1998). In watermelon, application of ET clearly enhanced maleness. That the same phyto-hormone application had different effects on two different plants is quite common, but in cucumber and watermelon we are dealing with plants that belong to the very same family (Cucurbitaceae) with a rather similar flower architecture.

The team of T. Trebitsh (Salman-Minkov *et al.*, 2008) studied the ACC synthase genes of watermelon and related *Citrullus* species to further explore the hormonal impact on the differentiation of stamens and pistils in cucurbits. Watermelons are either monoecious or andromonoecious, but in either sex type, the number of staminate flower-bearing leaf axils is much greater than the number of female or hermaphrodite flowers (respectively) in the leaf axils. The investigators isolated and characterized four genomic sequences of ACS encoding genes from watermelons: *CitACS1*, *CitACS2*, *CitACS3*, and *CitACS4*. The genes *CitACS1* and *CitACS3* were found to be expressed in floral tissues, though the exact location and timing (during floral development) were not documented. The *CitACS1* is also expressed in vegetative tissues. The expression of *CitACS1* was found to be up-regulated by treatment with auxin, gibberellin, or ACC. This gene had apparently no effect on the sex expression. The *CitACS3* gene was expressed in open flowers and in young staminate floral buds but, again, the exact localization of the expression, whether focused on stamen meristem or in other members of the floral buds, was not specified. The latter gene was not expressed in female flowers. The *CitACS3* gene is also up-regulated by ACC and the authors suggested that this gene is likely to be involved in ethylene-regulated anther development (possibly meaning promotion or suppression of stamen meristems). The expression of *CitACS2* was not detected in vegetative or reproductive organs. The *CitACS4* transcript was also not detected by the investigators. The *CitACS* genes showed polymorphism among and within different *Citrullus* groups, such as watermelon cultivars, *Citrullus lunatus* var. *citroides*, and the desert species *Citrullus colocynthis*. It is expected that further studies with the *CitACS* genes (especially *CitACS3*) will shed light on the involvement of ethylene in the differentiation of floral members of the *Citrullus* species.

A team of 13 investigators, most of them from France and headed by Abdelhafid Bendahmane of INRA (Boualem *et al.*, 2008), recently returned to the andromonoecious sex expression. This sex expression is the common one in regular (sweet) melon cultivars. The main shoot of andromonoecious melons mainly has staminate flowers that are followed in upper nodes with a few hermaphrodite flowers. The genetic difference between andromonoecious cultivars and monoecious melons is by only one allele, which in my publications was termed $M/m$. $St^+/-$, $M/-$, are monoecious while $St^+/-$, $m/m$ are andromonoecious. The Boualem *et al.* team termed the allele that causes andromonoecious sex expression: *a*. One may claim that *aa* converts pistillate flowers to hermaphrodite flowers. According to my model (Galun, 1961) the *a* allele eliminates the mutual antagonism between pistil and anther so that in andromonoecious melon floral meristems, the initiation of stamen does not cause the abortion of the pistil, and the initiation of the pistil does not cause the abortion of stamens. The French team developed another model for andromonoecious sex expression. They found in andromonoecious plants a mutation in the ACS and termed it *CmACS-7*. Expression of the enzyme that is encoded by *CmACS-7* apparently inhibits the development of male organs (in a way that they no longer cause the abortion of the female organs). The investigators found that an active CmACS-7 enzyme is required for the development of female flowers in monoecious melon plants, whereas a reduction of the enzymatic activity results in hermaphrodite flowers in andromonoecious melons. The investigators used *in situ* hybridization to follow the location of *CmACS-7* mRNA at various stages in the floral development and revealed spatial and temporal changes at the levels of transcription. For example, the early accumulation of this mRNA was strongly localized in carpel primordia. On the other hand, because *CmACS-7* expression levels and patterns were not different between female and hermaphrodite flowers, and because the loss of *CmACS-7* activity accounts for the functional variation, the authors concluded that *CmACs-7*-mediated ethylene production in the carpel primordia affects the development of stamens in female flowers but it is not required for carpel development. Due to the fact that other phytohormones are also involved in the sex expression of melons, we should wait until *in situ* hybridization of additional transcripts involved in the biosynthesis of these other hormones becomes available, in order to fully understand the impact of phytohormones on the sex expression of melon.

## Recent Studies on the Patterning of Specific Floral Organs

The understanding of the differentiation of flower organs started with the studies of the four whorls (sepals, petals, stamens, and carpel) in *Antirrhinum* and Arabidopsis. The early studies in this field were summarized by Schwarz-Sommer *et al.* (1990), and Coen and Meyerowitz (1991). These studies were followed by a great number of subsequent studies that were detailed in Chapter 12 of Galun (2007) and will not be repeated here.

Balanza *et al.* (2006) continued the study of the patterning of floral organs in Arabidopsis. They focused on the female floral organ, the gynoecium. In Arabidopsis, the gynoecium is composed of a carpel (the fourth and terminal whorl) that is built of two modified "leaves" (in line with the old Wolff/Goethe hypothesis of leaf metamorphosis). The approach of Balanza *et al.* (2006) was to regard the patterning of the carpel as a lateral outgrowth from the shoot, meaning, as modified leaves. Therefore, they started their update on carpel patterning with a rather detailed account of the molecular genetic information of the development of leaves from the shoot apical meristem (SAM). They thus reviewed the roles of a great number of genes that are involved in leaf development, such as those that encode class I KNOX homeobox transcription factors, *CUP-SHAPED COTYLEDON* (*CUC*), *ASYMMETRIC LEAVES1* (*AS1*), *AS2*, as well as genes that are involved in leaf symmetry (i.e. adaxial/abaxial sides of leaves) such as genes encoding the transcription factors *PHABULOSA*, *PHAVOLUTA*, and *REVOLUTA*. They also reviewed the role of the gene that encodes the transcription factor *AINTEGUMENT* (*ANT*) (see Table 4 above for acronyms of the various genes that are active in the SAM).

The Balanza *et al.* update also mentioned the role of auxin in leaf patterning and noted the higher presence of the auxin influx facilitator, AUX1, in the abaxial epidermis of the leaves. This could contribute to the formation of an abaxial–adaxial gradient of auxin concentration in the leaf. There is now a fair knowledge of the establishment of the abaxial–adaxial symmetry in leaves while the proximal distal axis is less understood. Only after summarizing the molecular genetics of leaf patterning, did Balanza *et al.* (2006) move to the gynoecium of Arabidopsis. The Arabidopsis gynoecium is derived from two congenitally fused carpels that emerge as a single primordium in the center (and terminal end) of the flower. Subsequently, a central invagination forms and the primordium elongates as an open hollow cylinder. Later, two opposing internal meristematic outgrowths form at a medial position, which in turn produce the placentae and ovules. The center is fused to form a septum. With the formation of the stigma (a single cell layer of papillar cells), the gynoecium is ready for fertilization. The authors found that most of the genes that affect the gynoecium development also play a role in leaf patterning, again verifying the centuries' old hypothesis of leaf metamorphosis. A leaf can be transformed into a carpel by expressing the corresponding organ identity genes, especially *AGAMOUS* (*AG*) and one or more of the *SEPALLATA* (*SEP*) genes. Conversely, simultaneous loss-of-function of the redundance *SEP* genes leads to complete absence of carpel development. The *ag* mutants lack carpels. However, there are hints that additional factors could specify, at least partially, carpel identity in an AG-independent pathway, such as *SPATULA* (*SPT*) and *CRABS CLAW* (*CRC*). The precise hierarchy leading to carpel identity is not yet completely understood, although there is evidence that AG is activated in the floral meristem by the joint action of the products of the floral meristem identity gene LEAFY and the meristem maintenance homeobox gene WUSCHEL. Another gene that plays a role in carpel development is *SHATTERPROOF* (*SHP*). Moreover, Balanza *et al.* (2006) placed *SHP* at the top of the carpel identity, AG-independent pathway. As in regular leaves, adaxial–abaxial patterning is established in the carpels at an early stage, and *KAN* genes initially become expressed in abaxial domains, together with Auxin Response

Factors *ETT* and *ARF4*. As for the role of auxin in carpel patterning, what is clear is that auxin is related to all the major pathways directing polarity in the gynoecium. However, how auxin signaling is translated into gene regulation and the specific nature of its targets (directing specific tissue differentiation) is still not understood. The future use of advanced methods, such as the use of immunolocalization of auxin with specific monoclonal antibodies, will surely provide further interesting information. Clearly, in addition to the prominent role of auxin in gynoecium differentiation, GAs and CKs probably also play a role in this differentiation. Moreover, there is a discrepancy between the numerous data on the effects of ET on the pistils of plants (e.g. in cucumber, as noted above) and the lack of information on the interaction of ET with the many genes that take place in flower (and leaf) patterning. Again, future studies will probably provide the missing information.

## Organogenesis of the Arabidopsis Petal

Vivian Irish (2008) recently reviewed the organogenesis of the petals of Arabidopsis. She advocated the petal of Arabidopsis as a preferred organ for studies that will reveal the regulation of cell division, cell expanding, cell and tissue type differentiation and patterning of a plant organ. As petals are dispensable for growth and reproduction, one can obtain various mutants that affect petals and maintain these mutants. In Arabidopsis, the petals have a simple laminar structure with only a small number of cell types, making the analyses of organogenesis relatively simple. Vivian Irish from Yale University therefore reviewed the recent studies on the patterning of Arabidopsis petals. The simple structure and dispensability of the Arabidopsis petals render them a proper model for the studies that intend to reveal how gene activities are translated into morphological forms. Irish claims that the examination of cell proliferation patterns in developing Arabidopsis petals indicates that directional growth is regulated at the whole organ level, implying that long-range signaling is necessary to coordinate the patterning processes. Typically, the genes that have roles in the patterning of the petals also have roles in the patterning of other "lateral" organs and the general "strategy" of petal patterning is the same as in these other organs: each gene is expressed at a certain stage of the petal differentiation but is "mute" at other stages and there are elaborated regulators that control the site and developmental stage of these gene expressions.

The mature Arabidopsis petal consists of a greenish claw and a distal white blade. There is a single layer of epidermal cells overlaying the mesophyll tissue and the vasculature. There is a difference between the adaxial and the abaxial cells of the epidermal layers. The petals grow relatively slowly and are derived from the L1 and L2 layers of the SAM. During the early stage of petal growth there is mainly cell division, while in later stages there is mainly cell elongation; then the petal reaches a spoon shape. The very young petals are green because of the presence of chloroplasts but later the plastids lose their chloroplasts and redifferentiate as leucoplasts. The petal blade then turns white.

There are several types of genes that play major roles in establishing aspects of petal organogenesis: genes involved in establishing the second whorl domain, genes required for specifying petal identity, and genes required for the differentiation of petal-specific cell types.

There are feedback loops between, and within, these gene types. Some of the above were schematized by Irish (2008) and are shown in Fig. 85. In Fig. 85A the genes involved in the petal organogenesis networks are shown. The expression patterns of key regulatory genes (such as *RBE, UFO, PROXY1, AG, PTL, AP3,* and *EEP1*).

An important issue in any patterning, and thus also of relevance in patterning of petals, is the establishment of boundaries. In these boundaries, specific cells fail to proliferate across a given region such as between the four main floral whorls (e.g. between the first and the second whorls, the sepals, and the petals, respectively; and between the second and third whorls, the petals and the stamens, respectively). Several genes have roles in establishing boundaries that specify the petal domain. One of these is *AGAMOUS* (*AG*) that encodes a MADS box transcription factor. MADS box is a conserved motif of 56 residues in a DNA-binding protein of transcription factors

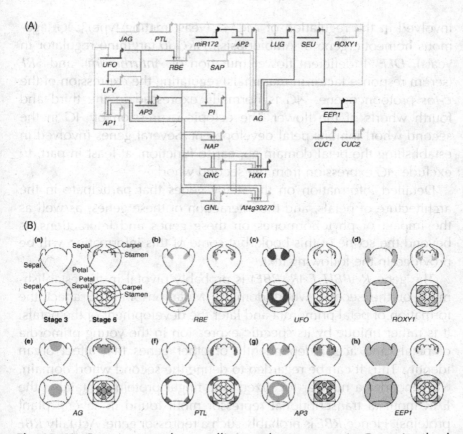

**Fig. 85.** (A) Genetic network controlling petal organogenesis. Genes involved in petal organogenesis networks are shown; positive regulatory interactions are denoted by a vertical arrow and negative regulatory interactions are denoted by a vertical line ending in a bar. Many of these regulatory interactions have been inferred from genetic data and so have not yet been proven to be direct. Putative transcription factors (AG, AP1, AP2, AP3, CUC1, CUC2, GNC, GNL, JAG, LFY, NAP, PI, PTL, RBE) are indicated in shades of blue and green; putative microRNAs (EEP, miR172) in black; putative transcriptional cofactors (HXK1, LUG, SEU, UFO) in shades of red; all other gene products (At4g30270, ROXY1, SAP1) are in shades of yellow. (B) Expression patterns of several key regulatory genes involved in petal organogenesis. (a) Diagram showing the positions of organ primordia at stages 3 and 6 of flower development; cartoons of lateral sections shown at top; transverse sections at bottom. (b–h) Cartoons of transcript distribution of key regulatory genes at stages 3 (left) and 6 (right), as indicated below each panel. Color coding in (B) are as in (A). (From Irish, 2008.)

involved in the regulation of *MCM1* (yeast mating type), *AG* (aga-mous homeotic gene of Arabidopsis), *ARG80* (arginine regulator in yeast), *DEF A* (deficient flower mutation of *Antirrhinum*), and *SRF* (serum response factor in mammals) regulating the expression of the *c-fos* protooncogene. *AG* is normally expressed in the third and fourth whorls of the flower. The ectopic expression of *AG* in the second whorl inhibits petal development. Several genes involved in establishing the petal domain appear to function, at least in part, to exclude *AG* expression from the second whorl.

Detailed information on the many genes that participate in the architecture of petals, and the interaction of these genes, as well as the impact of phytohormones on these genes and interactions, is beyond the scope of this book, but some of this information will be provided in the following text.

The gene *RABBIT EAR* (*RBE*) is probably pivotal in the establish-ment of the second whorl domain. Mutations in *RBE* affect the formation of petal primordia and later the development of the petals. It is rather unique by its specific expression in the young primordia of petals and acts independently of other genes that affect organ identity. Thus it can be regarded to define the second whorl domain. *RBE* encodes a nuclear localized zinc finger protein containing the EAR motif (a transcriptional repressor motif found in several plant proteins). Hence *RBE* is probably such a repressor gene. Actually *RBE* was found to restrict *AG* expression in the early phases of flower development. One possibility is that *RBE* might be responsible for defining the first/second whorl boundary in a non-cell-autonomous fashion. *RBE* expression itself is dependent on the action of two genes: *UNUSUAL FLORAL ORGANS* (*UFO*) and *PETALLOSS* (*PTL*). The *UFO* is expressed mainly in young petal and stamen primordia and encodes an F-box protein; thus it is probably involved in ubiq-uitination, so that $SCF^{UFO}$ can be formed and target a negative regulator in the second whorl, causing derepression of the latter regulator. *UFO* is only required during the very early differentiation of petals. In con-trast, *PTL* is commonly expressed in the first whorl (sepals) but *ptl* mutants have defects in the second whorl. *PTL* encodes a trihelix tran-scription factor and positively regulates the expression of *RBE*.

There are several genes that have also been implicated in establishing the petal domain by restricting *AG* expression. These include *APETALA2* (*AP2*), *AINTEGUMENTA* (*ANT*), *SEUSS* (*SEU*), *LEUNIG* (*LUG*), and *STERILE APETALA* (*SAP*). These affect not only petals; the *ROXY1* gene also negatively regulates *AG* and is mainly expressed in petal primordia. There is genetic evidence for the suggestion that *ROXY1* also acts to post-translationally modify *AG* repressors, such as *AP2*, *LUG*, *UFO*, and *RBE*.

MicroRNAs have a further role in defining boundaries of the second-whorl petal domain. *CUP-SHAPED COTYLEDON1* (*CUC1*) and *CUC2* encode a *NAC* family transcription factor (*NAC* stands for a family of transcription factors named after the first three founding members: *NAM*, *ATAF1*, and *CUC2*) that are expressed at the organ boundaries and repress growth. *CUC1* and *CUC2* are both regulated by the miR164 family of microRNAs. This regulatory cascade is essential for establishing the petal domain, since a mutation in *EARLY EXTRA PETALS1* (*EEP1*) that disrupts the miR164c gene causes extra or enlarged petals. The miR164c, that is expressed in several floral tissues — including young petal primordial — appears to down-regulate the expression of *CUC1* and *CUC2* transcripts, specifically in the second whorl. MiR172 has also been implicated in regulating the inner boundary of the second whorl domain by the repression of *AP2*, which, in turn regulates *AG*. Still, there are several "unknowns" in the regulation of boundaries. It nevertheless seems likely that the overlap in function of regional specifiers — such as UFO, with general organogenic processes controlling outgrowth, like those mediated by auxin flux — is required to establish separate petals in the second whorl.

## The Specification of Petal Identity

As already summarized in Galun (2007) and shown schematically in Fig. 72 of that book, a complex of *APETALA3* (*AP3*), *PISTILLATA* (*P1*), that encode two respective MADS box transcription factors, are essential for petal identity. These two factors probably act in conjugation with two other MADS box transcription factors, that are encoded by *AP1* and *SEPALLATA (SEP)*, to regulate petal-specific differentiation. The genes *AP3* and *P1* are expressed early in petal development and also continue their expression during further development. This continuous expression is probably required for the maintenance of petal identity throughout the petal organogenesis.

The expression of *AP3* seems to be correlated to the expressions of *UFO* and *LEAFY (LFY)*. LFY provides a temporal specificity and *UFO* defines the spatial domain of *AP3* activation. *UFO* probably acts as a transcription cofactor of *FLY* in *AP3* activation. This and other observations suggest that *UFO* might be pivotal in coordinating the different pathways that are required to establish petal organogenesis.

It should be noted that *AP3* binds to DNA only when complexed with the P1 protein. There are additional loops in this system: *AP3* directly promotes its own expression through binding of an *AP3-P1* heteromeric complex to *AP3* promoter sequences. *AP3* and *P1* also both positively regulate *P1* expression, although this regulation is probably indirect; *AP3* negatively regulates the *AP1* MADS box gene. *AP1*, in turn, positively regulates the expression of UFO in petal primordia, providing additional feedback control of *AP3* expression.

There is also an involvement of GA signaling. This signaling is required to maintain *AP3* expression at later stages of floral development. This signaling suppresses the function of at least two DELLA transcription regulators. Together, these processes suggest that complex homeostatic feedback controls act to maintain the appropriate *AP3* expression throughout petal development. It was found by mosaic analysis in which wild-type and mutant tissues are juxtaposed, that *AP3*-expressing cells are required for the intercellular signaling events responsible for establishing normal petal shape and size. However, AP3 protein does not traffic between cells, implying that downstream transcriptional targets of *AP3* act to coordinate these intracellular signaling events. The down-regulation of chlorophyll levels in petals probably reflects the *AP3*-mediated down-regulation of *GNC* (*GATA, nitrate-inducible, carbon-metabolism-involved*) and *GNL*-dependent (*GNC-like*) responses. Furthermore, *AP3* appears to function, in part, to regulate several transcriptional cascades that culminate in the regulation of plastid identity and cell expansion, providing one possible explanation for how *AP3* might act to coordinate the event required for shaping the petal.

Some genes have been shown to regulate the somewhat unique aspects of petal growth. The genes *JAGGED* (*JAG*), *ANT*, and *BIG BROTHER* (*BB*) regulate distinct aspects of petal proliferation. For example, *JAG* encodes a transcription factor with a $C_2H_2$ zinc-finger domain and is expressed in the distal petal blade during later phases of development, where it appears to promote both cell cycle progression and cell expansion. *JAG* also acts earlier in petal development to negatively regulate the expression of *PTL*. *ANT* encodes an *AP2*-domain-containing transcription factor and also acts to maintain a proliferative state in developing petals. *BB* appears to act independently of *JAG* and *ANT* in limiting petal growth.

The abovementioned review by Irish (2008) provided ample information on the patterning of petals in Arabidopsis (e.g. Fig. 85), but it is still unclear how the various processes are coordinated in space and time to finally result in a mature petal with specific size and shape. Moreover, in addition to the reported roles of auxin and gibberellins, other phytohormones may also have roles in petal patterning. One may ask a naive question: how is it that a simple plant organ requires such elaborated processes for its patterning? There is no simple answer to such a question but one should remember that plant patterning is not a simple engineering process. In plants, there is a very long evolutionary "history" and the controlling elements are built one above another so that complications accumulate during many millions of years.

In an unexpected manner, a member of the plant glutaredoxin family, *ROXY1*, was discovered by Xing *et al.* (2005) to be required for the normal development of petals in Arabidopsis. This discovery was made by a team from the Max Planck Institute (MPI) for Plant Breeding in Cologne, Germany (headed for many years by H. Saedler), where the main plant model for flower research was *Antirrhinum*. The role of *ROXY1* was unexpected because *ROXY1* encodes a glutaredoxin (GRX). GRXs are small disulfide oxidoreductases that posses a typical glutathione-reducible dithiol CXXC or CXXS active site that is required for the reduction of target protein disulfides. Such reactions were known in the defense of plants against stress but not in floral member patterning. Xing *et al.* (2005) isolated three alleles of the Arabidopsis gene *ROXY1* that initiate a reduced number of petal primordia and exhibit abnormalities during further petal development. The defects are restricted to the second whorl of the flower (the petals) and are independent of organ identity. *ROXY1* was found to be predominantly expressed in tissues that give rise to new flower primordia including petal precursor cells and petal primordia. Occasionally, filamentous organs with stigmatic structures were formed in the second whorl of the *roxy1* mutants, indicating an ectopic function of the Class C gene *AG* (according to the classic ABC model of floral patterning). The function of *ROXY1* in the negative regulation of *AG* is corroborated by premature and ectopic AG expression in the *roxy1-3 ap1-10* double mutants, as well as in double mutants of *roxy1* with repressors of *AG*, such as *ap2* (Figs. 71 and 72 of Galun, 2007, provide schemes for the updated ABC model). The authors noted that their data demonstrate that unexpectedly a plant glutareduxin is involved in flower development, probably by mediating post-transcriptional modifications in target proteins that are required for normal petal organ initiation and patterning.

In a further study of the MPI team with investigators from the University of Osnabrück (Germany) and from Australia (Li *et al.*, 2009), the theme of *ROXY1* effects on petal differentiation in Arabidopsis was further pursued. The authors noted that in the past, GRXs had been mainly associated with the redox-regulated processes, participating in stress responses, but Xing *et al.* (2005) showed that ROXY1 encodes a GRX that was found to affect petal development. *ROXY1* belongs to a land plant-specific class of GRXs that have a CC-type class active site motif. Plant GRXs fall into three classes of small ubiquitous glutathione-dependent oxidoreductases that play crucial roles in response to oxidative stress: CC, CPYC, and CGFS. Of these three groups, CPYC and CGFS are common to all prokaryotes and eukaryotes, while the CC-type class is specific for land plants. Out of the 31 *GRX* genes identified in Arabidopsis, 21 belong to the CC-type class. The investigators performed expression studies of yellow fluorescent protein-*ROXY1* fusion genes derived by the cauliflower mosaic virus 35S promoter, and revealed a nucleo-cytoplasmic distribution of *ROXY1*. The nuclear localization of ROXY was found to be indispensable and crucial for its activity in flower development.

Yeast two-hybrid screens identified TGA-motif-containing transcription factors as interacting proteins, which was confirmed by bimolecular fluorescence complementation experiments, showing their nuclear interaction *in planta*. Overlapping expression patterns of *ROXY1* and TGA genes during flower development demonstrated that ROXY1/TGA protein interactions can occur *in vivo* and support their biological relevance in petal development. Deletion analyses of ROXY1 demonstrated the importance of the C-terminus for its functionality and for mediating ROXY1/TGA protein interactions. *PERIANTHIA* (*PAN*) is a member of the TGA gene family. The investigators also found that phenotypic analysis of the *roxy1-2 pan* double mutant and an engineered chimeric repressor mutant from *PAN* supported a dual role for *ROXY1* in petal development: petal primordia initiation and (later) petal morphogenesis probably act by modulating other TGA factors and could act redundantly during differentiation of the second whorl of floral members.

It is noteworthy that Xing and Zachgo (2008) also found that *ROXY1* is not only involved in petal initiation and differentiation, but that this gene as well as *ROXY2* — the two glutaredoxin genes — are required for the development of anthers, the floral members of the third whorl.

## Patterning in the Third Whorl of the Arabidopsis Flower

I noted earlier that the involvement of genes, that were considered to be required in stress response in plants (*ROXY1*), was also required for the initiation of the floral members of the second whorl of flowers: the petals. However, then the same team found that the two Arabidopsis genes *ROXY1* and *ROXY2* (both of them glutaredoxin genes) are required for the correct differentiation of members of the third whorl (Xing and Zachgo, 2008); although in the latter case, the role of the glutaredoxin genes was not in the initiation and development of stamen but rather in the cell differentiation in the anthers. *ROXY1* and *ROXY2* are two CC-type GRXs encoding genes. Single mutants of either *roxy1* or *roxy2* produce normal and fertile anthers (i.e. with functional pollen grains), however, the double mutants *roxy1 roxy2* are sterile and do not produce pollen. Normal anthers are bilateral, symmetrical, four-lobed structures that produce the pollen. Each lobe develops from successive divisions of sub-epidermal archesporial cells formed in the anther's primordium that gives rise to three morphologically distinct cell layers. The endothecium, middle layer, and tapetum surround the pollen mother cells (PMCs). The latter will undergo meiosis and finally form the haploid microspores (the pollen). The functional tapetum is a source of nutrients and is thus essential for microspore maturation. A few anther genes were previously reported that control the normal differentiation of the tapetum (see Zhang *et al.*, 2006, for details). It appears that the loss of functional *ROXY1* and *ROXY2* results in defects in sporogenous cell formation in the adaxial anther lobes, whereas later stages, such as PMC and tapetum differentiation, are affected in abaxial lobes. Xing and Zachgo (2008) claimed, on the basis of their experimental work, that *ROXY1* and *ROXY2* are expressed with overlapping patterns during anther development. Lack of *ROXY1* and *ROXY2* function affects a large number of anther genes at the transcription level. Their genetic and RT-PCR data imply that *ROXY1/2* function downstream of the early-acting anther gene *SPOROCYTELESS/NOZZLE* and upstream of *DYSFUNCTIONAL TAPETUM1*, controlling tapetum development. Mutagenesis of a conserved glutathione-binding glycine in the ROXY1 protein indicated to the authors that CC-type GRXs need to interact with gluthatione to catalyze essential biosynthetic reactions. The analysis of these two "novel" genes indicated that redox regulation, as well as participating in plant stress defense mechanisms, might play a major role in the control of male gametogenesis. Clearly, since the tapetum tissue is a relatively "new" invention that happened with the emergence of angiosperms (probably less than 150 MYA), we are faced with the abovementioned "Pillars of the Mosque of Acre" in which "old" genes were modified to serve newer developments.

# Novel Studies on the Fourth Whorl of the Arabidopsis Flower

According to the classic ABC model, the fourth whorl of the Arabidopsis flower is determined by a combination of *AG* and *Sep* encoded proteins that combine into a complex. The carpel and the enclosed structure in the flower are termed *gynoecium*. The petal that was discussed earlier is considered the simplest floral organ, while the gynoecium is the most complex floral organ. Until recently, only little molecular information was available on the patterning of the gynoecium, but then two extensive studies were published consecutively in *Plant Cell*: Alvarez *et al.* (2009) and Trigueros *et al.* (2009) that deal with the molecular processes involved in the development of the Arabidopsis gynoecium. The Alvarez *et al.* publication was on a study performed in the laboratory of Yuval Eshed, at the Weizmann Institute of Science in Rehovot. The latter authors noted that the C Class *AGAMOUS* (*AG*)-clade MADs box genes are primary promoters of the gynoecium. The gynoecium is divided into a distal style (topped by the stigma) and a suspending ovary along the apical basal axis. The investigators showed that members of a clade of B3 domain transcription factors *NGATHA1* (*NGA1*) to *NGA4* are expressed distally in all lateral organs and all four have a redundant, but essential role, in style development. When all four genes were lost, it caused the formation of a gynoecium, where the style is replaced by valve-like projections and a reduction in style-specific *SHATTER-PROOF1* (*SHP1*) expression. In agreement, floral misexpression of NGA1 promotes ectopic style and *SHP1* expression. It was also found that *STYLISH1*, an auxin biosynthesis inducer, conditionally activates *NGA* genes, which in turn promotes distal expression of other *STY* genes in a putative positive feedback loop. Inhibited auxin transport, or lack of *YABBY1* gene activity, resulted in a basally expanded style domain and a broader expression of *NGA* genes. The authors speculated that early active gynoecium factors delimit *NGA* gene response to an auxin-based signal elicited by *STY* gene activity to restrict the activation of style program to a late and distal carpel domain. In brief, the *NGA* activity in Arabidopsis flowers is intimately linked to style formation and to the carpel genes. A genetic model for *NGA* activation and function in promoting style development is presented in Fig. 86.

The publication by Trigueros *et al.* (2009) which follows the Alvarez *et al.* (2009) article in the same *Plant Cell* issue deals with the same subject but resulted from independent research. The similarity of the two publications is amazing even with respect to trivial issues. Trigueros *et al.* wrote that "The gynoecium is the most complex floral organ ..." while Alvarez *et al.* wrote "The *Arabidopsis thaliana* gynoecium is one of the most complex organs of the plant ..." However, there are also small nuances between the two publications. The authors of Trigueros *et al.* (2009) noted that the characterization of the *NGA* gene family is based on an analysis of the activation-tagged mutant named *tower-of-pisa1* (*top1*) which was found to overexpress *NGA3*. The latter authors claimed that their data suggest that the NGA and STY factors act cooperatively to promote style specification, in part by directing *YUCCA*-mediated auxin synthesis in the apical gynoecium domain, and that their data point toward a complex landscape of regulatory activities and hormonal pathways established in the developing gynoecium.

## Grasses versus Arabidopsis

In recent years, since Arabidopsis became a favorite plant model for molecular differentiation studies, many more studies were performed on Arabidopsis than on grasses, such as rice and maize. Moreover, with respect to flowering, most grass research was focused on the differentiation of inflorescences (see Chapter 12 and Fig. 76). As for inflorescences, Barazesh and McSteen listed maize inflorescence developmental mutants, with similar genes from Arabidopsis and rice (Table 5), but no such comparisons were yet provided for floral organs of Arabidopsis vs floral organs of rice and maize. The floral organs of grasses were also given different names; only the terms stamen and carpel were identical in maize, rice, and Arabidopsis. The terms of the flower organs in grasses (rice and maize) are provided in Fig. 87.

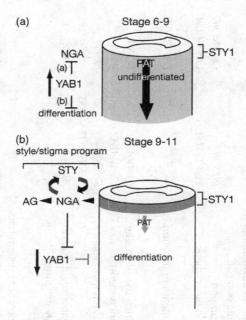

**Fig. 86.** Genetic models for *NGA* activation and function in promoting style development and differentiation in aerial organs. (a) During stages 6–9 of gynoecium development, early ovary factors, such as the *YABBY1* genes (*FIL* and *YAB3*), function to either/or (i) suppress *NGA* gene activity or (ii) suppress *NGA* gene activity passively by preventing the accumulation of an activating signal through an effect that has been defined here simplistically as delayed differentiation. It is suggested that a distal, auxin-based signal promoted by the *STY* genes, including *STY1*, acts to induce *NGA* gene activity at a threshold level. This level is not reached (faint shading) because of efficient PAT (large arrow) maintained in gynoecium at that stage. (b) In stages 9–11, gynoecium *YAB1* efficacy is reduced as is its ability to negatively regulate *NGA* gene activity or passively prevent *NGA* response to an activating signal. Under the auxin-based activator scenario, the polar transport of auxin from the distal site of *STY1*-mediated synthesis becomes inefficient and auxin accumulates to an *NGA*-activating threshold. Upon activation, *NGA* gene activity suppresses *YAB1* while promoting *AG* clade and *SHI/STY* gene family members. These in turn maintain *NGA* gene activity and their own expression through positive feedback. Together, the *NGA*, *STY*, and *AG* clade genes constitute a developmental module for style/stigma morphogenesis. (From Alvarez *et al.*, 2009.)

**Table 5.  Maize Inflorescence Development Mutants, with Similar Genes from Arabidopsis and Rice**

| Maize mutant | Protein | Function | Arabidopsis | Rice |
|---|---|---|---|---|
| **Auxillary meristem initiation** | | | | |
| Barren inflorescence1 (Bif1) | Serine/threonine protein kinase | Regulates auxin transport | — | — |
| barren inflorescence2 (bif2) | Serine/threonine protein kinase | Regulates auxin transport | PINOID | OsPINOID |
| sparse inflorescence1 (spi1) | Flavin mono-oxygenase | Auxin biosynthesis | YUCCA1/4 | OsYUCCA1 |
| barren stalk1 (ba1) | bHLH TF | Ax. meristem initiation | — | LAX PANICLE1 |
| **Meristem size** | | | | |
| thick tassel dwarf1 (td1) | Receptor-like kinase | Restricts meristem size | CLAVATA1 | FON1 |
| fasciated ear2 (fea2) | Receptor-like protein | Restricts meristem size | CLAVATA2 | — |
| knotted1 (kn1) | KNOX TF | Meristem maintenance | SHOOT-MERISTEMLESS | OSH1 |
| abphyl1 (abph1) | Type-A response regulator | Negative regulation of CK signaling | ARR | OsRR |
| **Meristem determinacy** | | | | |
| ramosa1 (ra1) | Zinc finger TF | Confers SPM determinacy | — | — |
| ramosa2 (ra2) | LOB-domain TF | Confers SPM determinacy | ASL4 | OsRA2 |
| ramosa3 (ra3) | Trehalose-6-phosphate phosphatase | Confers SPM determinacy | TPP | SISTER OF RA3 |

(Continued)

**Table 5.** *(Continued)*

| Maize mutant | Protein | Function | Arabidopsis | Rice |
|---|---|---|---|---|
| **Meristem determinacy and/or sex determination** | | | | |
| tasselseed4 (ts4) | miR172 microRNA | Confers SM determinacy | miR172 | — |
| indeterminate spikelet (ids1) | AP2-like TF | Confers SM determinacy | APETALA2 | OsIDS1 |
| Tasselseed6 (Ts6) | | | | OsTS2 |
| tasselseed2 (ts2) | Short chain dehydrogenase/ reductase | Floral organ abortion | ATA1 | OsTS2 |
| anther ear1 (an1) | Ent-kaurene synthase | GA biosynthesis | GA1 | OsCPS |
| Dwarf8 (D8) | DELLA TF | Negative regulator of GA response | GIBBERELLIC ACID INSENSITIVE | SLENDER1 |

TF — transcription factor. From Barazesh and McSteen (2008), where references were provided.

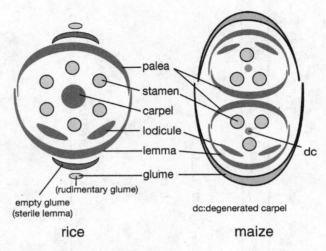

**Fig. 87.**    Flower organs of rice and maize. (From Bommert *et al.*, 2005.)

Barazesh and McSteen (2008) mentioned in their review mutations that affect floral organs in maize. This review and later research reports handle the impacts of genes and phytohormones on the sex expression of grasses, as will be detailed below. However, all the reports together are still far from being able to provide an overall model of the patterning of floret members in grasses.

Interestingly, branching of the inflorescence of grasses may be related to the patterning of floret organs. For example, in the masculinization mutants of maize *anther ear1* (*an1*) and *Dwarf8* (*D8*), the stamens and the lower floret in the ear do not abort, and some of the mutants have fewer long branches in the tassel. Some of these mutants have defects in GA metabolism.

In the feminizing mutants of plants that received the common name *tasselseed* (*ts*) mutants (composed actually of very different genes that encode different proteins), the sex in the florets of the tassel is reversed: it is the stamens rather than the pistils that are aborted. Some of these feminizing mutants, such as *ts4* and *Ts6*, have determinacy defects with increased branching in the tassel and the ear. Other *ts* mutants, such as *ts1* and *ts2*, do not have branching defects by themselves, but their synergistic genetic interaction with *ts4* or *Ts6* indicates that they too function in determinacy. The *ts2* encodes an enzyme with similarities to short chain dehydrogenase/reductase (SDR) enzymes. Biochemical analyses showed that the substracts of TS2 include hydroxysteroid hormones, but it is not yet clear if these include brassinosteroids.

*Ts4* encodes a microRNA (miR172) that is expressed at the boundary of auxillary meristems, where it promotes determinacy and sex determination, in part, by negatively regulating the *AP2*-like transcription factor *indeterminate spikelet1 (ids1)/tasselseed6 (Ts6)*.

Recently, three genes in the *ramosa* pathway were isolated. This pathway promotes the determinacy of the spikelet pair meristem (SPM) in maize. *Ra1* encodes a zinc-finger putative transcription factor and is expressed at the boundary of the SPM and the main axis, implying that it interacts with a mobile signal to impose determinacy on the SPM. It has been proposed that this signal could be a GA because exogenous application of GA has been reported to suppress the *ra1* phenotype. Auxin has also been implicated in the *ramosa* pathway because *ra2*, which encodes LATERAL ORGAN BOUNDARY (LOB) domain transcription factor, is proposed to act downstream of *bif2* (*barren inflorescence2*, see Table 5, a gene having the function of *PINOID* in Arabidopsis), which is involved in auxin transport.

As noted above, in *tasselseed2* (*TS2*) mutant plants (of maize) floret structures in the tassel adopt a female developmental program: they fail to abort pistils while arresting stamen development. Recent studies (Wu *et al.*, 2007) on the *tasselseed2* gene (*TS2*) hint to interesting relationships between this gene, brassinosteroids, and the induction of cell death involved in antagonizing the abortion of the pistil in florets of the tassel. By molecular cloning of *TS2*, it was revealed that it encodes a member of an evolutionary conserved superfamily of short-chain dehydrogenases/reductases. The TS2 is a tetrameric enzyme and it binds NAD(H) and NAD(P)(H). TS2 was established as a plant $3\beta/17\beta$-hydroxysteroid dehydrogenase and carbonyl/quinone reductase. As such it may involve brassinosteroids in the C6 oxidation pathway. Brassinosteroids themselves have not been shown to promote plant cell death. However, these phytohormones display a complex regulatory interaction with other plant hormones, such as abscisic acid, jasmonic acid, and GAs, shown to be essential in plant apoptosis.

Another recent publication (Pineda Rodo *et al.*, 2008) involved zeatin O-glucosylation with a phenocopy of the tasselseed phenotype of maize. These investigators transformed maize to obtain plants with a construct that contained the *ZOG1* gene (encoding a zeatin O-glucosyltransferase from *Phaseolus lunatus*) under the control of the constitutive ubiquitin (Ubi) promoter. They found that roots and leaves of the transformants had greatly increased levels of zeatin-O-glucoside. The vegetative characteristic of the hemizygous and homozygous *Ubi:ZOG1* plants resembled plants that are cytokinin-deficient: shorter statue, thinner stems, narrower leaves, smaller meristems, and strongly branching roots with increased mass. Transformants' leaves had a higher than normal content of active cytokinins compared to those of non-transformed sites. The *Ubi:ZOG1* plants also exhibited delayed senescence when grown during the spring/summer season. The latter phenomenon brings us more than 50 years back, when Richmond and Lang (1957) found that kinetin (a kind of synthetic cytokinin) extended the survival of detached *Xanthium* leaves. The most impressive and probably also surprising change in patterning of the *Ubi:ZOG1* transformed plants was the change in sex expression. The hemizygous transformants had reduced tassels with fewer spikelets but normal and viable pollen, while the homozygous transformants had very small tassels and feminized tassel florets that resembled the tasselseed phenotype. This change in phenotype clearly indicated that there is a link between cytokinins and sex-specific floral development in maize. The authors suggested that this link exists in all monocots. The future will tell...

The involvement of phytohormones in the development of maize floret organ emerges from another recent publication of Acosta *et al.* (2009), by the team of Stephen Dellaporta from Yale University and seven associates from various laboratories. This publication links *tasselseed1* to jasmonic acid signaling. The authors first reminded the readers that in maize, sex determination is controlled by a developmental cascade that leads to the formation of unisexual florets that are derived from an initially bisexual floral meristem. Abortion of pistil primordia in staminate florets is controlled (as narrated above) by a tasselseed-mediated cell death process. The investigators positionally cloned and characterized the function of the sex determination gene *tasselseed1* (*ts1*). They found a sequence (between 5' and 3' untranslated regions) that contains 7 exons and 6 introns. They also located eight *ts1* mutations in the exons. The translated TS1 protein is headed with a predicted chloroplast transit peptide and a lipoxygenase. Lipoxygenases are nonheme iron-containing fatty acid dioxygenases that catalyze the peroxidation of poly-unsaturated fatty acids, such as linoleic acid, α-linolenic acid, and arachidonic acid (see Chapter 6). The first dedicated step in jasmonate biosynthesis is the peroxidation of α-linolenic acid (18:3) by 13-lipogenase, to form hydroperoxyoctadecadienoic acid. The latter conversion was found to be the putative function of TS1. This conversion takes place in the plastids (as detailed earlier in Chapter 6). Thus, TS1 is probably involved in the biosynthesis of the phytohormone jasmonic acid. This supposition was supported by the investigators after the application of 1 mM JA to *ts1,ts1* or *ts2,ts2* mutants: this application restored the phenotype of the florets to staminate [ones].

# Epilogue

In Plato's *Symposium*, there is a discussion on Love — what it is and what its origin is. There were several speakers with very different proposals. For this Epilogue, the claim made by Aristophanes is very relevant. Aristophanes elaborated on the old myth that humans were originally created in "doubles": they had two faces looking "outside," four hands, four legs, and a round body. In some of them, one side was female, while the other side was male; in others there were two female sides, and in a third group there were two male sides. These were very naughty creatures causing great trouble to the Gods. Zeus assembled the Gods, and their decision was to cut these creatures vertically and make some surgical changes so that the resulting

**Table 6.   Plant Hormone Receptors**

| Hormone | Receptor Type | Receptors |
|---|---|---|
| Auxin | F-box protein | TIR1, AFBs |
| Abscisic acid | G-protein, chelatase | GTG1, GTG2, GCR2*, CHLH* |
| Cytokinin | Two-component regulators | CRE1, AHK2, AHK3 |
| Gibberellins | Hormone-sensitive lipase like | GID1 |
| Ethylene | Two-component regulators | ETR1, ERS1, ETR2, EIN4, ERS2 |
| Brassinosteroids | Leucine-rich repeat receptor-like kinases | BRI1 |
| Jasmonic acid | F-box protein | COI1 |
| Salicylic acid | Unknown | |
| Nitric oxide | Unknown | |

From Santner and Estelle (2009), where references were cited.

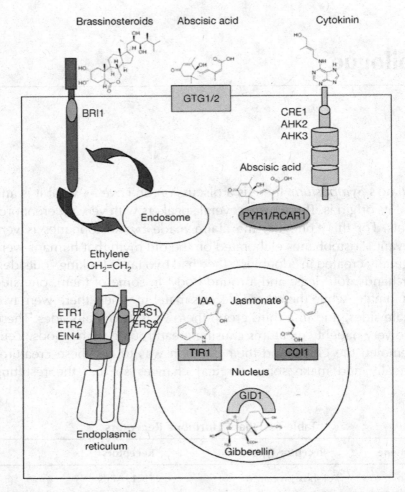

**Fig. 88.** Sites of plant hormone perception. BRI1 is a membrane-associated receptor that cycles between the plasma membrane and endosomal compartments. The extracellular leucine-rich repeat domain binds brassinosteroids and transduces the signal through an intracellular kinase domain. GTG1 and GTG2 are GDCR-type G proteins that bind abscisic acid. They have inherent GTPase activity but also interact with the only canonical G subunit in Arabidopsis. PYR1/RCAR1 is a soluble ABA receptor that represses PP2C phosphatases in the presence of ABA. The cytokinin receptors CRE1, AHK2, and AHK3 are plasma-associated and perceive cytokinin through their extracellular domains. Cytokinin binding triggers a phosphorylation

individuals would have two legs, two hands, one face, and would either be female or male. However, the individuals were then longing for their previous half and were attempting to re-fuse. This lust to re-fuse with one's original other-half is Love. Only the re-fusing of a male and a female could propagate normally.

Bisexuality, in the form of hermaphrodites, is the common form of sexuality in angiosperms: in most flowers, that are at maturity unisexual, there is a very early state of flower meristems where both stamen and pistils can be recognized. Only during later stages of flower differentiation the stamens or the pistils will stop developing and will abort, leading, respectively, to female (pistillate) or male (staminate) mature flowers. If neither of these floral members aborts, the flower will mature as bisexual. Changes in sex expression during lifetime also happen in vertebrate animals (as in some fishes), and even the young mammalian embryos start as hermaphrodites, but then the female organs may be suppressed to lead into unisexual male organisms.

In this book, I first examined a number of phytohormone types that differed vastly in respect to chemical structure, biosynthesis, receptors, and mode of inducing ubiquitination of specific factors involved in plant hormone action. In Part I of this book, I looked at each of the types of the presently-known phytohormone types, while in Part II, I focused on plant organs and tissue, and elaborated on the roles of the different plant hormones on the differentiation of the various organs and tissues, as well as on the interactions between phytohormones in fulfilling these roles. The latter aspect of interaction is a very active

---

**Fig. 88. (*Continued*)** cascade that is ultimately transmitted to response regulators in the nucleus. Like the cytokinin receptors, the known ethylene receptors are two-component regulators. All five receptors are active in the endoplasmic reticulum and transmit their signal through a common downstream component called CTR1. TIR1 and COI1 are F-box proteins that are integral components of SCF-type E2 ligases and recognize the plant hormones auxin and jasmonic acid, respectively. GID1 is a nuclear-localized receptor for gibberellins. Gibberellin binding to GID1 results in the enhanced degradation of DELLA proteins. (From Santner and Estelle, 2009).

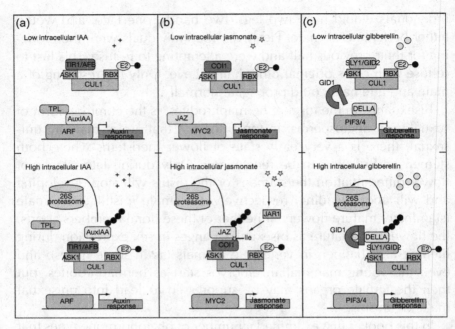

**Fig. 89.** SCFs are required for auxin, jasmonate, and gibberellin signaling. (a) The TIR1/AFB family of F-box proteins are auxin receptors. TIR1 is a component of the SCF complex that also consists of ASK, CUL, and RBX. Auxin binding stabilizes the TIR1-AUX/IAA complex, resulting in degradation of the AUX/IAAs, which in turn releases TPL and permits ARF-dependent transcription. (b) Binding of JA-Ile to COI1 JAZ binding and ubiquitination. This results in the derepression of MYC2-dependent transcription of jasmonate-responsive genes. (c) Gibberellin binding to the GID1 receptor promotes GID1–DELLA complex formation. GID1–DELLA binding promotes the interaction between the C terminus of the DELLA proteins and SCF^SLY1/GID2. Degradation of DELLA promotes the release of PIFs, thus permitting DNA-binding gibberellin responses. In each panel, the hormone receptor is colored red, the substrate protein is light blue, symbols representing hormones are yellow and components of the ubiquitin proteasome pathway are in grey. Black circles represent ubiquitin. (From Santner and Estelle, 2009).

issue in current studies. This is a rather complicated issue that is still mostly unknown but will likely be clarified in the coming years.

A very good summarizing review was recently published by Santner and Estelle (2009). These authors covered the recent advances in the understanding of plant hormones. They included nitric oxide (which I did not include) but avoided the review on peptide hormones, although the latter type of phytohormones have major roles in tissue differentiation, such as the control of the size of the apical shoot meristem.

To provide a general view of the present state of understanding of the plant hormone signaling, Santner and Estelle presented a list of hormone receptors, as is shown in Table 6.

In seven of the listed phytohormones, the receptor types are known, while for salicylic acid, nitric oxide, and strigolactones the receptors were not yet verified. These authors also provided a scheme for the sites of plant hormone perception of brassinosteroids, abscisic acid, cytokinins, ethylene, jasmonates, auxin, and gibberellins (Fig. 88). The SCFs that are required for the ubiquitination that leads to the activation of the auxin response genes, jasmonate response genes, and the gibberellin response genes, were provided in an additional scheme (see Fig. 89). The schemes indicate that the perception of jasmonate is similar to that of auxin. It is also clear now that the ubiquitin-proteosome pathways are important in hormone signaling. It is expected that within the coming years, additional information will be provided on the signaling of plant hormones that affect the architecture of plant organs and tissues, since several talented investigators are now engaged in this field of endeavor.

A book like the present one cannot be updated due to plant patterning being a field of very active endeavor, and it takes several months from the end of writing until the book is printed. Thus, for example, after the text of this book was written, three very good reviews appeared in the same issue of *Current Opinion in Genetics and Development*: "Apical Shoot Meristem Boundary" (by Rast and Simon, 2008), "Transcriptional Factor Interaction in DELLA" (by Daviere *et al.*, 2008), and "Evolution of Developmental Mechanisms in Plants" (by Langdale, 2008).

Readers who wish to be updated in these subjects should refer to these publications.

# References

Abdala, G., G. Castro, M. M. Guinazu, et al. (1996). "Occurrence of jasmonic acid in organs of Solanum tuberosum L and its effect on tuberization." *Plant Growth Regul* **19**(2): 139–143.

Acosta, I. F., H. Laparra, S. P. Romero, et al. (2009). "tasselseed1 is a lipoxy-genase affecting jasmonic acid signaling in sex determination of maize." *Science* **323**: 262–265.

Adams, D. O. and S. F. Yang (1979). "Ethylene biosynthesis — Identification of 1-aminocyclopropane-1-carboxylic acid as an intermediate in the conversion of methionine to ethylene." *Proc Natl Acad Sci USA* **76**(1): 170–174.

Addicott, F. T., J. L. Lyon, K. Ohkuma, et al. (1968). "Abscisic acid: a new name for Abscisin II." *Science* **159**: 1493.

Aitken, A. (2006). "14-3-3 proteins: A historic overview." *Semin Cancer Biol* **16**(3): 162–172.

Alonso, J. M. and A. N. Stepanova (2004). "The ethylene signaling pathway." *Science* **306**(5701): 1513–1515.

Alvarez, J. P., A. Goldshmidt, I. Efroni, et al. (2009). "The NGATHA distal organ development genes are essential for style specification in arabidopsis." *Plant Cell* **21**(5): 1373–1393.

Amasino, R. (2005). "1955: kinetin arrives: The 50th anniversary of a new plant hormone." *Plant Physiol* **138**(3): 1177–1184.

Anastasiou, E., S. Kenz, M. Gerstung, et al. (2007). "Control of plant organ size by KLUH/CYP78A5-dependent intercellular signaling." *Dev Cell* **13**(6): 843–856.

Anastasiou, E. and M. Lenhard (2007). "Growing up to one's standard." *Curr Opin Plant Biol* **10**(1): 63–69.

Angenent, G. C., J. Stuurman, K. C. Snowden and R. Koes (2005). "Use of Petunia to unravel plant meristem functioning." *Trends Plant Sci* **10**(5): 243–250.

Argueso, C. T., M. Hansen and J. J. Kieber (2007). "Regulation of ethylene biosynthesis." *J Plant Growth Regul* **26**(2): 92–105.

Arteca, R. N., D. S. Tsai, C. Schlagnhaufer and N. B. Mandava (1983). "The effect of brassinosteroid on auxin-induced ethylene production by etiolated mung bean segments." *Physiol Plant* **59**(4): 539–544.

Atsmon, D. and E. Galun (1960). "A morphogenetic study os staminate, pistillate and hermaphrodite flowers in Cucumis sativus." *Phytomorphol* **10**: 110–115.

Atsmon, D. and E. Galun (1962). "Physiology of sex in Cucumis sativus: Leaf age patterns and sexual differentiation of Cucumis sativus floral buds." *Ann Bot (Lond)* **26**: 137–146.

Atsmon, D., E. Galun and K. M. Jakob (1965). "Relative time of anthesis in pistillate and staminate cucumber flowers." *Ann Bot (Lond)* **29**: 277–282.

Auldridge, M. E., A. Block, J. T. Vogel, *et al.* (2006). "Characterization of three members of the Arabidopsis carotenoid cleavage dioxygenase family demonstrates the divergent roles of this multifunctional enzyme family." *Plant J* **45**(6): 982–993.

Bai, M. Y., L. Y. Zhang, S. S. Gampala, *et al.* (2007). "Functions of OsBZR1 and 14-3-3 proteins in brassinosteroid signaling in rice." *Proc Natl Acad Sci USA* **104**(34): 13839–13844.

Bainbridge, K., K. Sorefan, S. Ward and O. Leyser (2005). "Hormonally controlled expression of the Arabidopsis MAX4 shoot branching regulatory gene." *Plant J* **44**(4): 569–580.

Baker, A., I. A. Graham, M. Holdsworth, *et al.* (2006). "Chewing the fat: Beta-oxidation in signalling and development." *Trends Plant Sci* **11**(3): 124–132.

Balanza, V., M. Navarrete, M. Trigueros and C. Ferrandiz (2006). "Patterning the female side of Arabidopsis: The importance of hormones." *J Exp Bot* **57**(13): 3457–3469.

Bandyopadhyay, A., J. J. Blakeslee, O. R. Lee, *et al.* (2007). "Interactions of PIN and PGP auxin transport mechanisms." *Biochem Soc Trans* **35**(Pt 1): 137–141.

Bao, F., J. Shen, S. R. Brady, *et al.* (2004). "Brassinosteroids interact with auxin to promote lateral root development in Arabidopsis." *Plant Physiol* **135**(3): 1864–1864.

Barak, M. and T. Trebitsh (2007). "A developmentally regulated GTP binding tyrosine phosphorylated protein A-like cDNA in cucumber (Cucumis sativus L.)." *Plant Mol Biol* **65**(6): 829–837.

Barazesh, S. and P. McSteen (2008). "Hormonal control of grass inflorescence development." *Trends Plant Sci* **13**(12): 656–662.

Barkoulas, M., C. Galinha, S. P. Grigg and M. Tsiantis (2007). "From genes to shape: Regulatory interactions in leaf development." *Curr Opin Plant Biol* **10**(6): 660–666.

Barkoulas, M., A. Hay, E. Kougioumoutzi and M. Tsiantis (2008). "A developmental framework for dissected leaf formation in the Arabidopsis relative Cardamine hirsuta." *Nat Genet* **40**: 1136–1141.

Bartel, B., S. LeClere, M. Magidin and B. K. Zolman (2001). "Inputs to the active indole-3-acetic acid pool: De novo synthesis, conjugate hydrolysis, and indole-3-butyric acid beta-oxidation." *J Plant Growth Regul* **20**(3): 198–216.

Beerling, D. J. and A. J. Fleming (2007). "Zimmermann's telome theory of megaphyll leaf evolution: A molecular and cellular critique." *Curr Opin Plant Biol* **10**(1): 4–12.

Benjamins, R. and B. Scheres (2008). "Auxin: The looping star in plant development." *Annu Rev Plant Biol* **59**: 443–465.

Bennett, T., T. Sieberer, B. Willett, *et al.* (2006). "The Arabidopsis MAX pathway controls shoot branching by regulating auxin transport." *Curr Biol* **16**(6): 553–563.

Berger, Y., S. Harpaz-Saad, A. Brand, *et al.* (2009). "The NAC-domain transcription factor GOBLET specifies leaflet boundaries in compound tomato leaves." *Development* **136**(5): 823–832.

Bergmann, D. C. and F. D. Sack (2007). "Stomatal development." *Ann Rev Plant Biol* **58**: 163–181.

Berleth, T. and T. Sachs (2001). "Plant morphogenesis: Long-distance coordination and local patterning." *Curr Opin Plant Biol* **4**(1): 57–62.

Besserer, A., V. Puech-Pages, P. Kiefer, *et al.* (2006). "Strigolactones stimulate arbuscular mycorrhizal fungi by activating mitochondria." *Plos Biol* **4**(7): 1239–1247.

Beveridge, C. A. (2000). "Long-distance signalling and a mutational analysis of branching in pea." *Plant Growth Regul* **32**: 193–203.

Binder, B. M. (2007). "Rapid kinetic analysis of ethylene growth responses in seedlings: New insights into ethylene signal transduction." *J Plant Growth Regul* **26**(2): 131–142.

Binder, B. M. (2008). "The ethylene receptors: Complex perception for a simple gas." *Plant Sci* **175**(1–2): 8–17.

Binder, B. M., C. O'Malley R, W. Wang, *et al.* (2004). "Arabidopsis seedling growth response and recovery to ethylene. A kinetic analysis." *Plant Physiol* **136**(2): 2913–2920.

Binder, B. M., J. M. Walker, J. M. Gagne, *et al.* (2007). "The Arabidopsis EIN3 binding F-Box proteins EBF1 and EBF2 have distinct but overlapping roles in ethylene signaling." *Plant Cell* **19**(2): 509–523.

Bishop, G. J. (2007). "Refining the plant steroid hormone biosynthesis pathway." *Trends Plant Sci* **12**(9): 377–380.

Bjorklund, S., H. Antti, I. Uddestrand, *et al.* (2007). "Cross-talk between gibberellin and auxin in development of Populus wood: Gibberellin stimulates polar auxin transport and has a common transcriptome with auxin." *Plant J* **52**(3): 499–511.

Blakeslee, J. J., W. A. Peer and A. S. Murphy (2005). "Auxin transport." *Curr Opin Plant Biol* **8**(5): 494–500.

Blanchard, M. G. and E. S. Runkle (2008). "Benzyladenine promotes flowering in doritaenopsis and phalaenopsis orchids." *J Plant Growth Regul* **27**(2): 141–150.

Bleecker, A. B. (1999). "Ethylene perception and signaling: An evolutionary perspective." *Trends Plant Sci* **4**(7): 269–274.

Blein, T., A. Pulido, A. Vialette-Guiraud, *et al.* (2008). "A conserved molecular framework for compound leaf development." *Science* **322**(5909): 1835–1839.

Blilou, I., J. Xu, M. Wildwater, *et al.* (2005). "The PIN auxin efflux facilitator network controls growth and patterning in Arabidopsis roots." *Nature* **433**(7021): 39–44.

Bommert, N., N. Satoh-Nagasawa, D. Jackson and H.-Y. Hirano (2005). "Genetics and evolution of inflorescence and flower development in grasses." *Plant Cell Physiol* **46**: 69–78.

Booker, J., M. Auldridge, S. Wills, *et al.* (2004). "MAX3/CCD7 is a carotenoid cleavage dioxgenase requred for the synthesis of a novel plant signaling molecule." *Curr Biol* **14**: 1232–1238.

Booker, J., T. Sieberer, W. Wright, *et al.* (2005). "MAX1 encodes a cytochrome P450 family member that acts downstream of MAX3/4 to produce a carotenoid-derived branch-inhibiting hormone." *Dev Cell* **8**(3): 443–449.

Boualem, A., M. Fergany, R. Fernandez, *et al.* (2008). "A conserved mutation in an ethylene biosynthesis enzyme leads to andromonoecy in melons." *Science* **321**(5890): 836–838.

Boutte, Y., Y. Ikeda and M. Grebe (2007). "Mechanisms of auxin-dependent cell and tissue polarity." *Curr Opin Plant Biol* **10**(6): 616–623.

Braun, N., J. Wyrzykowska, P. Muller, *et al.* (2008). "Conditional Repression of AUXIN BINDING PROTEIN1 Reveals That It Coordinates Cell Division and Cell Expansion during Postembryonic Shoot Development in Arabidopsis and Tobacco." *Plant Cell* **20**(10): 2746–2762.

Braybrook, S. A. and J. J. Harada (2008). "LECs go crazy in embryo development." *Trends Plant Sci* **13**(12): 624–630.

Buchanan, B. B., W. Gruissem and J. D. Jones (2000). *Biochem Mol Biol Plants.* Rockville, MD, Amer Soc Plant Physiol.

Callos, J. D. and J. I. Medford (1994). "Organ positions and pattern-formation in the shoot apex." *Plant J* **6**(1): 1–7.

Casson, S. and J. E. Gray (2008). "Influence of environmental factors on stomatal development." *New Phytol* **178**(1): 9–23.

Chailakhyan, M. K. (1936). "New facts in support of the hormonal theory of plant development." *Comp Rend Acad Sci USSR* **13**: 79–83.

Chang, C. (2007). "Ethylene biosynthesis,perception and response." *J Plant Growth Regul* **26**: 89–91.

Chang, C., S. F. Kwok, A. B. Bleecker and E. M. Meyerowitz (1993). "Arabidopsis ethylene-response gene ETR1: Similarity of product to two-component regulators." *Science* **262**(5133): 539–544.

Chen, M., J. E. Markham, C. R. Dietrich, *et al.* (2008). "Sphingolipid long-chain base hydroxylation is important for growth and regulation of sphingolipid content and composition in Arabidopsis." *Plant Cell* **20**(7): 1862–1878.

Chen, Y. F., N. Etheridge and G. E. Schaller (2005). "Ethylene signal transduction." *Ann Bot (Lond)* **95**(6): 901–915.

Chen, Y. F., M. D. Randlett, J. L. Findell and G. E. Schaller (2002). "Localization of the ethylene receptor ETR1 to the endoplasmic reticulum of Arabidopsis." *J Biol Chem* **277**(22): 19861–19866.

Chen, Y. F., S. N. Shakeel, J. Bowers, *et al.* (2007). "Ligand-induced degradation of the ethylene receptor ETR2 through a proteasome-dependent pathway in Arabidopsis." *J Biol Chem* **282**(34): 24752–24758.

Cheng, Y., X. Dai and Y. Zhao (2006). "Auxin biosynthesis by the YUCCA flavin monooxygenases controls the formation of floral organs and vascular tissues in Arabidopsis." *Genes Dev* **20**(13): 1790–1799.

Cheng, Y. F., X. H. Dai and Y. D. Zhao (2007). "Auxin synthesized by the YUCCA flavin Monooxygenases is essential for embryogenesis and leaf formation in Arabidopsis." *Plant Cell* **19**(8): 2430–2439.

Chico, J. M., A. Chini, S. Fonseca and R. Solano (2008). "JAZ repressors set the rhythm in jasmonate signaling." *Curr Opin Plant Biol* **11**(5): 486–494.

Chini, A., S. Fonseca, G. Fernandez, *et al.* (2007). "The JAZ family of repressors is the missing link in jasmonate signalling." *Nature* **448**(7154): 666–671.

Chitwood, D. H., M. J. Guo, F. T. S. Nogueira and M. C. P. Timmermans (2007). "Establishing leaf polarity: The role of small RNAs and positional signals in the shoot apex." *Development* **134**(5): 813–823.

Chow, B. and P. McCourt (2006). "Plant hormone receptors: Perception is everything." *Genes Dev* **20**(15): 1998–2008.

Clouse, S. D. (2002). "Brassinosteroid signal transduction: Clarifying the pathway from ligand perception to gene expression." *Mol Cell* **10**(5): 973–982.

Clouse, S. D., M. Langford and T. C. McMorris (1996). "A brassinosteroid-insensitive mutant in Arabidopsis thaliana exhibits multiple defects in growth and development." *Plant Physiol* **111**(3): 671–678.

Clouse, S. D. and J. M. Sasse (1998). "Brassinosteroids: Essential regulators of plant growth and development." *Annu Rev Plant Physiol Plant Mol Biol* **49**: 427–451.

Coen, E. (1999). *The Art of the Genes*. Oxford, Oxford University Press.

Coen, E. S. and E. M. Meyerowitz (1991). "The war of the whorls — Genetic Interactions controlling flower development." *Nature* **353**(6339): 31–37.

Cohen, J. D. and W. J. Meudt (1983). "Investigations on the mechanism of the brassinosteroid response .1. Indole-3-acetic-acid metabolism and transport." *Plant Physiol* **72**(3): 691–694.

Consonni, G., C. Aspesi, A. Barbante, *et al.* (2003). "Analysis of four maize mutants arrested in early embryogenesis reveals an irregular pattern of cell division." *Sex Plant Reprod* **15**(6): 281–290.

Creelman, R. A. and J. E. Mullet (1997). "Biosynthesis and action of jasmonates in plants." *Annu Rev Plant Physiol Plant Mol Biol* **48**: 355–381.

Crozier, A., Y. Kanmiya, G. Bishop and T. Yokota (2000). "Biosynthesis of hormones and elicitor molecules" *Biochem Biol Plants*. B. B. Buchanan, W. Cruissem and R. L. Jones. Rockville, *Amer Soc Plant Physiologists*: 850–929.

Darwin, C. (1875). *Movements and Habits of Climbing Plants*. London, Joun Murray.

Dathe, W., H. Romnsch, A. Preiss, *et al.* (1981). "Endogenous plant hormones of the bean,Vicia faba, (-)-jasmonic acid, a plant growth inhibitor in pericarp." *Planta* **155**: 530–535.

Daviere, J. M., M. de Lucas and S. Prat (2008). "Transcriptional factor interaction: A central step in DELLA function." *Curr Opin Genet Dev* **18**(4): 295–303.

de Lucas, M., J. M. Daviere, M. Rodriguez-Falcon, *et al.* (2008). "A molecular framework for light and gibberellin control of cell elongation." *Nature* **451**(7177): 480–U411.

De Smet, I., S. Vanneste, D. Inze and T. Beeckman (2006). "Lateral root initiation or the birth of a new meristem." *Plant Mol Biol* **60**(6): 871–887.

Delker, C., B. K. Zolman, O. Miersch and C. Wasternack (2007). "Jasmonate biosynthesis in Arabidopsis thaliana requires peroxisomal beta-oxidation enzymes — additional proof by properties of pex6 and aim1." *Phytochemistry* **68**(12): 1642–1650.

Demole, E., E. Lederer and D. Mercier (1962). "Isolement et determination de la structure du jasmonate de methyle, constituant ordorant caracterisitique de l'essence de jasmin." *Helvetica Chimica Acta* **45**: 675–685.

Deprost, D., L. Yao, R. Sormani, *et al.* (2007). "The Arabidopsis TOR kinase links plant growth, yield, stress resistance and mRNA translation." *EMBO Reports* **8**(9): 864–870.

Devoto, A., M. Nieto-Rostro, D. Xie, *et al.* (2002). "COI1 links jasmonate signalling and fertility to the SCF ubiquitin-ligase complex in Arabidopsis." *Plant J* **32**(4): 457–466.

Dharmasiri, N., S. Dharmasiri and M. Estelle (2005a). "The F-box protein TIR1 is an auxin receptor." *Nature* **435**(7041): 441–445.

Dharmasiri, N., S. Dharmasiri, D. Weijers, *et al.* (2005b). "Plant development is regulated by a family of auxin receptor F box proteins." *Dev Cell* **9**(1): 109–119.

Dikic, I., S. Wakatsuki and K. J. Walters (2009). "Ubiquitin-binding domains — from structures to functions." *Nat Rev Mol Cell Biol* **10**(10): 659–671.

Dolfini, S., G. Consonni, C. Viotti, *et al.* (2007). "A mutational approach to the study of seed development in maize." *J Exp Bot* **58**(5): 1197–1205.

Dreher, K. and J. Callis (2007). "Ubiquitin, hormones and biotic stress in plants." *Ann Bot (Lond)* **99**(5): 787–822.

Du, L., F. Jiao, J. Chu, *et al.* (2007). "The two-component signal system in rice (Oryza sativa L.): A genome-wide study of cytokinin signal perception and transduction." *Genomics* **89**(6): 697–707.

Dubrovsky, J. G., M. Sauer, S. Napsucialy-Mendivil, *et al.* (2008). "Auxin acts as a local morphogenetic trigger to specify lateral root founder cells." *Proc Natl Acad Sci USA* **105**(25): 8790–8794.

Edlund, A., S. Eklof, B. Sundberg, *et al.* (1995). "A Microscale Technique for gas chromatography-mass spectrometry measurements of picogram amounts of indole-3-acetic acid in plant tissues." *Plant Physiol* **108**(3): 1043–1047.

Efroni, I., E. Blum, A. Goldshmidt and Y. Eshed (2008). "A protracted and dynamic maturation schedule underlies Arabidopsis leaf development." *Plant Cell* **20**(9): 2293–2306.

Emery, J. F., S. K. Floyd, J. Alvarez, *et al.* (2003). "Radial patterning of Arabidopsis shoots by class IIIHD-ZIP and KANADI genes." *Curr Biol* **13**(20): 1768–1774.

Esau, K. (1977). *Anatomy of Plants, 2nd Edition.* New York, John Wiley & Sons.

Eshed, Y., A. Izhaki, S. F. Baum, *et al.* (2004). "Asymmetric leaf development and blade expansion in Arabidopsis are mediated by KANADI and YABBY activities." *Development* **131**(12): 2997–3006.

Estelle, M. (1996). "Plant tropisms: the ins and outs of auxin." *Current Biology* **6**(12): 1589–1591.

Etheridge, N., B. P. Hall and G. E. Schaller (2006). "Progress report: Ethylene signaling and responses." *Planta* **223**(3): 387–391.

Evans, M. M. S. and M. K. Barton (1997). "Genetics of angiosperm shoot apical meristem development." *Annu Rev Plant Physiol Plant Mol Biol* **48**: 673–701.

Felix, G. and T. Boller (1995). "Systemin induces rapid ion fluxes and ethylene biosyntyhesis in Lycopersicon peruvianum cells." *Plant J* **7**: 381–389.

Ferreira, F. J. and J. J. Kieber (2005). "Cytokinin signaling." *Curr Opin Plant Biol* **8**(5): 518–525.

Fiers, M., E. Golemiec, J. Xu, *et al.* (2005). "The 14-amino acid CLV3, CLE19, and CLE40 peptides trigger consumption of the root meristem in Arabidopsis through a CLAVATA2-dependent pathway." *Plant Cell* **17**(9): 2542–2553.

Fiers, M., G. Hause, K. Boutilier, *et al.* (2004). "Mis-expression of the CLV3/ESR-like gene CLE19 in Arabidopsis leads to a consumption of root meristem." *Gene* **327**(1): 37–49.

Fischer, U., Y. Ikeda, K. Ljung, *et al.* (2006). "Vectorial information for Arabidopsis planar polarity is mediated by combined AUX1, EIN2, and GNOM activity." *Curr Biol* **16**(21): 2143–2149.

Fleming, A. J. (2005). "Formation of primordia and phyllotaxy." *Curr Opin Plant Biol* **8**(1): 53–58.

Fontanet, P. and C. M. Vicient (2008). Maize Embryogenesis. *Plant Embryogenesis (Vol. 427 of Methods in Molecular Biology)*. M. F. Suarez and P. V. Bozhkov. Totowa, N.J., Humana Press (Springer Verlag): 17–29.

Foo, E., E. Bullier, M. Goussot, *et al.* (2005). "The branching gene RAMOSUS1 mediates interactions among two novel signals and auxin in pea." *Plant Cell* **17**(2): 464–474.

Foster, A. S. and E. M. Gifford (1959). *Comparative Morphology of Vascular Plants*. San Francisco, W.H.Freeman & Comp.

Frankel, R. and A. Galun (1977). *Pollination Mechanisms, Reproduction and Plant Breeding*. Heidelberg, Springer Verlag.

Freedberg, D. (2002). *The Eye of the Lynx*. Chicago, University of Chicago Press.

Friml, J. (2003). "Auxin transport — shaping the plant." *Curr Opin Plant Biol* **6**(1): 7–12.

Friml, J., E. Benkova, I. Blilou, *et al.* (2002). "AtPIN4 mediates sink-driven auxin gradients and root patterning in Arabidopsis." *Cell* **108**(5): 661–673.

Friml, J. and M. Sauer (2008). "In their neighbours shadow." *Nature* **453**: 298–299.

Fujioka, S., Y. H. Choi, S. Takatsuto, *et al.* (1996). "Identification of castasterone, 6-deoxocastasterone, typhasterol and 6-deoxotyphasterol from the shoots of Arabidopsis thaliana." *Plant Cell Physiol* **37**(8): 1201–1203.

Fujioka, S. and A. Sakurai (1997). "Conversion of lysine to L-pipecolic acid induces flowering in Lemna paucicostata 151." *Plant Cell Physiol* **38**(11): 1278–1280.

Fukuda, H., Y. Hirakawa and S. Sawa (2007). "Peptide signaling in vascular development." *Curr Opin Plant Biol* **10**: 477–482.

Galun, E. (1959a). "Effect of seed treatment on sex expression in the cucumber." *Experientia* **12**: 218–219.

Galun, E. (1959b). "Effect of gibberellic acid and naphthaleneacetic acid on sex expression and some morphological characters in the cucumber plant." *Phyton* **13**: 1–8.

Galun, E. (1959c). "The role of auxins in the sex expression of the cucumber." *Physiol Plant* **12**: 48–61.

Galun, E. (1959d). "The cucumber tendril — a new test organ for gibberellic acid." *Experientia* **15**: 184–185.

Galun, E. (1961a). "Study of the inheritance of sex expression in the cucumber: The interaction of major genes with modifying genetic and non-genetic factors." *Genetica* **32**: 134–163.

Galun, E. (1961b). "Gibberellic acid as a tool for the estimation of the time interval between physiological bisexuality of cucumber floral buds." *Phyton* **16**: 57–62.

Galun, E. (2003). *Transposable Elements*. Dordrecht, Kluwer Academic Publishers (Now Springer Verlag).

Galun, E. (2005). *RNA Silensing*. Singapore, World Scientific.

Galun, E. (2007). "*Plant Patterning — Structural and Molecular Genetic Aspects.*", Singapore, World Scientific.

Galun, E. and D. Atsmon (1960). "The leaf floral buds relationship of genetic sexual types in the cucumber plant." *Bull Res Counc Israel* **9D**: 43–50.

Galun, E. and A. Breiman (1997). *Transgenic Plants*. London, Imperial College Press.

Galun, E. and E. Galun (2001). *The Manufacture of Medical and Health Products by Transgenic Plants*. London, Imperial College Press.

Galun, E., J. Gressel and A. Keynan (1964). "Suppression of floral induction by actinomycin D — aninhibitor of 'messenger' RNA synthesis." *Life Sci* **3**: 911–915.

Galun, E., S. Izhar and D. Atsmon (1965). "Determination of relative auxin content in hermaphrodite and andromonoecious Cucumis sativus L." *Plant Physiol* **40**(2): 321–326.

Galun, E., Y. Jung and A. Lang (1962). "Culture and sex modification of male cucumber floral buds in vitro." *Nature* **194**: 595–598.

Galun, E., Y. Jung and A. Lang (1963). "Morphogenesis of floral buds of cucumber cultured in vitro." *Dev Biol* **6**: 370–387.

Galun, E. and J. G. Torrey (1969). "Initiation and suppression of apical hairs of Fucus embryos." *Dev Biol* **19**: 447–459.

Galun, M., P. Keller, D. Malki, *et al.* (1983b). "Recovery of Uranium (Vi) from solution using precultured penicillium biomass." *Water Air Soil Poll* **20**(2): 221–232.

Galun, M., P. Keller, M. Malki, *et al.* (1983a). "Removal of uranium (VI) from solution by fungal biomass and fungal wall related biopolymers." *Science* **219**: 285–286.

Gampala, S. S., T. W. Kim, J. X. He, *et al.* (2007). "An essential role for 14-3-3 proteins in brassinosteroid signal transduction in Arabidopsis." *Dev Cell* **13**(2): 177–189.

Gan, Y., H. Yu, J. Peng and P. Broun (2007). "Genetic and molecular regulation by DELLA proteins of trichome development in Arabidopsis." *Plant Physiol* **145**(3): 1031–1042.

Geisler, M. and A. S. Murphy (2006). "The ABC of auxin transport: The role of p-glycoproteins in plant development." *FEBS Lett* **580**(4): 1094–1102.

Gendron, J. M. and Z.-Y. Wang (2007). "Multiple mechanisms modulate brassinosteroid signaling." *Curr Opin Plant Biol* **10**: 436–441.

Gifford, E. M. (1954). "The shoot apex in angiospems." *Bot Rev* **20**: 477–529.

Giulini, A., J. Wang and D. Jackson (2004). "Control of phyllotaxy by the cytokinin-inducible response regulator homologue ABPHYL1." *Nature* **430**(7003): 1031–1034.

Goldberg, R. B., G. Depaiva and R. Yadegari (1994). "Plant embryogenesis — zygote to seed." *Science* **266**(5185): 605–614.

Golovko, A., F. Sitbon, E. Tillberg and B. Nicander (2002). "Identification of a tRNA isopentenyltransferase gene from Arabidopsis thaliana." *Plant Mol Biol* **49**(2): 161–169.

Gomez-Mena, C. and R. Sablowski (2008). "Arabidopsis thaliana homeobox Gene1 establishes the basal boundaries of shoot organs and controls stem growth." *Plant Cell* **20**(8): 2059–2072.

Gomez-Roldan, V., S. Fermas, P. B. Brewer, *et al.* (2008). "Strigolactone inhibition of shoot branching." *Nature* **455**(7210): 189–194.

Graebe, J. E. (1987). "Gibberellin biosynthesis and control." *Annu Rev Plant Physiol* **38**: 419–465.

Griffiths, J., K. Murase, I. Rieu, *et al.* (2006). "Genetic characterization and functional analysis of the GID1 gibberellin receptors in Arabidopsis." *Plant Cell* **18**(12): 3399–3414.

Guilfoyle, T. J. and G. Hagen (2007). "Auxin response factors." *Curr Opin Plant Biol* **10**: 453–460.

Guo, H. and J. R. Ecker (2003). "Plant responses to ethylene gas are mediated by SCF(EBF1/EBF2)-dependent proteolysis of EIN3 transcription factor." *Cell* **115**(6): 667–677.

Guo, H. and J. R. Ecker (2004). "The ethylene signaling pathway: New insights." *Curr Opin Plant Biol* **7**(1): 40–49.

Hake, S. (2008). "Inflorescence architecture: The transition from branches to flowers." *Curr Biol* **18**: R1106–R1108.

Hall, B. P., S. N. Shakeel and G. E. Schaller (2007). "Ethylene receptors: Ethylene perception and signal transduction." *J Plant Growth Regul* **26**(2): 118–130.

Hamberg, M., A. Sanz and C. Castresana (1999). "Alpha-oxidation of fatty acids in higher plants — Identification of a pathogen-inducible oxygenase (PIOX) as an alpha-dioxygenase and biosynthesis of 2-hydroperoxylinolenic acid." *J Biol Chem* **274**(35): 24503–24513.

Hamilton, A. J., M. Bouzayen and D. Grierson (1991). "Identification of a tomato gene for the ethylene-forming enzyme by expression in yeast." *Proc Natl Acad Sci USA* **88**(16): 7434–7437.

Hao, S. G., C. B. Beck and D. M. Wang (2003). "Structure of the earliest leaves: Adaptations to high concentrations of atmospheric $CO_2$." *Int J Plant Sci* **164**(1): 71–75.

Hareven, D., T. Gutfinger, A. Parnis, *et al.* (1996). "The making of a compound leaf: Genetic manipulation of leaf architecture in tomato." *Cell* **84**(5): 735–744.

Hartweck, L. M. and N. E. Olszewski (2006). "Rice GIBBERELLIN INSENSITIVE DWARF1 is a gibberellin receptor that illuminates and raises questions about GA signaling." *Plant Cell* **18**(2): 278–282.

Haughn, G. W. and C. R. Somerville (1988). "Genetic control of morphogenesis in Arabidopsis." *Dev Genet* **9**: 73–89.

Hay, A. and M. Tsiantis (2006). "The genetic basis for differences in leaf form between Arabidopsis thaliana and its wild relative Cardamine hirsuta." *Nat Genet* **38**(8): 942–947.

Hayama, R., B. Agashe, E. Luley, *et al.* (2007). "A circadian rhythm set by dusk determines the expression of FT homologs and the short-day photoperiodic flowering response in Pharbitis." *Plant Cell* **19**(10): 2988–3000.

He, J. X., J. M. Gendron, Y. Sun, *et al.* (2005). "BZR1 is a transcriptional repressor with dual roles in brassinosteroid homeostasis and growth responses." *Science* **307**(5715): 1634–1638.

He, Y. and R. M. Amasino (2005) "The role of chromatin modification in flowering time control." *Trends Plant Sci* **10**(1): 30–35.

Hedden, P. and Y. Kamiya (1997). "Gibberellin biosynthesis: Enzymes, genes and their regulation." *Ann Rev Plant Physiol Plant Mol Biol* **48**: 431–460.

Hedden, P., A. L. Phillips, M. C. Rojas, *et al.* (2002). "Gibberellin biosynthesis in plants and fungi: A case of convergent evolution?" *J Plant Growth Regul* **20**(4): 319–331.

Hershko, A. and A. Ciechanover (1998). "The ubiquitin system." *Annu Rev Biochem* **67**: 425–479.

Heslop-Harrison, J. (1957). "The experimental modification of sex expression in flowering plants." *Bio Rev* **32**: 38–90.

Heslop-Harrison, J. (1961). "Experimental control of sexuality and inflorescence structure in Zea mays." *Proc Linn Soc London* **172**: 108–123.

Higuchi, M., M. S. Pischke, A. P. Mahonen, *et al.* (2004). "In planta functions of the Arabidopsis cytokinin receptor family." *Proc Natl Acad Sci USA* **101**(23): 8821–8826.

Himmelbach, A., Y. Yang and E. Grill (2003). "Relay and control of abscisic acid signaling." *Curr Opin in Plant Biol* **6**(5): 470–479.

Hirano, K., M. Ueguchi-Tanaka and M. Matsuoka (2008). "GID1-mediated gibberellin signaling in plants." *Trends Plant Sci* **13**(4): 192–199.

Hirayama, T. and K. Shinozaki (2007). "Perception and transduction of abscisic acid signals: Keys to the function of the versatile plant hormone ABA." *Trends Plant Sci* **12**(8): 343–351.

Hirose, N., K. Takei, T. Kuroha, *et al.* (2008). "Regulation of cytokinin biosynthesis, compartmentalization and translocation." *J Exp Bot* **59**(1): 75–83.

Hochholdinger, F., W. J. Park, M. Sauer and K. Woll (2004). "From weeds to crops: genetic analysis of root development in cereals." *Trends Plant Sci* **9**(1): 42–48.

Hofer, J., L. Turner, C. Moreau, *et al.* (2009). "Tendril-less regulates tendril formation in pea leaves." *Plant Cell* **21**(2): 420–428.

Hofmeister, W. (1868). *Allgemeine Morphologie der Gewachse*, Enggelmann.

Horiguchi, G., A. Ferjani, U. Fujikura and H. Tsukaya (2006). "Coordination of cell proliferation and cell expansion in the control of leaf size in Arabidopsis thaliana." *J Plant Res* **119**(1): 37–42.

Howe, G. A. (2004). "Jasmonates as signals in the wound response." *J Plant Growth Regul* **23**(3): 223–237.

Howell, S. H., S. Lall and P. Che (2003). "Cytokinins and shoot development." *Trends Plant Sci* **8**(9): 453–459.

Hu, J. H., M. G. Mitchum, N. Barnaby, *et al.* (2008). "Potential sites of bioactive gibberellin production during reproductive growth in Arabidopsis." *Plant Cell* **20**(2): 320–336.

Hua, J. and E. M. Meyerowitz (1998). "Ethylene responses are negatively regulated by a receptor gene family in Arabidopsis thaliana." *Cell* **94**(2): 261–271.

Huang, S., R. E. Cerny, Y. Qi, *et al.* (2003). "Transgenic studies on the involvement of cytokinin and gibberellin in male development." *Plant Physiol* **131**(3): 1270–1282.

Inoue, T., M. Higuchi, Y. Hashimoto, *et al.* (2001). "Identification of CRE1 as a cytokinin receptor from Arabidopsis." *Nature* **409**(6823): 1060–1063.

Irish, E. E. (1999). "Maize sex determination" *Sex Det Plants*. C. Ainsworth. Oxford (UK), BIOS Scientific Publishers.

Irish, V. F. (2008). "The Arabidopsis petal: A model for plant organogenesis." *Trends Plant Sci* **13**(8): 430–436.

Ito, Y., I. Nakanomyo, H. Motose, *et al.* (2006). "Dodeca-CLE peptides as suppressors of plant stem cell differentiation." *Science* **313**(5788): 842–845.

Ivanchenko, M. G., W. C. Coffeen, T. L. Lomax and J. G. Dubrovsky (2006). "Mutations in the Diageotropica (Dgt) gene uncouple patterned cell division during lateral root initiation from proliferative cell division in the pericycle." *Plant J* **46**(3): 436–447.

Jaeger, K. E., A. Graf and P. A. Wigge (2006). "The control of flowering in time and space." *J Exp Bot* **57**(13): 3415–3418.

Jaeger, K. E. and P. A. Wigge (2007). "FT protein acts as a long-range signal in Arabidopsis." *Curr Biol* **17**(12): 1050–1054.

Jaffe, M. J. and A. W. Galston (1968). "Physiology of tendrils." *Annu Rev Plant Physiol* **19**: 417–434.

Jasinski, S., P. Piazza, J. Craft, *et al.* (2005). "KNOX action in Arabidopsis is mediated by coordinate regulation of cytokinin and gibberellin activities." *Curr Biol* **15**(17): 1560–1565.

Jasinski, S., A. Tattersall, P. Piazza, *et al.* (2008). "PROCERA encodes a DELLA protein that mediates control of dissected leaf form in tomato." *Plant J* **56**(4): 603–612.

Jiao, Y., O. S. Lau and X. W. Deng (2007). "Light-regulated transcriptional networks in higher plants." *Nat Rev Genet* **8**(3): 217–230.

Jonsson, H., M. G. Heisler, B. E. Shapiro, *et al.* (2006). "An auxin-driven polarized transport model for phyllotaxis." *Proc Natl Acad Sci USA* **103**(5): 1633–1638.

Kakimoto, T. (2001). "Identification of plant cytokinin biosynthetic enzymes as dimethylallyl diphosphate: ATP/ADP isopentenyltransferases." *Plant Cell Physiol* **42**(7): 677–685.

Kaparakis, G. and P. G. Alderson (2008). "Role for cytokinins in somatic embryogenesis of pepper (Capsicum annuum L.)." *J Plant Growth Regul* **27**(2): 110–114.

Kato-Noguchi, H. and Y. Tanaka (2008). "Effect of ABA-beta-D-glucopyranosyl ester and activity of ABA-beta-D-glucosidase in Arabidopsis thaliana." *J Plant Physiol* **165**(7): 788–790.

Katsir, L., H. S. Chung, A. J. Koo and G. A. Howe (2008a). "Jasmonate signaling: A conserved mechanism of hormone sensing." *Curr Opin Plant Biol* **11**(4): 428–435.

Katsir, L., A. L. Schilmiller, P. E. Staswick, *et al.* (2008b). "COI1 is a critical component of a receptor for jasmonate and the bacterial virulence factor coronatine." *Proc Natl Acad Sci USA* **105**(19): 7100–7105.

Kazan, K. and J. M. Manners (2008). "Jasmonate signaling: Towards an integrated view." *Plant Physiol* **146**: 1459–1468.

Kende, H. (1993). "Ethylene biosynthesis." *Ann Rev Plant Physiol Plant Mol Biol* **44**: 283–307.

Kende, H. and J. Zeevaart (1997). "The five "Classical" plant hormones." *Plant Cell* **9**(7): 1197–1210.

Kepinski, S. (2006). "Integrating hormone signaling and patterning mechanisms in plant development." *Curr Opin Plant Biol* **9**(1): 28–34.

Kepinski, S. and O. Leyser (2005). "The Arabidopsis F-box protein TIR1 is an auxin receptor." *Nature* **435**(7041): 446–451.

Kerstetter, R. A. and S. Hake (1997). "Shoot meristem formation in vegetative development." *Plant Cell* **9**: 1001–1010.

Kevany, B. M., D. M. Tieman, M. G. Taylor, *et al.* (2007). "Ethylene receptor degradation controls the timing of ripening in tomato fruit." *Plant J* **51**(3): 458–467.

Kidner, C. A. and M. C. P. Timmermans (2007). "Mixing and matching pathways in leaf polarity." *Curr Opin Plant Biol* **10**(1): 13–20.

Kieber, J. J. (1997). "The ethylene response pathway in Arabidopsis." *Annu Rev Plant Physiol Plant Mol Biol* **48**: 277–296.

Kieber, J. J. R., M. G. Roman, K. a. Feldmann and J. R. Ecker (1993). "Ctr1, a negative regulator of the ethylene response pathway in Arabidopsis, encodes a member of the raf family of protein-kinases." *Cell* **72**(3): 427–441.

Kim, I. and P. C. Zambryski (2008). Intercellular trafficking of macromolecules during embryogenesis. *Plant Embryogenesis (Vol. 427 of Methods in Molecular Biology)*. M. F. Suarez and P. V. Bozhkov. Totowa, N.J., Humana Press (Springer Verlag): 145–155.

Kimura, S., D. Koenig, J. Kang, *et al.* (2008). "Natural variation in leaf morphology results from mutation of a novel KNOX gene." *Curr Biol* **18**(9): 672–677.

Klee, H. J. (2004). "Ethylene signal transduction. Moving beyond Arabidopsis." *Plant Physiol* **135**(2): 660–667.

Kloosterman, B., C. Navarro, G. Bijsterbosch, *et al.* (2007). "StGA2ox1 is induced prior to stolon swelling and controls GA levels during potato tuber development." *Plant J* **52**(2): 362–373.

Knopf, R. R. and T. Trebitsh (2006). "The female-specific Cs-ACS1G gene of cucumber. A case of gene duplication and recombination between the non-sex-specific 1-aminocyclopropane-1-carboxylate synthase gene and a branched-chain amino acid transaminase gene." *Plant Cell Physiol* **47**(9): 1217–1228.

Knott, J. E. (1934). "Effect of a localized photoperiod on spinach." *Pro Am Soc Hort Sci* **31**: 152–154.

Koes, R. (2008). "Evolution and development of virtual inflorescences." *Trends Plant Sci* **13**(1): 1–3.

Kondo, T., S. Sawa, A. Kinoshita, *et al.* (2006). "A plant peptide encoded by CLV3 identified by *in situ* MALDI-TOF MS analysis." *Science* **313**(5788): 845–848.

Kramer, E. M. and M. J. Bennett (2006). "Auxin transport: A field in flux." *Trends Plant Sci* **11**(8): 382–386.

Krizek, B. A. (2009). "Making bigger plants: Key regulators of final organ size." *Curr Opin Plant Biol* **12**: 17–22.

Kubicki, B. (1970). "Cucumber hybrid seed production based on gynoecious lines multiplied with the aid of complementary hermaphroditic lines." *Genet Polonicum* **11**: 181–186.

Kuhlemeier, C. (2007). "Phyllotaxis." *Trends Plant Sci* **12**(4): 143–150.

Kurakawa, T., N. Ueda, M. Maekawa, *et al.* (2007). "Direct control of shoot meristem activity by a cytokinin-activating enzyme." *Nature* **445**(7128): 652–655.

Kutter, C., H. Schob, M. Stadler, *et al.* (2007). "MicroRNA-mediated regulation of stomatal development in Arabidopsis." *Plant Cell* **19**(11): 2417–2429.

Laibach, F. and F. J. Kribben (1949). "Der Einfluss von Wuchsstoff auf die Bildung mannnlicher und weiblicher Bluten bei einer monoziischen Pflanze (Cucumis sativus)." *Ber Deut Botan Gesell* **62**: 53–55.

Lang, A. (1965). Physiology of flower initiation. *Ency Plant Physiol* W. Ruhland. Berlin, Springer Verlag. **15/1**: 1380–1536.

Langdale, J. A. (2008). "Evolution of developmental mechanisms in plants." *Curr Opin Genet Dev* **18**(4): 368–373.

Lange, T. (1998). "Molecular biology of gibberellin synthesis." *Planta* **204**: 409–419.

Lau, S., G. Jurgens and I. De Smet (2008). "The evolving complexity of the auxin pathway." *Plant Cell* **20**(7): 1738–1746.

Laux, T., T. Wurschum and H. Breuninger (2004). "Genetic regulation of embryonic pattern formation." *Plant Cell* **16**: S190–S202.

Lebel-Hardenack, S., D. Ye, H. Koutnikova, *et al.* (1997). "Conserved expression of a TASSELSEED2 homolog in the tapetum of the dioecious Silene latifolia and Arabidopsis thaliana." *Plant J* **12**(3): 515–526.

Leevers, S. J. and H. McNeill (2005). "Controlling the size of organs and organisms." *Curr Opin Cell Biol* **17**(6): 604–609.

Letham, D. S. (1963). "Zeatin, a factor inducing cell division isolated from Zea mays." *Life Sci* **2**: 569–573.

Leyser, O. (2005). "Auxin distribution and plant pattern formation: How many angels can dance on the point of PIN?" *Cell* **121**(6): 819–822.

Leyser, O. (2006). "Dynamic integration of auxin transport and signalling." *Curr Biol* **16**(11): R424–433.

Li, H. and H. Guo (2007). "Molecular basis of the ethylene signaling and response pathway in Arabidopsis." *J Plant Growth Regul* **26**: 106–117.

Li, J. (2005). "Brassinosteroid signaling: From receptor kinases to transcription factors." *Curr Opin Plant Biol* **8**(5): 526–531.

Li, J. and J. Chory (1997). "A putative leucine-rich repeat receptor kinase involved in brassinosteroid signal transduction." *Cell* **90**(5): 929–938.

Li, J. and H. Jin (2007). "Regulation of brassinosteroid signaling." *Trends Plant Sci* **12**(1): 37–41.

Li, L., J. Xu, Z. H. Xu and H. W. Xue (2005). "Brassinosteroids stimulate plant tropisms through modulation of polar auxin transport in Brassica and Arabidopsis." *Plant Cell* **17**(10): 2738–2753.

Li, S., A. Lauri, M. Ziemann, *et al.* (2009). "Nuclear activity of ROXY1, a glutaredoxin interacting with TGA factors, is required for petal development in Arabidopsis thaliana." *Plant Cell* **21**(2): 429–441.

Lifschitz, E., T. Eviatar, A. Rozman, *et al.* (2006a). "The tomato FT gene triggers conserved systemic flowering signals that regulate termination and substitute for light and photoperiodic stimuli." *Comp Biochem Phys A* **143**(4): S166–S167.

Lifschitz, E., T. Eviatar, A. Rozman, *et al.* (2006b). "The tomato FT ortholog triggers systemic signals that regulate growth and flowering and substitute for diverse environmental stimuli." *Proc Natl Acad Sci USA* **103**(16): 6398–6403.

Lin, M. K., H. Belanger, Y. J. Lee, *et al.* (2007). "FLOWERING LOCUS T protein may act as the long-distance florigenic signal in the cucurbits." *Plant Cell* **19**(5): 1488–1506.

Lippman, Z. B., O. Cohen, J. P. Alvarez, *et al.* (2008). "The making of a compound inflorescence in tomato and related nightshades." *Plos Biol* **6**(11): 2424–2435.

Liu, J., K. F. Xia, J. C. Zhu, *et al.* (2006). "The nightshade proteinase inhibitor IIb gene is constitutively expressed in glandular trichomes." *Plant and Cell Physiol* **47**(9): 1274–1284.

Liu, X., Y. Yue, B. Li, *et al.* (2007). "A G protein-coupled receptor is a plasma membrane receptor for the plant hormone abscisic acid." *Science* **315**(5819): 1712–1716.

Ljung, K., A. K. Hull, J. Celenza, *et al.* (2005). "Sites and regulation of auxin biosynthesis in Arabidopsis roots." *Plant Cell* **17**(4): 1090–1104.

Lopez-Bucio, J., E. Hernandez-Abreu, L. Sanchez-Calderon, *et al.* (2005). "An auxin transport independent pathway is involved in phosphate stress-induced root architectural alterations in arabidopsis. Identification of BIG as a mediator of auxin in pericycle cell activation." *Plant Physiol* **137**(2): 681–691.

Lorenzo, O. and R. Solano (2005). "Molecular players regulating the jasmonate signalling network." *Curr Opin Plant Biol* **8**(5): 532–540.

Lurssen, K., K. Naumann and R. Schroder (1979). "1-Aminocyclopropane-1-carboxylic acid — intermediate of the ethylene biosynthesis in higher-plants." *Zeitschrift fur Pflanzenphysiologie* **92**(4): 285–294.

MacMillan, J. (1997). "Biosynthesis of the gibberellin plant hormones." *Nat Prod Rep* **14**: 221–243.

MacMillan, J. (2002). "Occurrence of gibberellins invascular plants, fungi, and bacteria." *J Plant Growth Regul* **20**(4): 387–442.

Mahonen, A. P., M. Higuchi, K. Tormakangas, *et al.* (2006). "Cytokinins regulate a bidirectional phosphorelay network in Arabidopsis." *Curr Biol* **16**(11): 1116–1122.

Malamy, J. E. and P. N. Benfey (1997a). "Organization and cell differentiation in lateral roots of Arabidopsis thaliana." *Development* **124**(1): 33–44.

Malamy, J. E. and P. N. Benfey (1997b). "Down and out in Arabidopsis: The formation of lateral roots." *Trends Plant Sci* **2**(10): 390–396.

Mandaokar, A., B. Thines, B. Shin, *et al.* (2006). "Transcriptional regulators of stamen development in Arabidopsis identified by transcriptional profiling." *Plant J* **46**(6): 984–1008.

Mathieu, J., N. Warthmann, F. Kuttner and M. Schmid (2007). "Export of FT protein from phloem companion cells is sufficient for floral induction in Arabidopsis." *Curr Biol* **17**(12): 1055–1060.

Matsubayashi, Y. and Y. Sakagami (2006). "Peptide hormones in plants." *Annu Rev Plant Biol* **57**: 649–674.

Mattsson, J., W. Ckurshumova and T. Berleth (2003). "Auxin signaling in Arabidopsis leaf vascular development." *Plant Physiol* **131**(3): 1327–1339.

Mattsson, J., Z. R. Sung and T. Berleth (1999). "Responses of plant vascular systems to auxin transport inhibition." *Development* **126**(13): 2979–2991.

McCourt, P. (1999). "Genetic analysis of hormone signaling." *Ann Rev Plant Physiol Plant Mol Biol* **50**: 219–243.

McCourt, P. and R. Creelman (2008). "The ABA receptors — we report you decide." *Curr Opin Plant Biol* **11**: 474–478.

McCully, M. E. (1975). The development of lateral roots. *Devel Funct Roots.* J. G. A. K. Torrey, D.T., Academic Press: 105–124.

McGrath, K. C., B. Dombrecht, J. M. Manners, *et al.* (2005). "Repressor- and activator-type ethylene response factors functioning in jasmonate signaling and disease resistance identified via a genome-wide screen of Arabidopsis transcription factor gene expression." *Plant Physiol* **139**(2): 949–959.

McSteen, P. and S. Hake (2001). "Barren inflorescence2 regulates axillary meristem development in the maize inflorescence." *Development* **128**(15): 2881–2891.

McSteen, P. and O. Leyser (2005). "Shoot branching." *Ann Rev Plant Biol* **56**: 353–374.

McSteen, P., S. Malcomber, A. Skirpan, *et al.* (2007). "Barren inflorescence2 encodes a co-ortholog of the PINOID serine/threonine kinase and is required for organogenesis during inflorescence and vegetative development in maize." *Plant Physiol* **144**(2): 1000–1011.

McSteen, P. and Y. Zhao (2008). "Plant hormones and signaling: Common themes and new developments." *Dev Cell* **14**(4): 467–473.

Meyer, A., O. Miersch, C. Buttner, *et al.* (1984). "Occurrence of the plant-growth regulator jasmonic acid in plants." *J Plant Growth Regul* **3**(1): 1–8.

Meyerowitz, E. M. (1987). "Arabidopsis thaliana." *Annu Rev Genet* **21**: 93–111.

Meyerowitz, E. M. (2001). "Prehistory and history of Arabidopsis research." *Plant Physiol* **125**(1): 15–19.

Miersch, O., H. Bohlmann and C. Wasternack (1999). "Jasmonates and related compounds from Fusarium oxysporum." *Phytochemistry* **50**(4): 517–523.

Miersch, O., a. Meyer, S. Vorkefeld and G. Sembdner (1986). "Occurrence of (+)-7-Iso-jasmonic acid in Vicia faba L and its biological-activity." *J Plant Growth Regul* **5**(2): 91–100.

Miller, C. O. (1961). "Kinetin-like compound in maize." *Proc Nat Acad Sci USA* **47**: 170–174.

Miller, C. O., F. Skoog, F. S. Okumura, *et al.* (1955). "Structure and synthesis of kinetin." *J Amer Chem Soc* **78**: 2662–2663.

Minina, E. G. (1938). "On the phenotipical modification of sexual characters in higher plants under the influence of nutrition and other external factors (in Russian)." *Comp Rend Acad Sci USSR (Doklady)* **21**: 302–305.

Mitchell, J. W., N. Mandava, J. F. Worley, *et al.* (1970). "Brassins — a new family of plant hormones from rape pollen." *Nature* **225**(5237): 1065–1066.

Miyawaki, K., P. Tarkowski, M. Matsumoto-Kitano, *et al.* (2006). "Roles of Arabidopsis ATP/ADP isopentenyltransferases and tRNA isopentenyl-transferases in cytokinin biosynthesis." *Proc Natl Acad Sci USA* **103**(44): 16598–16603.

Mok, D. W. and M. C. Mok (2001). "Cytokinin metabolism and action." *Annu Rev Plant Physiol Plant Mol Biol* **52**: 89–118.

Moon, J., G. Parry and M. Estelle (2004). "The ubiquitin-proteasome pathway and plant development." *Plant Cell* **16**(12): 3181–3195.

Mora-Garcia, S., G. Vert, Y. H. Yin, *et al.* (2004). "Nuclear protein phosphatases with Kelch-repeat domains modulate the response to bras sino steroids in Arabidopsis." *Genes Dev* **18**(4): 448–460.

Morris, S. E., M. C. H. Cox, J. J. Ross, *et al.* (2005). "Auxin dynamics after decapitation are not correlated with the initial growth of axillary buds." *Plant Physiol* **138**(3): 1665–1672.

Mouchel, C. F., K. S. Osmont and C. S. Hardtke (2006). "BRX mediates feedback between brassinosteroid levels and auxin signalling in root growth." *Nature* **443**(7110): 458–461.

Muller, B. and J. Sheen (2007). "Advances in cytokinin signaling." *Science* **318**(5847): 68–69.

Muller, B. and J. Sheen (2008). "Cytokinin and auxin interaction in root stem-cell specification during early embryogenesis." *Nature* **453**(7198): 1094–U1097.

Nacry, P., G. Canivenc, B. Muller, *et al.* (2005). "A role for auxin redistribution in the responses of the root system architecture to phosphate starvation in Arabidopsis." *Plant Physiol* **138**(4): 2061–2074.

Nadeau, J. A. (2009). "Stomatal development: New signals and fate determinants." *Curr Opin Plant Biol* **12**(1): 29–35.

Nagpal, P., C. M. Ellis, H. Weber, *et al.* (2005). "Auxin response factors ARF6 and ARF8 promote jasmonic acid production and flower maturation." *Development* **132**(18): 4107–4118.

Nakajima, M., A. Shimada, Y. Takashi, *et al.* (2006). "Identification and characterization of Arabidopsis gibberellin receptors." *Plant J* **46**(5): 880–889.

Nakatsuka, A., S. Murachi, H. Okunishi, *et al.* (1998). "Differential expression and internal feedback regulation of 1-aminocyclopropane-1-carboxylate synthase, 1-aminocyclopropane-1-carboxylate oxidase, and ethylene receptor genes in tomato fruit during development and ripening." *Plant Physiol* **118**(4): 1295–1305.

Nambara, E. and A. Marion-Poll (2005). "Abscisic acid biosynthesis and catabolism." *Annu Rev Plant Biol* **56**: 165–185.

Napoli, C. A., C. A. Beveridge and K. C. Snowden (1999). "Reevaluating concepts of apical dominance and the control of axillary bud outgrowth." *Curr Topics Dev Biol* **44**: 127–169.

Nawy, T., W. Lukowitz and M. Bayer (2008). "Talk global, act local — Patterning the Arabidopsis embryo." *Curr Opin Plant Biol* **11**(1): 28–33.

Nemhauser, J. L. (2008). "Dawning of a new era: Photomorphogenesis as an integrated molecular network." *Curr Opin Plant Biol* **11**(1): 4–8.

Nemhauser, J. L. and J. Chory (2004). "Bring it on: New insights into the mechanism of brassinosteroid action." *J Exp Bot* **55**(395): 265–270.

Nickell, L. G., Ed. (1983). *Plant Growth Reg Chemicals.* Boca Raton, CRC Press, Inc.

Notaguchi, M., M. Abe, T. Kimura, *et al.* (2008). "Long-distance, graft-transmissible action of Arabidopsis FLOWERING LOCUS T protein to promote flowering." *Plant Cell Physiol* **49**: 1645–1658.

Ohnishi, T., A. M. Szatmari, B. Watanabe, *et al.* (2006). "C-23 hydroxylation by Arabidopsis CYP90C1 and CYP90D1 reveals a novel shortcut in brassinosteroid biosynthesis." *Plant Cell* **18**(11): 3275–3288.

Okushima, Y., H. Fukaki, M. Onoda, *et al.* (2007). "ARF7 and ARF19 regulate lateral root formation via direct activation of LBD/ASL genes in Arabidopsis." *Plant Cell* **19**(1): 118–130.

Ongaro, V. and O. Leyser (2008). "Hormonal control of shoot branching." *J Exp Bot* **59**: 67–74.

Ori, N., A. R. Cohen, A. Etzioni, *et al.* (2007). "Regulation of LANCEOLATE by miR319 is required for compound-leaf development in tomato." *Nat Genet* **39**(6): 787–791.

Ori, N., Y. Eshed, G. Chuck, *et al.* (2000). "Mechanisms that control knox gene expression in the Arabidopsis shoot." *Development* **127**(24): 5523–5532.

Ortega-Martinez, O., M. Pernas, R. J. Carol and L. Dolan (2007). "Ethylene modulates stem cell division in the Arabidopsis thaliana root." *Science* **317**(5837): 507–510.

Paciorek, T., E. Zazimalova, N. Ruthardt, *et al.* (2005). "Auxin inhibits endocytosis and promotes its own efflux from cells." *Nature* **435**(7046): 1251–1256.

Paponov, I. A., W. D. Teale, M. Trebar, *et al.* (2005). "The PIN auxin efflux facilitators: Evolutionary and functional perspectives." *Trends Plant Sci* **10**(4): 170–177.

Park, S. and J. J. Harada (2008). Arabidopsis embryogenesis. *Plant Embryogenesis (Vol. 427 of Methods in Molecular Biology).* M. F. Suarez and P. V. Bozhkov. Totowa, N.J., Humana Press (Springer Verlag ): 3–16.

Parthier, B. (1990). "Jasmonates — Hormonal regulators or stress factors in leaf senescence." *J Plant Growth Regul* **9**(1): 57–63.

Pauw, B. and J. Memelink (2004). "Jasmonate-responsive gene expression." *J Plant Growth Regul* **23**(3): 200–210.

Pei, Z. M. and K. Kuchitsu (2005). "Early ABA signaling events in guard cells." *J Plant Growth Regul* **24**(4): 296–307.

Pekker, I., J. P. Alvarez and Y. Eshed (2005). "Auxin response factors mediate Arabidopsis organ asymmetry via modulation of KANADI activity." *Plant Cell* **17**(11): 2899–2910.

Peng, P. and J. Li (2003). "Brassinosteroid signal transduction: A mix of conservation and novelty." *J Plant Growth Regul* **22**(4): 298–312.

Perata, P. and L. A. Voesenek (2007). "Submergence tolerance in rice requires Sub1A, an ethylene-response-factor-like gene." *Trends Plant Sci* **12**(2): 43–46.

Perl-Treves, R. and E. Galun (1985). "The Cucumis plastome: Physical map, intrageneric varietion and phylogenetic relationships." *Theor Appl Genet* **71**: 417–429.

Perl-Treves, R. and P. A. Rajagopalan (2006). Close, yet separate: Patterns of male and female floral development in monoecious species. *Flower Dev Manipulation* C. C. Ainsworth, Blackwell: 1–23.

Perl-Treves, R., D. Zamir, N. Navot and E. Galun (1985). "Phylogeny of Cucumis based on isozyme variability and its comparison with plastome phylogeny." *Theor Appl Genet* **71**: 430–436.

Piazza, P., S. Jasinski and M. Tsiantis (2005). "Evolution of leaf developmental mechanisms." *New Phytol* **167**(3): 693–710.

Pineda Rodo, A., N. Brugiere, R. Vankova, *et al.* (2008). "Over-expression of a zeatin O-glucosylation gene in maize leads to growth retardation and tasselseed formation." *J Exp Bot* **59**(10): 2673–2686.

Pinon, V., J. P. Etchells, P. Rossignol, *et al.* (2008). "Three PIGGYBACK genes that specifically influence leaf patterning encode ribosomal proteins." *Development* **135**(7): 1315–1324.

Pringle, P. (2008). *The Murder of Nikolai Vavilov: The Story of Stalin's Persecution of one of the Great Scientists of the Twentieth Century*, Simon & Schuster.

Prusinkiewicz, P., Y. Erasmus, B. Lane, *et al.* (2007). "Evolution and development of inflorescence architectures." *Science* **316**(5830): 1452–1456.

Rampey, R. A., S. LeClere, M. Kowalczyk, *et al.* (2004). "A family of auxin-conjugate hydrolases that contributes to free indole-3-acetic acid levels during Arabidopsis germination." *Plant Physiol* **135**(2): 978–988.

Rast, M. I. and R. Simon (2008). "The meristem-to-organ boundary: More than an extremity of anything." *Curr Opin Genet Dev* **18**(4): 287–294.

Raveh, D., E. Huberman and E. Galun (1973). "In vitro culture of tobacco protoplasts: Use of feeder techniques to support division of cells plated at low densities." *In Vitro* **9**: 216–222.

Razem, F. A., A. El-Kereamy, S. R. Abrams and R. D. Hill (2006). "The RNA-binding protein FCA is an abscisic acid receptor." *Nature* **439**(7074): 290–294.

Rebocho, A. B., M. Bliek, E. Kusters, *et al.* (2008). "Role of EVERGREEN in the development of the cymose petunia inflorescence." *Dev Cell* **15**(3): 437–447.

Redei, G. P. (1970). "Arabidopsis thaliana — a review of the genetics and bilology." *Bibliographica Genetica* **20**: 1–151.

Redei, G. P. (1975). "Arabidopsis as a genetic tool." *Annu Rev Genet* **9**: 111–127.

Reinhardt, D. (2005). "Phyllotaxis — a new chapter in an old tale about beauty and magic numbers." *Curr Opin Plant Biol* **8**(5): 487–493.

Reinhardt, D., M. Frenz, T. Mandel and C. Kuhlemeier (2003). "Microsurgical and laser ablation analysis of interactions between the zones and layers of the tomato shoot apical meristern." *Development* **130**(17): 4073–4083.

Reinhardt, D., M. Frenz, T. Mandel and C. Kuhlemeier (2005). "Microsurgical and laser ablation analysis of leaf positioning and dorsoventral patterning in tomato." *Development* **132**(1): 15–26.

Reinhardt, D., T. Mandel and C. Kuhlemeier (2000). "Auxin regulates the initiation and radial position of plant lateral organs." *Plant Cell* **12**(4): 507–518.

Richmond, A. E. and A. Lang (1957). "Effect of kinetin on protein content and survival of detached Xanthim leaves." *Science* **125**: 650–651.

Rieu, I., O. Ruiz-Rivero, N. Fernandez-Garcia, *et al.* (2008). "The gibberellin biosynthetic genes AtGA20ox1 and AtGA20ox2 act, partially redundantly, to promote growth and development throughout the Arabidopsis life cycle." *Plant J* **53**(3): 488–504.

Rimon Knopf, R. and T. Trebitsh (2006). "Female-specific *Cs-ACSIG* gene of cucumber. A case of gene duplication and recombination between the non-sex specific l-aminocyclopropane-l-carboxylate synthase gene and a branched-chain amino acid transaminase gene." *Plant Cell Physiol* **47**: 1217–1228.

Risk, J. M., R. C. Macknight and C. L. Day (2008). "FCA does not bind abscisic acid." *Nature*: E5–E6.

Robles, L. M., J. S. Wampole, M. J. Christians and P. B. Larsen (2007). "Arabidopsis enhanced ethylene response 4 encodes an EIN3-interacting TFIID transcription factor required for proper ethylene response, including ERF1 induction." *J Exp Bot* **58**(10): 2627–2639.

Rodriguez-Concepcion, M. and A. Boronat (2002). "Elucidation of the methylerythritol phosphate pathway for isoprenoid biosynthesis in bacteria and plastids. A metabolic milestone achieved through genomics." *Plant Physiol* **130**(3): 1079–1089.

Rohmer, M. (1999). "The discovery of a mevalonate-independent pathway for isoprenoid biosynthesis in bacteria, algae and higher plants." *Nat Prod Rep* **16**(5): 565–574.

Roig-Villanova, I., J. Bou-Torrent, A. Galstyan, *et al.* (2007). "Interaction of shade avoidance and auxin responses: A role for two novel atypical bHLH proteins." *EMBO J* **26**(22): 4756–4767.

Rudich, J. and E. Zamski (1985). Citrullus lunatus. *Handbook of flowering* A. Halevy. Boca Ratom, FL, CRC Press. **2**: 272–274.

Sablowski, R. (2007). "The dynamic plant stem cell niches." *Curr Opin Plant Biol* **10**(6): 639–644.

Sachs, J. (1887). *Vorlesungen ueber Pflanzenphysiologie.* Liepzig, Engelmann.

Sachs, T. (1981). "The control of the patterned differentiation of vascular tissues." *Adv Bot Res* **9**: 151–262.

Sachs, T. (1991). The polarization of tissues. *Pattern Formation in Plant Tissues.* T. Sachs. Cambridge, Cambridge University Press: 52–69.

Sakakibara, H. (2006). "Cytokinins: Activity, biosynthesis, and translocation." *Annu Rev Plant Biol* **57**: 431–449.

Salman-Minkov, A., A. Levi, S. Wolf and T. Trebitsh (2008). "ACC synthase genes are polymorphic in watermelon (Citrullus spp.) and differentially expressed in flowers and in response to auxin and gibberellin." *Plant Cell Physiol* **49**(5): 740–750.

Santner, A. and M. Estelle (2009). "Recent advances and emerging trends in plant hormone signalling." *Nature* **459**: 1071–1078.

Sato, T. and A. Theologis (1989). "Cloning the mRNA encoding 1-aminocyclopropane-1-carboxylate synthase, the key enzyme for ethylene biosynthesis in plants." *Proc Natl Acad Sci USA* **86**(17): 6621–6625.

Sauer, M. and J. Friml (2008). Visualization of auxin gradients in embryogenesis. *Plant Embryogenesis (Vol. 427 of Methods in Nolecular Biology).*

M. F. Suarez and P. V. Bozhkov. Totowa, N.J., Humana Press (Springer Verlag): 135–144.

Scarpella, E., D. Marcos, J. Friml and T. Berleth (2006). "Control of leaf vascular patterning by polar auxin transport." *Gene Dev* **20**(8): 1015–1027.

Schaller, F., A. Schaller and A. Stintzi (2004). "Biosynthesis and metabolism of jasmonates." *J Plant Growth Regul* **23**(3): 179–199.

Scheres, B. and L. Xu (2006). "Polar auxin transport and patterning: Grow with the flow." *Gene Dev* **20**(8): 922–926.

Schilmiller, A. L. and G. A. Howe (2005). "Systemic signaling in the wound response." *Curr Opin Plant Biol* **8**(4): 369–377.

Schilmiller, A. L., A. J. K. Koo and G. A. Howe (2007). "Functional diversification of acyl-coenzyme a oxidases in jasmonic acid biosynthesis and action." *Plant Physiol* **143**(2): 812–824.

Schuetz, M., T. Berleth and J. Mattsson (2008). "Multiple MONOPTEROS-dependent pathways are involved in leaf initiation." *Plant Physiol* **148**(2): 870–880.

Schwartz, S. H., X. Q. Qin and J. A. D. Zeevaart (2003). "Elucidation of the indirect pathway of abscisic acid biosynthesis by mutants, genes, and enzymes." *Plant Physiol* **131**(4): 1591–1601.

Schwarz-Sommer, Z., P. Huijser, W. Nacken, *et al.* (1990). "Genetic control of flower development by homeotic genes in Antirrhinum majus." *Science* **250**: 931–936.

Schwechheimer, C. (2008). "Understanding gibberellic acid signaling — are we there yet?" *Curr Opin Plant Biol* **11**(1): 9–15.

Schwechheimer, C. and K. Schwager (2004). "Regulated proteolysis and plant development." *Plant Cell Rep* **23**(6): 353–364.

Schwechheimer, C. and B. C. Williger (2009). "Shedding light on gibberellic acid signalling." *Curr Opin Plant Biol* **12**: 57–62.

Scutt, C. P., M. Vinauger-Douard, C. Fourquin, *et al.* (2006). "An evolutionary perspective on the regulation of carpel development." *J Exp Bot* **57**(10): 2143–2152.

Seo, M. and T. Koshiba (2002). "Complex regulation of ABA biosynthesis in plants." *Trends Plant Sci* **7**(1): 41–48.

Serna, L. (2009). "Emerging parallels between stomatal and muscle cell lineages." *Plant Physiol* **149**(4): 1625–1631.

Shalit, A., A. Rozman, A. Goldshmidt, *et al.* (2009). "The flowering hormone florigen functions as a general systemic regulator of growth and termination." *Proc Natl Acad Sci USA* **106**(20): 8392–8397.

Shani, E., O. Yanai and N. Ori (2006). "The role of hormones in shoot apical meristem function." *Curr Opin Plant Biol* **9**(5): 484–489.

Shantz, E. M. and F. C. Steward (1955). "Identification of compound A from coconut milk as 1,3-diphenylurea." *J Amer Chem Soc* **77**: 6351–6353.

Shen, Y. Y., X. F. Wang, F. Q. Wu, *et al.* (2006). "The Mg-chelatase H subunit is an abscisic acid receptor." *Nature* **443**(7113): 823–826.

Sheridan, W. F. and J. K. Clark (1987). "Maize embryogeny — a promising experimental system." *Trends Genet* **3**(1): 3–6.

Shifriss, O. (1961). "Sex control in cucumbers." *J Heredity* **52**: 5–12.

Shifriss, O. and E. Galun (1956). "Sex expression in the cucumber." *Proc Amer Soc Hort Sci* **67**: 479–486.

Siegel, S., P. Keller, M. Galun, *et al.* (1986). "Biosorption of lead and chromium by Penicillium preparations." *Water Air Soil Poll* **27**(1–2): 69–75.

Simons, J. L., C. A. Napoli, B. J. Janssen, *et al.* (2007). "Analysis of the DECREASED APICAL DOMINANCE genes of petunia in the control of axillary branching." *Plant Physiol* **143**(2): 697–706.

Skoog, F. and D. J. Armstrong (1970). "Cytokinins." *Annu Rev Plant Physiol* **21**: 359–384.

Smalle, J., J. Kurepa, P. Yang, *et al.* (2003). "The pleiotropic role of the 26S proteasome subunit RPN10 in Arabidopsis growth and development supports a substrate-specific function in abscisic acid signaling." *Plant Cell* **15**(4): 965–980.

Smalle, J. and R. D. Vierstra (2004). "The ubiquitin 26S proteasome proteolytic pathway." *Annu Rev Plant Biol* **55**: 555–590.

Smith, R. S., S. Guyomarc'h, T. Mandel, *et al.* (2006). "A plausible model of phyllotaxis." *Proc Natl Acad Sci USA* **103**(5): 1301–1306.

Snow, M. and R. Snow (1931). "Experiments in phyllotaxis." *Philos Trans R Soc London Ser B* **221**: 1–43.

Snowden, K. C., A. J. Simkin, B. J. Janssen, *et al.* (2005). "The Decreased apical dominance1/Petunia hybrida CAROTENOID CLEAVAGE DIOXYGENASE8 gene affects branch production and plays a role in leaf senescence, root growth, and flower development." *Plant Cell* **17**(3): 746–759.

Somerville, C. R. and W. L. Ogren (1979). "Phosphoglycolate phosphatase-deficient mutant of Arabidopsis." *Nature* **280**(5725): 833–836.

Sorefan, K., J. Booker, K. Haurogne, *et al.* (2003). "MAX4 and RMS1 are orthologous dioxygenase-like genes that regulate shoot branching in Arabidopsis and pea." *Genes Dev* **17**(12): 1469–1474.

Souer, E., A. B. Rebocho, M. Bliek, *et al.* (2008). "Patterning of inflorescences and flowers by the F-box protein DOUBLE TOP and the LEAFY homolog ABERRANT LEAF AND FLOWER of petunia." *Plant Cell* **20**(8): 2033–2048.

Spanu, P., D. Reinhardt and T. Boller (1991). "Analysis and cloning of the ethylene-forming enzyme from tomato by functional expression of its mRNA in Xenopus laevis oocytes." *Embo J* **10**(8): 2007–2013.

Spartz, A. K. and W. M. Gray (2008). "Plant hormone receptors: New perceptions." *Gene Dev* **22**(16): 2139–2148.

Staswick, P. E. (2008). "JAZing up jasmonate signaling." *Trends Plant Sci* **13**(2): 66–71.

Steeves, T. A. and I. M. Sussex (1989). *Patterns in Plant Development*. New York, Cambridge University Press.

Stepanova, A. N., J. Robertson-Hoyt, J. Yun, *et al.* (2008). "TAA1-mediated auxin biosynthesis is essential for hormone crosstalk and plant development." *Cell* **133**(1): 177–191.

Stirnberg, P., I. J. Furner and H. M. Ottoline Leyser (2007). "MAX2 participates in an SCF complex which acts locally at the node to suppress shoot branching." *Plant J* **50**(1): 80–94.

Suarez, M. F. and P. V. Bozhkov (2008). *Plant Embryogenesis*. Totowa, N.J., Humana Press (Springer Verlag).

Sugimoto-Shirasu, K. and K. Roberts (2003). ""Big it up": endoreduplication and cell-size control in plants." *Curr Opin Plant Biol* **6**(6): 544–553.

Sugiyama, K., T. Kanno and M. Morishita (1998). "Evaluation method of female flower bearing ability in watermelon using silver thiosulfate (STS)." *J Japan Soc Hort Sci* **67**(2): 185–189.

Sussex, I. M. (1989). "Developmental programming of the shoot meristem." *Cell* **56**(2): 225–229.

Svistoonoff, S., A. Creff, M. Reymond, *et al.* (2007). "Root tip contact with low-phosphate media reprograms plant root architecture." *Nat Gene* **39**(6): 792–796.

Swain, S. M. and D. P. Singh (2005). "Tall tales from sly dwarves: Novel functions of gibberellins in plant development." *Trends Plant Sci* **10**(3): 123–129.

Swarup, R., J. Friml, A. Marchant, *et al.* (2001). "Localization of the auxin permease AUX1_ suggests two functionally distinct hormone transport pathways operate in the Arabidopsis root apex." *Genes Dev* **15**(20): 2648–2653.

Takei, K., N. Ueda, K. Aoki, *et al.* (2004a). "AtIPT3 is a key determinant of nitrate-dependent cytokinin biosynthesis in Arabidopsis." *Plant Cell Physiol* **45**(8): 1053–1062.

Takei, K., T. Yamaya and H. Sakakibara (2004b). "Arabidopsis CYP735A1 and CYP735A2 encode cytokinin hydroxylases that catalyze the biosynthesis of trans-Zeatin." *J Biol Chem* **279**(40): 41866–41872.

Takeno, K. and R. P. Pharis (1982). "Brassinosteroid-induced bending of the leaf lamina of dwarf rice seedlings — an auxin-mediated phenomenon." *Plant Cell Physiol* **23**(7): 1275–1281.

Tan, X., L. I. Calderon-Villalobos, M. Sharon, *et al.* (2007). "Mechanism of auxin perception by the TIR1 ubiquitin ligase." *Nature* **446**(7136): 640–645.

Tang, W., T. W. Kim, J. A. Oses-Prieto, *et al.* (2008). "BSKs mediate signal transduction from the receptor kinase BRI1 in Arabidopsis." *Science* **321**(5888): 557–560.

Tao, Y., J. L. Ferrer, K. Ljung, *et al.* (2008). "Rapid synthesis of auxin via a new tryptophan-dependent pathway is required for shade avoidance in plants." *Cell* **133**(1): 164–176.

Teaster, N. D., C. M. Motes, Y. Tang, *et al.* (2007). "N-Acylethanolamine metabolism interacts with abscisic acid signaling in Arabidopsis thaliana seedlings." *Plant Cell* **19**(8): 2454–2469.

Thimann, K. V. and F. Skoog (1933). "Studies on the growth hormone of plants. III. The inhibiting action of the growth substance on bud development." *Proc Nat Acad Sci (USA)* **19**: 714–716.

Thines, B., L. Katsir, M. Melotto, *et al.* (2007). "JAZ repressor proteins are targets of the SCF(COI1) complex during jasmonate signalling." *Nature* **448**(7154): 661–665.

To, J. P. and J. J. Kieber (2008). "Cytokinin signaling: Two-components and more." *Trends Plant Sci* **13**(2): 85–92.

Torii, K. U. (2004). "Leucine-rich repeat receptor kinases in plants: Structure, function, and signal transduction pathways." *Int Rev Cytol* **234**: 1–46.

Torrey, J. G. (1986). Endogenous and exogenous influences on the regulation of lateral root formation. *New Root Formation in Plants and Cuttings*. M. B. Jackson, Martinus Nijhoff: 32–66.

Torrey, J. G. and E. Galun (1970). "Apolar embryos of Fucus resulting from osmotic and chemical treatment." *Amer J Botany* **57**: 111–119.

Traw, M. B. and J. Bergelson (2003). "Interactive effects of jasmonic acid, salicylic acid, and gibberellin on induction of trichomes in Arabidopsis." *Plant Physiol* **133**(3): 1367–1375.

Trebitsh, T., J. Rudich and J. Riov (1987). "Auxin, biosynthesis of ethylene and sex expression in cucumber (Cucumis sativus)." *Plant Growth Regul* **5**(2): 105–113.

Trebitsh, T., J. E. Staub and S. D. ONeill (1997). "Identification of a 1-aminocyclopropane-1-carboxylic acid synthase gene linked to the female (F) locus that enhances female sex expression in cucumber." *Plant Physiol* **113**(3): 987–995.

Trevaskis, B., M. N. Hemming, E. S. Dennis and W. J. Peacock (2007). "The molecular basis of vernalization-induced flowering in cereals." *Trends Plant Sci* **12**(8): 352–357.

Trewavas, A. J. (1982). "Growth substance sensitivity: The limiting factor in plant development." *Physiol Plantarum* **55**: 60–72.

Trigueros, M., M. Navarrete-Gomez, S. Sato, *et al.* (2009). "The NGATHA genes direct style development in the Arabidopsis gynoecium." *Plant Cell* **21**: 1394–1409.

Tsiantis, M., M. I. N. Brown, G. Skibinski and J. A. Langdale (1999). "Disruption of auxin transport is associated with aberrant leaf development in maize." *Plant Physiol* **121**(4): 1163–1168.

Tsiantis, M. and A. Hay (2003). "Comparative plant development: The time of the leaf?" *Nat Rev Genet* **4**(3): 169–180.

Tucker, M. R., A. Hinze, E. J. Tucker, *et al.* (2008). "Vascular signalling mediated by ZWILLE potentiates WUSCHEL function during shoot meristem stem cell development in the Arabidopsis embryo." *Development* **135**(17): 2839–2843.

Tucker, M. R. and T. Laux (2007). "Connecting the paths in plant stem cell regulation." *Trends Cell Biol* **17**(8): 403–410.

Turck, F., F. Fornara and G. Coupland (2008). "Regulation and identity of florigen: FLOWERING LOCUS T moves center stage." *Annu Rev Plant Biol* **59**: 573–594.

Turing, A. (1952). "The chemical basis of differentiation." *Philos Trans R Soc London Ser B* **237**: 37–72.

Ubeda-Tomas, S., R. Swarup, J. Coates, *et al.* (2008). "Root growth in Arabidopsis requires gibberellin/DELLA signalling in the endodermis." *Nat Cell Biol* **10**(5): 625–628.

Ueda, J. and J. Kato (1980). "Isolation and identification of a senescence-promoting substance from wormwood (Artemisia absinthium L)." *Plant Physiol* **66**(2): 246–249.

Ueguchi, C., H. Koizumi, T. Suzuki and T. Mizuno (2001). "Novel family of sensor histidine kinase genes in Arabidopsis thaliana." *Plant Cell Physiol* **42**(2): 231–235.

Ueguchi-Tanaka, M., M. Ashikari, M. Nakajima, *et al.* (2005). "GIBBERELLIN INSENSITIVE DWARF1 encodes a soluble receptor for gibberellin." *Nature* **437**(7059): 693–698.

Ueguchi-Tanaka, M., M. Nakajima, E. Katoh, *et al.* (2007a). "Molecular interactions of a soluble gibberellin receptor, GID1, with a rice DELLA protein, SLR1, and gibberellin." *Plant Cell* **19**(7): 2140–2155.

Ueguchi-Tanaka, M., M. Nakajima, A. Motoyuki and M. Matsuoka (2007b). "Gibberellin receptor and its role in gibberellin signaling in plants." *Annu Rev Plant Biol* **58**: 183–198.

Umehara, M., A. Hanada, S. Yoshida, *et al.* (2008). "Inhibition of shoot branching by new terpenoid plant hormones." *Nature* **455**(7210): 195–200.

Van Norman, J. M., R. L. Frederick and L. E. Sieburth (2004). "BYPASS1 negatively regulates a root-derived signal that controls plant architecture." *Curr Biol* **14**(19): 1739–1746.

Van Norman, J. M. and L. E. Sieburth (2007). "Dissecting the biosynthetic pathway for the bypass1 root-derived signal." *Plant J* **49**(4): 619–628.

Vardi, A., P. Spiegelroy and E. Galun (1975). "Citrus cell culture — isolation of protoplasts, plating densities, effect of mutagens and regeneration of embryos." *Plant Sci Lett* **4**(4): 231–236.

Vert, G. and J. Chory (2006). "Downstream nuclear events in brassinosteroid signalling." *Nature* **441**(7089): 96–100.

Vert, G., J. L. Nemhauser, N. Geldner, *et al.* (2005). "Molecular mechanisms of steroid hormone signaling in plants." *Annu Rev Cell Dev Biol* **21**: 177–201.

Vert, G., C. L. Walcher, J. Chory and J. L. Nemhauser (2008). "Integration of auxin and brassinosteroid pathways by Auxin Response Factor 2." *Proc Natl Acad Sci USA* **105**(28): 9829–9834.

Vick, B. A. and D. C. Zimmerman (1984). "Biosynthesis of jasmonic acid by several plant species." *Plant Physiol* **75**(2): 458–461.

Vieten, A., M. Sauer, P. B. Brewer and J. Friml (2007). "Molecular and cellular aspects of auxin-transport-mediated development." *Trends Plant Sci* **12**(4): 160–168.

Voesenek, L. A., T. D. Colmer, R. Pierik, *et al.* (2006). "How plants cope with complete submergence." *New Phytol* **170**(2): 213–226.

von Arnold, S. and D. Claphan (2008). Spruce embryogenesis. *Plant Embryogenesis (Vol. 427 of Methods in Molecular Biology).* M. F. Suarez and P. V. Bozhkov. Totowa, N.J., Humana Press (Springer Verlag): 31–47.

Wang, K. L., H. Li and J. R. Ecker (2002). "Ethylene biosynthesis and signaling networks." *Plant Cell* **14** (Suppl): S131–151.

Wang, L., S. Allmann, J. S. Wu and I. T. Baldwin (2008b). "Comparisons of LIPOXYGENASE3- and JASMONATE-RESISTANT4/6-silenced plants reveal that jasmonic acid and jasmonic acid-amino acid conjugates play different roles in herbivore resistance of Nicotiana attenuata." *Plant Physiol* **146**(3): 904–915.

Wang, X. and J. Chory (2006). "Brassinosteroids regulate dissociation of BKI1, a negative regulator of BRI1 signaling, from the plasma membrane." *Science* **313**(5790): 1118–1122.

Wang, X., U. Kota, K. He, *et al.* (2008a). "Sequential transphosphorylation of the BRI1/BAK1 receptor kinase complex impacts early events in brassinosteroid signaling." *Dev Cell* **15**(2): 220–235.

Wang, Z. Y. and J. X. He (2004). "Brassinosteroid signal transduction — choices of signals and receptors." *Trends Plant Sci* **9**(2): 91–96.

Ward, J. T., B. Lahner, E. Yakubova, *et al.* (2008). "The effect of iron on the primary root elongation of Arabidopsis during phosphate deficiency." *Plant Physiol* **147**(3): 1181–1191.

Ward, S. P. and O. Leyser (2004). "Shoot branching." *Curr Opin Plant Biol* **7**(1): 73–78.

Wardlaw, C. W. (1957). "On the organization and reactivity of the shoot apex in vascular plants." *Amer J Botany* **44**: 176–185.

Wasternack, C. (2004). "Introductory remarks on biosynthesis and diversity in actions." *J Plant Growth Regul* **23**(3): 167–169.

Wasternack, C. (2006). Oxylipins: Biosynthesis, signal transduction and action. *Plant Hormone Signaling. Annu Plant Revs.* Haden, P. and S. Thomas. Oxford, Blackwell Publishing: 182–228.

Wasternack, C. (2007). "Jasmonates: An update on biosynthesis, signal transduction and action in plant stress response, growth and development." *Ann Bot (Lond)* **100**(4): 681–697.

Wasternack, C. and B. Hause (2002). "Jasmonates and octadecanoids: Signals in plant stress responses and development." *Prog Nucleic Acid Res Mol Biol* **72**: 165–221.

Weiler, E. W., T. Albrecht, B. Groth, *et al.* (1993). "Evidence for the involvement of jasmonates and their octadecanoid precursors in the tendril coiling response of Bryonia dioica." *Phytochemistry* **32**(3): 591–600.

Weiss, D. and N. Ori (2007). "Mechanisms of cross talk between gibberellin and other hormones." *Plant Physiol* **144**(3): 1240–1246.

Went, F. and K. V. Thimann (1937). *Phytohormones.* Mcmillan, N.Y.

West, M. A. L. and J. J. Harada (1993). "Embryogenesis in higher plants — an overview." *Plant Cell* **5**(10): 1361–1369.

Wolpert, L. (1998). *Principles of Development.* Oxford, Oxford University Press.

Woodward, A. W. and B. Bartel (2005). "Auxin: Regulation, action, and interaction." *Ann Bot (Lond)* **95**(5): 707–735.

Wu, G., W. C. Lin, T. B. Huang, *et al.* (2008). "KANADI1 regulates adaxial-abaxial polarity in Arabidopsis by directly repressing the transcription of ASYMMETRIC LEAVES2." *Proc Natl Acad Sci USA* **105**(42): 16392–16397.

Wu, X. Q., S. Knapp, A. Stamp, *et al.* (2007). "Biochemical characterization of TASSELSEED 2, an essential plant short-chain dehydrogenase/reductase with broad spectrum activities." *Febs J* **274**(5): 1172–1182.

Xing, S., M. G. Rosso and S. Zachgo (2005). "ROXY1, a member of the plant glutaredoxin family, is required for petal development in Arabidopsis thaliana." *Development* **132**: 1555–1565.

Xing, S. and S. Zachgo (2008). "ROXY1 and ROXY2, two Arabidopsis glutaredoxin genes, are required for anther development." *Plant J* **53**: 790–801.

Yadav, V., C. Mallappa, S. N. Gangappa, *et al.* (2005). "A basic helix-loop-helix transcription factor in Arabidopsis, MYC2, acts as a repressor of blue light-mediated photomorphogenic growth." *Plant Cell* **17**(7): 1953–1966.

Yamaguchi-Shinozaki, K. and K. Shinozaki (2006). "Transcriptional regulatory networks in cellular responses and tolerance to dehydration and cold stresses." *Annu Rev Plant Biol* **57**: 781–803.

Yamane, H., H. Takagi, H. Abe, *et al.* (1980). "Identification of jasmonic acid related compounds and their structure-activity relationships on growth of rice seedlings." *Agric Biol Chem* **44**: 2857–2864.

Yamasaki, S., N. Fujii and H. Takahashi (2003). "Characterization of ethylene effects on sex determination in cucumber plants." *Sex Plant Reprod* **16**(3): 103–111.

Yan, Y. X., S. Stolz, A. Chetelat, *et al.* (2007). "A downstream mediator in the growth repression limb of the jasmonate pathway." *Plant Cell* **19**(8): 2470–2483.

Yanai, O., E. Shani, K. Dolezal, *et al.* (2005). "Arabidopsis KNOXI proteins activate cytokinin biosynthesis." *Curr Biol* **15**(17): 1566–1571.

Yang, J. C., J. H. Zhang, K. Liu, *et al.* (2007). "Abscisic acid and ethylene interact in rice spikelets in response to water stress during meiosis." *J Plant Growth Regul* **26**(4): 318–328.

Yevdakova, N. A., V. Motyka, J. Malbeck, *et al.* (2008). "Evidence for importance of tRNA-dependent cytokinin biosynthetic pathway in the moss Physcomitrella patens." *J Plant Growth Regul* **27**(3): 271–281.

Yin, Y. H., Z. Y. Wang, S. Mora-Garcia, *et al.* (2002). "BES1 accumulates in the nucleus in response to brassinosteroids to regulate gene expression and promote stem elongation." *Cell* **109**(2): 181–191.

Yokota, T. (1997). "The structure, biosynthesis and function of brassinosteroids." *Trends Plant Sci* **2**: 137–142.

Yonekura-Sakakibara, K., M. Kojima, T. Yamaya and H. Sakakibara (2004). "Molecular characterization of cytokinin-responsive histidine kinases in maize. Differential ligand preferences and response to cis-zeatin." *Plant Physiol* **134**(4): 1654–1661.

Yopp, J. H., N. B. Mandava and J. M. Sasse (1981). "Brassinolide, a growth-promoting steroidal lactone. 1. Activity in selected auxin bioassays." *Physiol Plantarum* **53**(4): 445–452.

Young, T. E., J. Giesler-Lee and D. R. Gallie (2004). "Senescence-induced expression of cytokinin reverses pistil abortion during maize flower development." *Plant J* **38**(6): 910–922.

Zeevaart, J. A. D. (1976). "Physiology of flower formation." *Annu Rev Plant Physiol Plant Mol Biol* **27**: 321–348.

Zeevaart, J. A. D. (2008). "Leaf-produced floral signals." *Curr Opin Plant Biol* **11**: 541–547.

Zeevaart, J. A. D. and R. A. Creelman (1988). "Metabolism and physiology of abscisic acid." *Annu Rev Plant Physiol Plant Mol Biol* **39**: 439–473.

Zentella, R., Z. L. Zhang, M. Park, *et al.* (2007). "Global analysis of DELLA direct targets in early gibberellin signaling in Arabidopsis." *Plant Cell* **19**(10): 3037–3057.

Zhang, D. P., Z. Y. Wu, X. Y. Li and Z. X. Zhao (2002). "Purification and identification of a 42-kilodalton abscisic acid-specific binding protein from epidermis of broad bean leaves." *Plant Physiol* **128**(2): 714–725.

Zhang, W., Y. J. Sun, L. Timofejeva, *et al.* (2006). "Regulation of Arabidopsis tapetum development and function by dysfunctional tapetum1 (dyt1) encoding a putative bHLH transcription factor." *Development* **133**(16): 3085–3095.

Zhao, Y. (2008). "The role of local biosynthesis of auxin and cytokinin in plant development." *Curr Opin Plant Biol* **11**(1): 16–22.

Zhao, Y., Y. F. Hu, M. Q. Dai, *et al.* (2009). "The WUSCHEL-related homeobox gene WOX11 is required to activate shoot-borne crown root development in rice." *Plant Cell* **21**(3): 736–748.

Zhao, Y., A. K. Hull, N. R. Gupta, *et al.* (2002). "Trp-dependent auxin biosynthesis in Arabidopsis: Involvement of cytochrome P450s CYP79B2 and CYP79B3." *Genes Dev* **16**(23): 3100–3112.

Zimmermann, W. (1952). "Main results of the telome theory." *Paleobotanist* **1**: 456–470.

Zou, J., S. Zhang, W. Zhang, *et al.* (2006). "The rice HIGH-TILLERING DWARF1 encoding an ortholog of Arabidopsis MAX3 is required for negative regulation of the outgrowth of axillary buds." *Plant J* **48**(5): 687–698.

# Corpus Acknowledgment

The publisher and the author of this book are grateful for the permission given by the following publishers to reproduce their material in this book.

| S/No. | Publisher | Figures and Tables |
|---|---|---|
| 1 | American Association for the Advancement of Science | Fig. 16 |
| 2 | American Society of Plant Biologists | Figs. 11, 12, 14, 15, 21, 37, 43, 52, 54, 60, 61, 67, 78, 86 |
| 3 | Wiley-Blackwell Publishing Company | Figs. 38, 40, 49, 77 |
| 4 | Annual Reviews | Figs. 1, 17, 18, 20, 22, 23, 27, 29, 32, 33, 44, 45, 46, 66, 80 |
| 5 | Springer Verlag | Figs. 2, 19, 48, 56<br>Table 2 |
| 6 | Oxford University Press | Figs. 3, 5, 34, 35, 39, 51, 69, 87 |
| 7 | Elsevier Ltd. | Figs. 7, 8, 9, 13, 23, 24, 26, 28, 30, 36, 41, 47, 53, 55, 57, 58, 59, 62, 63, 64, 65, 68, 70, 71, 72, 73, 74, 75, 76, 79, 85<br>Table 5 |
| 8 | Macmillan Publishing Ltd. | Figs. 50, 88, 89 |

# Index

1 (Kn1) 302
1-aminocyclopropane-1-carboxylic
acid (ACC) 52, 339
2-aminoethoxyvinyl glycine (AVG)
234
2-chloroethylphosphonic acid
("ethephon") 50

ABA biosynthetic pathway 156
ABA catabolic pathway 157
ABA glucosyl ester (ABA-GE) 157
ABA receptor 368
ABC model of floral patterning 325
abscisic acid (ABA) 151, 227, 367,
368, 371
ACC synthase 52
ACC synthase genes 343
acronyms for jasmonates 136
adaxial/abaxial polarity 312
adaxialization 304
AGAMOUS (AG) 346, 348
*Agrobacterium tumefaciens* 14, 117
AINTEGUMENT (ANT) 203, 345, 351
aleurone cells 85
α-amylase 72, 85
alfalfa 29
*Alternaria* 136
*Alternaria bassicola* 142
*Amborella trichopoda* 190
aminoethoxyvinyl glycine (AVG) 50,
330, 339
Amos 13
andromonoecious 340

andromonoecious sex expression 344
antipodals 208
*Antirrhinum* 306, 353
*Antirrhinum majus* 190, 289
APETALA1 (AP1) 277
APETALA2 (AP2) 217, 351
APETALA3 (AP3) 351
apical dominance 173, 273
apoptosis 363
apple 59
*Aquilegia caerulea* 310
Arabidopsis vii, 10, 29, 193, 275, 280
Arabidopsis embryogenesis 207
*Arabidopsis gynoecium* 357
*Arabidopsis thaliana* 2, 190
arbuscular symbiotic fungi 275
archaebacteria 153
Archesporial cells 356
ARGONAUTE (AGO) 257
ARGOS 203
Aristophanes 367
aromatic CKs 116
*Artemisia absinthium* 133
*Asparagus officinalis* 335
ASYMMETRIC LEAVES1 (AS1) 311,
345
ASYMMETRIC LEAVES2 (AS2) 311
auxillary bud (AB) 273
auxin 11, 227, 367, 369, 370, 371
AUXIN BINDING PROTEIN1 (ABP1)
265
AUXIN RESISTANT1 (AUX1) 272
auxin responsive factors (ARF) 41

auxin responsive genes (ARG)  41
auxin signal transduction  44

*Baba Batra* of the Bavli Talmud  111
barley  282
*Baruch spinoza*  227
*Bauhinia*  315
BELL-type protein  260
β-inhibitor complex  151
BIG BROTHER (BB)  205, 353
biosynthesis of ABA  152
biosynthesis of BRs  93
biosynthesis of CKs  118
bonsai trees  1
*Botrytis*  153
BR signal transduction  101
brace roots (BR)  231
branching-inhibitory factors  175
*Brassica napus*  91, 232
Brassicaceae  283
brassinolide (BL)  235
brassinolide (BR₁)  92
brassinosteroids (BR)  36, 195, 227,
     235, 363, 367, 368, 371
brefeldin A  243
*BREVIPEDICELLUS (BP)*  260
*BREVIS RADIX (BRX)*  198, 235
broomrapes (Orobanche)  297
bundle sheath cells,  31

C₁₉-Gibberellins  75
C₂₀-Gibberellins  75
*Cabomba aquatica*  190
Caesalpinioideae  314
campesterol  93
canalization hypothesis  320
canalization model  318
cane  59
*Cannabis sativa*  335
*Capsella*  225
Carboniferous eras  298
Cardamine hirsuta  299, 308, 310
*Carica papaya*  335
carotenoid cleavage by dioxygenases
     (CCDs)  176

carotenoid-derived hormone  182
carpel  325, 358
carpel development  189
Catasterone,  93
*Catharanthus roseus*  93
causality  v
*Cercospora*  153
cherubim  33
Cheshire Cat  13
circadian clock  196, 197
CitACS3  343
*Citrullus lanatus*  334
*Citrus*  157
*Citrus junos*  157
*Citrus* protoplasts  167
*Citrus unshiu*  96
Class III HD-ZIP genes  304
CLAVATA3 (CLV3)  255
CLAVATA3 hormones  165
*CLAVATA3/ESR* related genes (*CLE*)
     170
cleavage dioxygenase (CCD)  275
*CLV* genes  213
coconut fractions  111
coffee  59
Columella  33
compound leaves  309
conjugation of IAA and IBA  20
control of branching in monocots
     180
coronatine  147
coronatine-insensitive 1  144
cortex  32, 231
cotyledons  32
crown gall  93, 117
crown roots  237
cryptochromes  196, 278
cucumber  59, 296, 330, 339, 342,
     347
Cucumis  334
*Cucumis melo*  72, 331, 335
*Cucumis sativa*  72
Cucumis sativus  335
*Cucurbita*  281
Cucurbitaceae  341

cultured tobacco cells (BY2) 266
CUP-SHAPED COTYLEDON (CUC)
  259, 345
CUP-SHAPED COTYLEDON1
  (CUC1) 214, 255, 351
cyanobacteria 60
cyclophilin 240
Cymose 285
cytokinin (CK) 116, 227, 367, 368,
  371
cytokinin receptor 118, 369
cytokinin signaling 123
cytoskeleton 28
c-zeatin (cZ) 115

Da Xue 1
decussate 270, 272
DELLA 81, 192–194, 196, 308,
  313, 352, 370
DELLA proteins 199, 369
DELLA repressor proteins 237
DELLA repressors 267
determinated inflorescence 285
deubiquitylating enzyme (DUB)
  11
Devonian era 297, 298
dichlorophenoxyacetic acid
  (2,4-D) 13
dichotomized stems 298
diphenylurea 126
dissected leaves 307
Distichous 270, 272
*Distylium racemosum* 91
diterpenes 152
diterpenoid compounds 71
dodder (Cuscuta) 297
*Drosophila melanogaster* 2

egg cell 208
embryo research 207
embryo sac 207
embryogenesis 207
endodermal cells 237
endodermis 32, 231
endoplasic reticulum (ER) 58, 202

endosome 28
endosymbiosis 58
endothecium 356
*ENHANCER OF PINOID* 214
ent-copalyl diphosphate synthase (CPS)
  74
ent-Kaurene 75
enzyme (E2) 7
EPIDERMAL PATTERNING 321
epidermis 231
*Escherichia coli* 121
ET perception 58
ET signal transduction 51, 57
ethephon 329, 340, 342
ethylene (ET) 227, 367, 371
ethylene receptor (ETR) 58, 369
ethylene-responsive genes 64
evolutionary 289

FACTOR1 (EPF1) 321
far-red light 268
F-box proteins 183, 369
Fe toxicity 243
female gametophyte 207, 208
floral organs 179
floret 294, 295, 335
*Florigen* 279, 292
flower patterning 325
*FLOWERIN LOCUS C (FLC)* 260
*Fucus* 5
*Fusarium oxysporum* 136

GA12-aldehyde 75
Galilei 2
Galileo 1
*Galium odoratum* 270
gametophytes 207
GAs biosynthesis 73
geotropism 15
geranylgeranyl diphosphate (GGDP)
  74, 75
*Gibberella fujikuroi* 71
Gibberellins 367, 369–371
GIBBERRELIC ACID INSENSITIVE
  DWARF1 (GID) 267

globular embryo  210
Glomeromycota  185
glumes  181, 294, 295
Glutaredoxin (GRX)  353–355
*GNOM*  38
Goethe, J.W.  2, 247
*Gramineae*  180
gravitropic curvature  15
Great Learning  1
guard mother cell (GMC)  322
gymnosperms  189
gynoecious  340
gynoecium  305, 335, 336, 346,
    358

HD-ZIPIII genes  311
heart-shaped embryo  210
hermaphrodite flower  329
hermaphrodite plants  329
hermaphrodites  369
herring sperm DNA  112
Hillel the Elder  114
history of cytokinins (CKs)  111
*HOBBIT* (*HBT*)  217
hormone receptors  371
human brain  108
Hume, David  v
*Humulus lupulus*  335
hybrid-seed cucumber  326
Hydra  25, 317
hypocotyl  25, 32
hypophysis  222

IAA biosynthesis in Arabidopsis  16
IAA influx  25
immunolocalization  27
indeterminated inflorescence  285
indole butyric acid, IBA  15
indole-3-butyric acid (IBA)  237
indoleacetic acid (IAA)  15
inflorescence architectures  288, 289
inflorescences  294
insects  205
*Ipomoea nil*  282
IPT (isopentenyl transferase)  115

isopentenyl diphospate (IPP)  152
Isoprenoid biosynthesis  154
isprenoid CKs  116

JA conjugate synthase (JAR1)  143
JAGGED (JAG)  353
JA-isoleucine  131
jasmonate  370, 371
jasmonate biosynthesis  136
jasmonate responses  144
jasmonic acid (JA)  131, 364, 365,
    367, 369
JA-valine  131
Jerusalem artichoke  72

*Kalanchoe daigermontiana*  219
KANADI  304
KANADI gene  305
kinetin  112
KNAT1  303
KNOTTED  302
KNOTTED1-type homeodomain  303
KNOTTED-like homeobox (KNOXI)
    proteins  253
KNOX genes  299
KNOXI  192, 255, 307, 310, 311
KNOXI protein  255, 259, 305

laminae  297
LANCEOLATE (LA)  308
lanceolate leaves  302
*Lasiodiploida* (*Botryodiploidi*)
    *theobromae*  133
lateral root initiation  239
Lathyrus odorantus  315
leaf venation  35
LEAFY  346
leaves  325
lemma (l)  294
lettuce  157
Leucoplasts  348
LEUNIG (LUG)  351
light photoreceptor PHYB  199
*Lilium elegans*  96
lodicules  294

long day (LD) plants 279
lovastatin 222
love 367
*Lycopersicum esculen* 55
lycophyte 267
Lysenko, Trofim 113
*Lysimachia dethroides* 270

MADS box 348, 351
maize 19, 21, 59, 180, 190
maize embryogenesis 207
male gametophyte 207
mammals 205
megaspore mother cell 208
melon 330, 342
meristemoid mother cells (MMCs) 322
meristemoids 322
metamorphogenesis 297
metamorphosis 325, 345
methionine cycle 52
methyl jasmonate (MeJA) 131
methylerythriol 4-phosphate (MEP) 153
mevalonic acid (MVA) 152
microphyle 208
Mimosoideae 314
MIR 165/166 304
miR164 351
miR172 351, 363
miR319 308, 309
miR824 323
miRNA159 193
monoecious 326, 340
monoecious species 334
*MONOPTEROS* 215
monoterpenes 152
*morphogen* 25

naphthalene-1-acetic acid (NAA) 13
NGATHA1 (NGA1) 357
*Nicotiana* 167
nitric oxide (NO) 242, 367
nitrilases 19
nursing cultures 167

octadecanoids 131
*Oncidium* 59
organ growth 205
*Orobanche* 184, 275
*Oryza* 181
ovary 357
overtopping 298
ovules 346

palea 294
Papilionoideae 314
parthenocarpic 330
pea 193, 194, 275, 314
peach 59
pear 59
*Pelargonium* 59
peptide hormones 165
perception 40
perception and signal transduction of GA 81
perception of brassinosteroids 103
PERIANTHIA (PAN) 355
pericycle 32, 125, 231, 241
pericycle cells 239
pericycle endodermis 232
persimmon 59
petal primordia 350
PETALLOSS (PTL) 350
petals 200, 325, 347
petunia 25, 59, 177, 274, 275, 285
*Petunia hybrida* 190
PHABULOSA (PHB) 304, 345
*Pharbitis nil* 277, 279
*Phaseolus lunatus* 364
*Phaseolus vulgaris* 91, 96
PHAVOLUTA (PHV) 304, 345
phloem companion cells 125
phosphates (Pi) 242
photomorphogenesis 277
photoperiod pathway 278
photoperiods 72
phototropism 15, 277
phyllotaxis 248, 260, 265, 270
phylogeny of angiosperms 189
*Physcomitrella* 267

*Physcomitrella patens* 89, 142
phytochromes 196, 197, 278
phytomer 178, 273, 291
phytosulfokine peptides 167
PIGGYBACK1 (PGY1) 311
Pillars of the Mosque of Acre 5,
	356
PINFORMED (PIN) 272
PINOID (PID) 38, 214
PISTILLATA (P1) 351
pistillate flowers 326, 329
*Pisum sativum* 310, 315
placentae 346
planation 298, 299
plasma membrane (PM) 40, 127
*Plasmodium falciparum* 153
PLETHORA (*PLT*) 28
PLETHORA genes 217
plum 59
Plumbaginaceae family 209
*Plumbago zeylanica* 209
polar flow of IAA 229
pollen grain 207
pollen mother cells (PMCs) 356
pollen tube 208
polyamine synthesis 56
polypeptides 195
*position* 4
potato 29
pre-precambrial cells 317, 319, 320
primary root 232
procambrium 316
proembryo 210
proteasome 7
proteasome-mediated proteolysis
	186
protoderm 210
PROTODERMAL FACTOR2 (PDF2)
	218
protophloem 33
provasculature 33
*Pseudomonas syringae* 147

quiescent center (QC) 40, 216, 231,
	232

RABBIT EAR (RBE) 350
raceme 285, 335
radio immunoassay, 133
rapid alkalinization factor 171
receptors of CKs 126
red light 268
repression of branching 275
response to jasmonate 133, 135
retinoblastoma protein 258
REVOLUTA (REV) 304, 345
rice 10, 29, 190, 281
*Ricinus communis* 334, 335
root apex 165
root cap 242
root hair 39, 228, 241
root system architecture (RSA) 244
rosetta leaves 199
*Rosmarinus officinalis* 133

*Saccharomyces cerevisiae* 121
S-adenosyl-L-methionine (SAM) 52
salicylic acid 312, 367
SAM synthetase 52
SCARECROW (*SCR*) 217
SCFTIR1 complex 45
SCR/SP11 peptides 168
Scrophulariaceae 283
*Selaginella* 267
*Selaginella moellendorffi* 89
self-incompatibility 191
seminal roots 229, 232
sepals 325
septum 346
Seraphim 33
sesquiterpenes 152
SEUSS (SEU) 351
sex expression 296, 326, 340
shade avoidance 277
shade avoidance syndrome (SAS) 5
SHATTERPROOF (SHP) 346
SHATTERPROOF1 (SHP1) 357
shoot apex 32, 165
shoot branching 274, 276
SHOOT MERISTEMLESS (STM)
	260

shoot-apex meristem (SAM) 247
Short Day (SD) plants 279
*SHORTROOT* (*SHR*) 217
signal transduction vi, 36, 143
signal transduction of BR 100
Silene latifolia 339
silkless (sk) 338
Silurian era 297
silverthiosulphate 332, 342
single letter designation 81
Sir Charles and Sir Francis Darwin
    15
Skoog, Folke 111
snapdragon 285, 289
Solanaceae 283
*Solanum lycopersicum* 148, 310
*Solanum toberosum* 310
somatic embryogenesis 219
soybean 29
sperm cells 207
spikelet 181, 294, 295, 335, 364
spinach 279
*Spinacia oleracea* 335
spiral 270
spruce embryogenesis 207
stamens 294, 325
staminate 365
staminate flowers 326, 329
statolithes 28
stem apical meristem 195
*Stenophragma thalianum* vii
STERILE APETALA (SAP) 351
Steward F.C. 111
stigma 207, 346, 357
stomata formation 321
*Striga* 184, 275
strigolactones 184, 195, 274, 275
STYLISH1 357
sugar 59
sulphates ($SO_4^{2-}$) 242
synergids 208
systemines 165

tapetum 356
tassel 294, 335, 364

tassel seed 337, 339, 364
Telome Theory 298
tendrils 314
teosinte 181
tetraterpenes 152
the *bakanae* disease 71
the *CLAVATA* genes 169
the Prophet Amos 49
the quiescent center (QC) 234
*Thea sinensis* 96
tillering in rice 177
tillering 274
TIR1 proteins 261
tobacco 193, 280
tomato 59, 280
tomato (*Solanum lycopersicum*) 291
TomSys 166
TOO MANY MOUTHS (TMMs) 323
*Torenia fournieri* 209
torpedo-stage of the embryo 219
tracheary elements 170
tracheophytes 297
transcription factors (TFs) 135, 144
transcription regulators 352
transduction of ET 50
transition to flowering 194
transmission of the flowering signal
    277
TRANSPORT INHIBITOR RESPONSE
    PROTEIN (TIR1) 265
trichlorophenoxybutyric acid
    (2,4,5 TB) 15
trichome 41, 312, 314
triple-response 57
triploid endosperm tissue 208
*Tripsacum* 339
*Trisetum distichophyllum* 270
*Triticum aestivum* 279
Turing, Alan M. 25
type A response regulators 118
type B response regulators 118
t-zeatin (tZ) 115

Ub activator (E1) 7
Ub conjugating 7

Ub ligase (E3) 7
ubiquitin (Ub) 7
ubiquitination vi, 6, 65, 84, 350, 369, 370
ubiquitination of repressors 133
ubiquitin-proteasome system (UPS) 6
United States Department of Agriculture (USDA) 91
UNUSUAL FLORAL ORGANS (UFO) 350
UPS 6
*Urtica dioica* 270

vascular parenchyma 31
Vavilov, Nicolei 113
vegetative nucleus 207
venation in angiosperm leaves 315
vernalization 278, 282, 284
*Vicia faba* 133, 159

watermelon (Citrullus lanatus) 342, 343
webbing 298, 300
wheat 21, 29, 59, 282

whorled 270
Wolff, Caspar Friedrich 2
Wolff, C.F. 247
*WOODEN LEG (WOL)* 218
wormholes 289
WUSCHEL (WUS) 255, 346

Xanthium 364
*Xenopus* 39

YABBY 304, 305
YABBY1 357
yeast 205
yeast extract 112
*YUC* genes 25
YUCCA flavin monooxygenase 25
YUCCA protein 268

*Zea mays* 335
zeatin 114
Zeus 367
Zinnia elegans cells 320
*Zinnia* 168
*ZWILLE (ZLL)* 257
zygote 208